# COMPUTER ARITHMETIC IN THEORY AND PRACTICE

*This is a volume in*
COMPUTER SCIENCE AND APPLIED MATHEMATICS
*A Series of Monographs and Textbooks*

*Editor:* WERNER RHEINBOLDT

A complete list of titles in this series appears at the end of this volume.

# COMPUTER ARITHMETIC IN THEORY AND PRACTICE

*Ulrich W. Kulisch*

FAKULTÄT FÜR MATHEMATIK
UNIVERSITÄT KARLSRUHE
KARLSRUHE, WEST GERMANY

*Willard L. Miranker*

MATHEMATICAL SCIENCES DEPARTMENT
IBM THOMAS J. WATSON RESEARCH CENTER
YORKTOWN HEIGHTS, NEW YORK

1981

ACADEMIC PRESS
A Subsidiary of Harcourt Brace Jovanovich, Publishers
New York London Toronto Sydney San Francisco

ACADEMIC PRESS, INC.
111 Fifth Avenue, New York, New York 10003

*United Kingdom Edition published by*
ACADEMIC PRESS, INC. (LONDON) LTD.
24/28 Oval Road, London NW1 7DX

**Library of Congress Cataloging in Publication Data**

Kulisch, Ulrich.
Computer arithmetic in theory and practice.

(Computer science and applied mathematics)
Bibliography: p.
Includes index.
1. Algebra, Abstract. 2. Interval analysis (Mathematics) 3. Floating–point arithmetic.
I. Miranker, Willard L., joint author. II. Title.
QA162.K84 512'.02 80–765
ISBN 0–12–428650–X

PRINTED IN THE UNITED STATES OF AMERICA

81 82 83 84 9 8 7 6 5 4 3 2 1

*To the memory of*
*Oskar Perron*
*1880–1975*
*whose fundamental work on numbers*
*so clarifying in its form*
*was an inspiring model for us*
*to develop the mathematics of*
*Computer Arithmetic*

Oskar Perron

# CONTENTS

## Chapter 4. **Interval Arithmetic**



# Part 2. **IMPLEMENTATION OF ARITHMETIC ON COMPUTERS**

## Chapter 5. **Floating-Point Arithmetic**



## Chapter 6. **Implementation of Floating-Point Arithmetic on a Computer**

## Chapter 7. **Computer Arithmetic and Programming Languages**

# PREFACE

This book deals with computer arithmetic and treats it in a more general sense than usual. This results in more far-reaching concepts and a more uniform theory. These in turn serve to simplify even the restricted traditional notions of computer arithmetic. They also clarify and otherwise facilitate the process of implementation of computer arithmetic. The text summarizes both an extensive research activity that went on during the past fifteen years and the experience gained through various implementations of the entire arithmetic package on diverse processors, including even microprocessors. These implementations require an extension of existing higher programming languages to accommodate the entire arithmetic package. The text is also based on lectures held at the Universität Karlsruhe during the preceding decade.

While the collection of research articles that contribute to this text is not too large in number, we refrain from a detailed review of them and refer to the list of references. Since our text synthesizes and organizes diverse contributions from these sources into a coherent and global setting, we found it best not to interrupt the continuous flow of exposition with detailed citations to original sources.

The text consists of two parts. Part 1, comprising the first four chapters, deals with the theory of computer arithmetic, while Part 2, comprising the last three chapters, treats the implementation of arithmetic on computers. Our development shows that a sound implementation of arithmetic on computers depends critically on the theoretical development. Such an implementation requires the establishment of various isomorphisms between different definitions of arithmetic operations. These isomorphisms are to be established for the spaces in which the actual computer operations reside. Therefore we begin with a development of these spaces.

Central to this treatise is the concept of semimorphism. Properties of this concept are directly derivable from the diverse models that we study here. The concept of semimorphism supplies an ordering principle that permits the whole treatise to be understood and developed in close analogy to principles of geometry, whereby the mapping properties of a semimorphism replace those of a group of transformations in a geometry. Following this idea, we define the structures of the spaces occurring in computations on computers as invariants with respect to semimorphisms. The algorithms for the arithmetic operations in various spaces, which are given in Part 2, describe the implementation of these semimorphisms on computers. The result is an arithmetic with many sound and desirable properties (optimal accuracy, theoretical describability, closedness of the theory, applicability, etc.). The similarity to geometric principles just referred to is the guideline that leads—as we see it—to this closed and well-rounded presentation of computer arithmetic.

Implementation of the algorithms is made along natural lines with which the nonspecialist views computers. This principle helps to avoid those misunderstandings of the computer by its users that are caused by tricky implementations.

We mention incidentally that a sound implementation of traditional computer arithmetic is captured by what we call the vertical definition of arithmetic.

This book, of course, can be used as a textbook for lectures on the subject of computer arithmetic. If one is interested only in the more practical aspects of implementation of arithmetic on computers, Part 2 along with acceptance, a priori, of some results of Part 1, also suffices as a textbook for lectures.

We regard the availability of the directed roundings and interval operations as essential tools on the computer. Apart from the great impact on the insight and theoretical understanding of computer arithmetic exerted by the development of interval arithmetic, these tools are necessary if a computer is to be used to control rounding errors. Employment of such techniques provides computation with guarantees, and thus permits use of the computer for verification and *decidability*. Such questions cannot be studied by using arithmetic which rounds in the traditional sense. By taking appropriate measures, interval operations can be made as fast as the corresponding ordinary floating-point operations.

The theory developed in Part 1 permits additional applications to numerical analysis, applications that are beyond the scope of this book. We mention a few of them:

A clear and precise definition of all arithmetic operations on a computer is a fundamental and necessary condition for reliable error analysis of numerical algorithms. The definition of the arithmetic operations on a computer by semimorphisms is also sufficient for complete control of the computational error due to the approximate arithmetic performed by the computer for many standard problems of numerical analysis. Examples of such problems are linear systems of equations, inversion of matrices, eigenvalue problems, zeros of polynomials, nonlinear systems of equations, evaluation of arithmetic expressions and mathematical functions, linear programming problems, numerical quadrature, and initial and boundary value problems for ordinary differential equations.

The concept of semimorphism can be used directly for an axiomatic definition of arithmetic within the syntax and semantics of programming languages.

This in turn should be useful for correctness proofs of numerical programs.

The similarity of the concept of a semimorphism to the Kuratowski axioms of a topology is clear. This has been used to redefine the concept of stability of algorithms in numerical mathematics [1, 76, 82, 83].

The concept of semimorphism can be applied to study and describe the approximation of function spaces by certain subspaces.

The spaces that we later call ringoids and vectoids and in which computations on computers are performed provide a framework for the study of the cyclic termination of iterative methods on computers [28, 29, 67, 86].

Last, but hardly least, we note that the spaces of interval arithmetic can be developed much further. It is well known that the intervals over $\boldsymbol{R}$ and $\boldsymbol{C}$ form regular semigroups with respect to addition and multiplication. Such semigroups can be imbedded into groups by a well-known algebraic procedure. The extended spaces then permit the definition of concepts of metric and norm in a manner quite similar to their definition in a field or a linear space. This greatly simplifies the analysis of interval algorithms [1, 24, 25, 26, 80].

Many of these additional applications are subjects of current research, and we defer their exposition to a follow-up volume.

We are grateful to all those who have contributed through their research to this treatment. We once more refer to the bibliography at the end of this text. We especially owe thanks to Dr. P. Schweitzer of IBM Germany and to Dr. R. A. Toupin of the IBM Research Center. Their support and their interest, so congenial in its nature, have been critical for the completion of this text. We are also grateful to J. Genzano. Her virtuosity and devotion to the physical preparation of the text provided a constant force accelerating the work.

# COMPUTER ARITHMETIC
# IN THEORY AND PRACTICE

# INTRODUCTION AND PRELIMINARY DEFINITION OF COMPUTER ARITHMETIC

In this treatise we present a study of computer arithmetic. Central to this is the specification of the spaces that form the natural setting for arithmetic when it is performed in a computer. In earlier days of computer development the properties of computer arithmetic were influenced by such features as simplification of computer architecture, hardware and software considerations, and even the cost of the technology employed. Experience has shown that these influences led to economies that frequently resulted in added costs and burdens to the user. A typical example of the latter is the frequent production of an inexplicable, even incorrect computation. The arithmetical properties of computer operations should be precisely specified by mathematical methods. These properties should be simple to describe and, we hope, easily understood. This would enable both designers and users of computers to work with a more complete knowledge of the computational process.

To L. Kronecker we owe the remark, "*God created the natural numbers, all else is man made.*" This difference is reflected in computers, where arithmetic on the natural numbers may be performed exactly, while for all else, only approximately. In fact, there are some interesting mathematical questions involved even in the implementation of integer arithmetic on computers. However, we postpone discussing these and assume that the hardware designer has satisfactorily solved this particular problem.

In addition to the integers, numerical algorithms are usually defined in the space $\boldsymbol{R}$ of real numbers and the vectors $V\boldsymbol{R}$ or matrices $M\boldsymbol{R}$ over the real numbers. Additionally, the corresponding complex spaces $\boldsymbol{C}$, $V\boldsymbol{C}$, and $M\boldsymbol{C}$

also occur ocassionally. All these spaces are ordered with respect to the order relation $\leq$. (In all product sets the order relation $\leq$ is defined componentwise.) Recently, numerical analysts have begun to define and study algorithms for intervals defined over these spaces. If we denote the set of intervals over an ordered set $\{M, \leq\}$ by $IM$, we obtain the spaces ***IR***, ***IVR***, ***IMR***, and ***IC***, ***IVC***, ***IMC***.

In Fig. 1, to which we shall repeatedly refer, we present a table of spaces and operations. For example, in the second column of this figure we list the various spaces in which arithmetic is performed and which we have introduced in the previous paragraph.

| | I | | II | | III | | IV | V |
|---|---|---|---|---|---|---|---|---|
| 1 | | | ***R*** | $\supset$ | *D* | $\supset$ | *S* | $+ - \cdot /$ |
| 2 | | | ***VR*** | $\supset$ | *VD* | $\supset$ | *VS* | $\times$<br>$+ -$ |
| 3 | | | ***MR*** | $\supset$ | *MD* | $\supset$ | *MS* | $\times$<br>$+ - \cdot$ |
| 4 | ***PR*** | $\supset$ | ***IR*** | $\supset$ | ***ID*** | $\supset$ | ***IS*** | $+ - \cdot /$ |
| 5 | ***PVR*** | $\supset$ | ***IVR*** | $\supset$ | ***IVD*** | $\supset$ | ***IVS*** | $\times$<br>$+ -$ |
| 6 | ***PMR*** | $\supset$ | ***IMR*** | $\supset$ | ***IMD*** | $\supset$ | ***IMS*** | $\times$<br>$+ - \cdot$ |
| 7 | | | ***C*** | $\supset$ | ***CD*** | $\supset$ | ***CS*** | $+ - \cdot /$ |
| 8 | | | ***VC*** | $\supset$ | ***VCD*** | $\supset$ | ***VCS*** | $\times$<br>$+ -$ |
| 9 | | | ***MC*** | $\supset$ | ***MCD*** | $\supset$ | ***MCS*** | $\times$<br>$+ - \cdot$ |
| 10 | ***PC*** | $\supset$ | ***IC*** | $\supset$ | ***ICD*** | $\supset$ | ***ICS*** | $+ - \cdot /$ |
| 11 | ***PVC*** | $\supset$ | ***IVC*** | $\supset$ | ***IVCD*** | $\supset$ | ***IVCS*** | $\times$<br>$+ -$ |
| 12 | ***PMC*** | $\supset$ | ***IMC*** | $\supset$ | ***IMCD*** | $\supset$ | ***IMCS*** | $\times$<br>$+ - \cdot$ |

FIGURE 1. Table of the spaces occurring in numerical computations.

For arithmetic purposes, a real number is usually represented by an infinite $b$-adic expansion, with operations performed on these expansions defined as the limit of the sequence of results obtained by operating on finite portions of the expansions. In principle, a computer could approximate this limiting process, but the apparent inefficiency of such an approach eliminates its serious implementation even on the fastest computers. In fact, for arithmetic purposes, the real numbers are approximated by a subset $S$ in which all operations are simple and rapidly performable. The most common choice for this subset $S$ is the so-called floating-point system with a fixed number of digits in the mantissa. If a prescribed accuracy for the computation cannot be achieved by operating within $S$, we use a larger subset $D$ of $\boldsymbol{R}$ with the property $\boldsymbol{R} \supset D \supset S$. For arithmetic purposes we define vectors, matrices, intervals, and so forth as well as the corresponding complexifications over $S$ and $D$. So doing, we obtain the spaces $VS$, $MS$, $IS$, $IVS$, $IMS$, $CS$, $VCS$, $MCS$, $ICS$, $IVCS$, $IMCS$ and the corresponding spaces over $D$. These two collections of spaces are listed in the third and fourth columns of Fig. 1. For example, $CS$ is the set of all pairs of elements of $S$, $VCS$ the set of all $n$-tuples of such pairs, $ICS$ the set of all intervals over the ordered set $\{CS, \leq\}$, and so forth.

Characteristically, $S$ and $D$ are chosen as the sets of floating-point numbers of single and double length, respectively. However, in Fig. 1, $S$ and $D$ are generic symbols for a whole system of subsets of $\boldsymbol{R}$ with arithmetic properties that we shall subsequently define.

Having defined the sets listed in the third and fourth columns of Fig. 1, we turn to the question of defining operations within these sets. These operations are supposed to approximate operations that are defined on the corresponding sets listed in the second column of Fig. 1. Figure 1 has four blocks, and in every set in each of the first lines of each block, we are to define an addition, a subtraction, a multiplication, and a division. For the sets in the last line of each such block, for instance, we need to define an addition, a subtraction, and a multiplication. These required operations are listed in the fifth column of Fig. 1. Furthermore, the lines in Fig. 1 are not mutually independent arithmetically. By this we mean, for instance, that a vector can be multiplied by a scalar as well as by a matrix; an interval vector can be multiplied by an interval as well as by an interval matrix. These latter multiplication types are indicated in Fig. 1 by means of a $\times$ sign between lines in the fifth column therein.

As a preliminary and informal definition of computer arithmetic, we say the following: *By computer arithmetic, we understand all operations that have to be defined in all of the sets listed in the third and fourth columns of Fig. 1 as well as in certain combinations of these sets.* The sets $S$ and $D$ may, for instance, be thought of as floating-point numbers of single and double mantissa length. In a good programming system, these operations should be available as operators for all admissible combinations of data types.

We interpret this definition in somewhat more detail. First we make a count of the number of multiplications that occur in the computer arithmetic as defined above. Later on we make the analogous count for the other operations. But the multiplication count itself is enough to show that it is too much to expect the average computer user to define the system of operations by himself.

If $\boldsymbol{Z}$ denotes the set of integers on the computer, we have the *five basic data types*: $\boldsymbol{Z}, S, \boldsymbol{CS}, \boldsymbol{IS}, \boldsymbol{ICS}$, which in an appropriate programming language may be called integer, real, complex, real interval, and complex interval. If $a$ and $b$ are operands, each a possible one of these five data types, the table in Fig. 2 shows the resulting type of the product $a * b$.

| $a * b$ | $Z$ | $S$ | $CS$ | $IS$ | $ICS$ |
|---|---|---|---|---|---|
| $Z$ | $Z$ | $S$ | $CS$ | $IS$ | $ICS$ |
| $S$ | $S$ | $S$ | $CS$ | — | — |
| $CS$ | $CS$ | $CS$ | $CS$ | — | — |
| $IS$ | $IS$ | — | — | $IS$ | $ICS$ |
| $ICS$ | $ICS$ | — | — | $ICS$ | $ICS$ |

FIGURE 2. Multiplication table for the basic numerical data types $a * b$.

A dash in the table of Fig. 2 means that the product $a * b$ is not defined a priori. Indeed a floating-point number is an approximate representation of a real, while an interval is a precisely defined object. The product of the two, which ought to be an interval, may then not be precisely specified. If the user is obliged to multiply a floating-point number and a floating-point interval, he has to employ a transfer function that transforms this floating-point operand into a floating-point interval. (This implicitly endows a precision to the floating-point number.) At this point ordinary multiplication of floating-point intervals may be employed.

If the programming language used has a so-called strong typing concept, the table in Fig. 2 shows that for multiplication among pairs of the five basic data types, 17 multiplication routines are required. We shall see that the multiplication corresponding to the entries in the framed part of the table of Fig. 2 must be supplied with three different roundings if the multiplications corresponding to interval types are to be correctly defined. This requires 16 additional multiplications or a total of 33 multiplication routines for the five basic data types.

Now let $T_1$, $T_2$, and $T_3$ denote one of the basic data types $\boldsymbol{Z}$, $S$, $\boldsymbol{CS}$, $\boldsymbol{IS}$, $\boldsymbol{ICS}$. We consider the sets of matrices $MT_i$, vectors $VT_i$, and transposed vectors $V^T T_i$, $i = 1(1)3$, whose components are chosen from among the basic data types. Elements of these sets can also be multiplied. Figure 3 displays the types of such products. In the figure, $T_3$ is to be replaced by

| $a \cdot b$ | $T_2$ | $MT_2$ | $VT_2$ | $V^{T}T_2$ |
|---|---|---|---|---|
| $T_1$ | $T_3$ | $MT_3$ | $VT_3$ | — |
| $MT_1$ | — | $MT_3$ | $VT_3$ | — |
| $VT_1$ | — | — | — | $MT_3$ |
| $V^{T}T_1$ | — | $V^{T}T_3$ | $T_3$ | — |

FIGURE 3. Table for the multiplication $a \cdot b$ of matrices and vectors over the basic data types.

the resulting type of Fig. 2 if the components of the operands are of the type $T_1$ and $T_2$, respectively. The products with operands of type $T_1$ and $T_2$ were already counted (in relation to Fig. 2). In addition to these products, we see seven essential {matrix, vector} multiplications† in Fig. 3. This leads in principle to $33 \times 8 = 264$ different multiplications.

The table for the addition and subtraction of the basic data types is identical to that of Fig. 2. The division table is likewise identical except for the single entry that corresponds to the quotient of two operands of type $Z$. Thus for each of the three cases of addition, subtraction, and division of the five basic data types, we require, as before, 33 different routines.

The table for matrix and vector addition and subtraction is given in Fig. 4.

| $+\ -$ | $T_2$ | $MT_2$ | $VT_2$ | $V^{T}T_2$ |
|---|---|---|---|---|
| $T_1$ | $T_3$ | — | — | — |
| $MT_1$ | — | $MT_3$ | — | — |
| $VT_1$ | — | — | $VT_3$ | — |
| $V^{T}T_1$ | — | — | — | — |

FIGURE 4. Addition and subtraction table of matrices and vectors over the basic data types.

Summarizing, we can say that we have 99 ($=3 \times 33$) different additions and subtractions, 264 different multiplications, and 33 different divisions.

When we deal with interval spaces, we are obliged to append to the arithmetic operations the operations of intersection ($\cap$) and taking the

† A subtle point concerns the occurrence of the set $IMS$ in Fig. 1, while for $T_i = IS$ in Fig. 3 the set $MIS$ is listed. Later on we shall prove that they are isomorphic with respect to the order relation $\leq$ and all arithmetic operations.

convex hull ($\bar{\cup}$) of pairs of intervals. Figure 5, which lists the different possibilities, shows that there are $12 (= 3 \times 4)$ different intersections and $12 (= 3 \times 4)$ different takings of the convex hull. In Fig. 5, $T_1$ resp. $T_2$ denote ***IS*** or ***ICS***. The intersection and convex hull are to be taken componentwise.

| $\cap\ \bar{\cup}$ | ***IS*** | ***ICS*** |
|---|---|---|
| ***IS*** | ***IS*** | ***ICS*** |
| ***ICS*** | ***ICS*** | ***ICS*** |

| $\cap\ \bar{\cup}$ | $T_2$ | $MT_2$ | $VT_2$ | $V^T T_2$ |
|---|---|---|---|---|
| $T_1$ | $T_3$ | — | — | — |
| $MT_1$ | — | $MT_3$ | — | — |
| $VT_1$ | — | — | $VT_3$ | — |
| $V^T T_1$ | — | — | — | — |

FIGURE 5. Tables for intersection and convex hull.

The subject of the first part of this text concerns the development of definitions for these many operations and also specifies simple natural structures that form the settings of these operations. The second part then deals with the implementation of these operations on computers. In particular, we shall see that the large number of operations that we counted above can be reduced and built up from a relatively small number of fundamental algorithms and routines.

We shall see that there are in principle two different basic methods for defining computer arithmetic. These are called the vertical method and the horizontal method. For the horizontal method, the arithmetic is assumed to be known in the leftmost set of each row of Fig. 1. On the other hand, for the vertical method, the arithmetic is assumed to be defined by some means in each set of the first row of Fig. 1. Both methods then define the arithmetic in all of the sets in Fig. 1 by appropriate extension procedures relevant to each method. By way of illustration of these two possibilities, we consider a simple detail of Fig. 1 that concerns the sets ***R***, *D*, and *S* as well as the spaces of matrices *M**R***, *MD*, and *MS*. We assume that an arithmetic in *D* and *S* is defined. By the vertical definition of the arithmetic in *MD* and *MS*, we mean that the operations in *MD* and *MS* are defined by the operations in *D* and *S* and the usual formulas for the addition and multiplication of real matrices (see the following figure). On most computers this is precisely the method of definition of addition and multiplication for floating-point matrices. While the horizontal method to be developed has many advantages, it turns out that both methods lead to the same abstract structures as settings for the arithmetics. It will develop that the structures derived from $\boldsymbol{R}$ can be described as ordered ringoids resp. as ordered vectoids, while those derived from $\boldsymbol{C}$ are weakly ordered ringoids resp. weakly ordered vectoids. (For definitions, see below.)

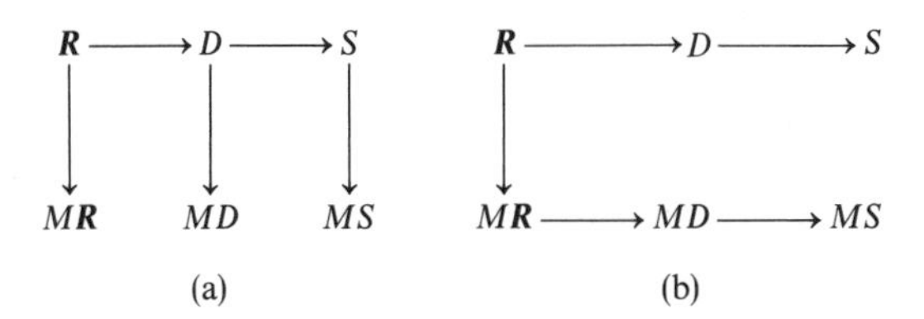


We now turn to a more detailed description of the horizontal method for defining the arithmetic operations. Let $M$ be one of the sets listed in Fig. 1, and let us assume that certain operations and relations are defined for its elements. Further, let $\bar{M}$ be a set of rules (axioms) given for the elements of $M$. By way of example, the commutative law of addition might be one such rule. Then we call the pair $\{M, \bar{M}\}$ a structure. We shall also refer to $\{M, \bar{M}\}$ as the structure in the set $M$. The structure is well known in the sets $\boldsymbol{R}$, $V\boldsymbol{R}$, $M\boldsymbol{R}$, $\boldsymbol{C}$, $V\boldsymbol{C}$, and $M\boldsymbol{C}$ listed in Fig. 1. Now let $M$ be one of these sets and $*$ be one of the operations defined in $M$. Then in the power set $\boldsymbol{P}M$, which is the set of all subsets of $M$, an operation $*$ can be defined by

$$\bigwedge_{A,B \in \boldsymbol{P}M} A * B := \{a * b \,|\, a \in A \wedge b \in B\}.$$

If we apply this definition to all operations $*$ of $M$, the structure of the power set $\{\boldsymbol{P}M, \overline{\boldsymbol{P}M}\}$ can be derived from the structure $\{M, \bar{M}\}$.

Summarizing, we can say that the structure $\{M, \bar{M}\}$ in the sets listed in the leftmost element of every row in Fig. 1 is always known. We use these structures in order to derive the structures of the subsets of each of the corresponding rows. We do this by means of a general principle that we are now going to describe.

First, we can state that in the leftmost element of every row in Fig. 1, a minus operator is defined. We shall see later that if $D$ and $S$ are floating-point systems, all the subsets $N \subseteq M$ of Fig. 1 have the property

(S) $\bigwedge_{a \in N} -a \in N \wedge o, e \in N.$

Here $o$ denotes the neutral element of addition and $e$ the neutral element of multiplication if it exists. Further, we assume that the elements of $M$ are mapped into the subset $N \subseteq M$ by a *rounding*, which we assume to have the basic property

(R1) $\bigwedge_{a \in N} \square a = a.$

If one has to approximate a given structure $\{M, \bar{M}\}$ by a structure $\{N, \bar{N}\}$, one is initially tempted to try it in such a manner that the mapping has properties like isomorphism or homomorphism. In our applications, however, $N$ is a subset of $M$ in general of different cardinality. There does not exist a one-to-one mapping, and therefore there exists no isomorphism between sets of different cardinality. But even a homomorphism would preserve the structure of a group, a ring, or a vector space. It can be shown by means of simple examples even in the case of floating-point numbers that neither can a homomorphism be realized in a sensible way. It seems, however, desirable to stay as close to a homomorphism as possible. We shall see later that in addition to (S) and (R1), the following three properties can be realized in all cases of Fig. 1:

(RG) $\bigwedge_{a,b \in M} a \boxast b := \square(a * b)$ for all $* \in \{+, -, \cdot, /\}$ in $M$ and corresponding $\boxast$ in $N$,

(R2) $\bigwedge_{a,b \in M} (a \leq b \Rightarrow \square a \leq \square b)$,

(R4) $\bigwedge_{a \in M} \square(-a) = -\square a.$

These properties can be shown to be necessary conditions for a homomorphism between $M$ and $N$. We therefore call a mapping with the properties (S), (RG), (R1), (R2), and (R4) a semimorphism. In a semimorphism, (RG) defines the operations $\boxast, * \in \{+, -, \cdot, /\}$, in the subset $N \subseteq M$, (R2) means that the rounding is monotone, and (R4) that it is antisymmetric.

Although the properties (R1), (R2), and (R4) do not define the rounding uniquely, the rounding $\square: M \to N$ is responsible not only for the mapping of the elements but also for the resulting structure $\{N, \bar{N}\}$. This means that the set of rules $\bar{N}$ valid in $N$ depends essentially on the properties of the rounding function. More precisely, $\bar{N}$ can be defined as the set of rules that is invariant with respect to a semimorphism i.e., $\bar{N} \subseteq \bar{M}$. This means that the structure $\{N, \bar{N}\}$ becomes a generalization of $\{M, \bar{M}\}$. If we consider the mapping between the second and the third columns in any row of Fig. 1, we get a proper generalization $\bar{N} \subset \bar{M}$. Proceeding from the third to the fourth column—and possibly proceeding with additional such steps—we always have $\bar{N} = \bar{M}$. We shall see later that the structures of an ordered or weakly ordered ringoid or vectoid represent the properties that are invariant with respect to semimorphisms. That this assertion can be proved in all cases of Fig. 1 establishes the horizontal definition of computer arithmetic.

A method that defines computer arithmetic is a reasonable one if it can be implemented by algorithms that are comparable in speed to those of other definitions of computer arithmetic. At first sight it seems doubtful that formula (RG) in particular can be implemented on computers at all. In order to determine the approximation $a \boxast b$, the exact result $a * b$ seems to be necessary in (RG). If, for instance, in the case of addition in a decimal floating-point system, $a$ is of the magnitude $10^{50}$ and $b$ of the magnitude $10^{-50}$, about 100 decimal digits in the mantissa would be necessary in order to represent $a + b$. Not even the largest computers have such long accumulators. Still more difficult situations for implementation of (RG) arise in the case of floating-point matrix multiplication or in the case of the division of two complex floating-point numbers.

Nevertheless, we shall prove by means of special algorithms that the formula system (S), (RG), (R1), (R2), and (R4) can be realized in all cases of Fig. 1. These algorithms show that whenever $a * b$ is not reasonably representable on the computer, it is sufficient to replace it by an appropriate and representable value $a \tilde{*} b$ with the property $\square(a * b) = \square(a \tilde{*} b)$. Then $a \tilde{*} b$ can be used to define $a \boxast b := \square(a * b) := \square(a \tilde{*} b)$. These algorithms are comparable in speed with the corresponding algorithms of the vertical definition of the arithmetic operations.

The horizontal definition generates certain desirable properties. The formulas (RG), (R1), and (R2), in particular, used in the definition, result in a computer arithmetic with an optimal accuracy in a sense to be made precise later. Coupled with certain other reasons, this in turn allows for a simplification of the error analyses of numerical algorithms. Further, we also obtain the following reasonable compatibility properties between the structures $\{M, \bar{M}\}$ and $\{N, \bar{N}\}$, which are central to our subject:

(RG1) $\bigwedge_{a,b \in N} (a * b \in N \Rightarrow a \boxast b = a * b)$ for all $* \in \{+, -, \cdot, /\}$,

(RG2) $\bigwedge_{a,b,c,d \in N} (a * b \leq c * d \Rightarrow a \boxast b \leq c \boxast d)$ for all $* \in \{+, -, \cdot, /\}$,

(RG4) $\bigwedge_{a \in N} -a = \boxminus a := (-e) \boxdot a.$

For these many reasons, we strongly recommend the horizontal method for defining computer arithmetic. Part 1 of this text explores both the horizontal and the vertical method theoretically; in Part 2, which deals with computer implementation of arithmetic, the horizontal method is employed extensively.

Finally we remark that the formulas (S), (RG), (R1), (R2), and (R4) of a semimorphism can also be used as an axiomatic definition of computer arithmetic in the context of programming languages.

# Part 1 / THEORY OF COMPUTER ARITHMETIC

# Chapter 1 / **FIRST CONCEPTS**

**Summary:** In this chapter we give an axiomatic characterization of the essential properties of the sets and subsets displayed in Fig. 1. We then define the notion of rounding from a set $M$ into a subset $N$ and study the key properties of certain special roundings and their interactions with simple arithmetic operations. To accomplish this, we employ several lattice theoretic concepts that are developed at the beginning of this chapter.

## 1. ORDERED SETS

We begin with a few well-known concepts and properties of ordered sets.

**Definition 1.1:** A relation $\leq$ in a set $M$ is called an *order relation*, and $\{M, \leq\}$ is called an *ordered set* if we have

(O1) $\bigwedge_{a \in M} a \leq a$ (reflexivity),

(O2) $\bigwedge_{a,b,c \in M} (a \leq b \wedge b \leq c \Rightarrow a \leq c)$ (transitivity),

(O3) $\bigwedge_{a,b \in M} (a \leq b \wedge b \leq a \Rightarrow a = b)$ (antisymmetry).

An ordered set is called *linearly* or *totally* ordered if in addition

(O4) $\bigwedge_{a,b \in M} a \leq b \vee b \leq a.$ ■

$\{M, \leq\}$ being an ordered set just means that there is an order relation defined in $M$. It does not mean that for any two elements $a$, $b \in M$ either $a \leq b$ or $b \leq a$ holds. The latter is valid in linearly ordered sets. If for two elements $a$, $b \in M$ neither $a \leq b$ nor $b \leq a$, then $a$ and $b$ are called incomparable (in notation $a \| b$). Ordered sets are sometimes also called partially ordered sets. Since we shall consider many specially ordered sets, we suppress the modifier, partially, to avoid bulky expressions. This practice is also quite common in the literature.

If $\{M, \leq\}$ is an ordered (resp. linearly ordered) set and $T \subseteq M$, then $\{T, \leq\}$ is also an ordered (resp. linearly ordered) set. To each pair $\{a, b\}$ of elements in an ordered set $\{M, \leq)$, where $a \leq b$, the *interval* $[a, b]$ is defined by

$$[a, b] := \{x \in M \mid a \leq x \leq b\}.$$

A subset $T \subseteq M$ is called *convex* if with any two elements $a$, $b \in T$, the whole interval $[a, b] \in T$. The smallest convex superset of a set is called its *convex hull*.

**Definition 1.2:** A relation $<$ in a set $M$ is called *antireflexive* and the pair $\{M, <\}$ an *antireflexively ordered set* if we have

(AO1) $\bigwedge_{a \in M} \neg(a < a)$ (antireflexivity),

(AO2) $\bigwedge_{a,b,c \in M} (a < b \wedge b < c \Rightarrow a < c)$ (transitivity). ■

The following lemma describes a well-known relation between the orderings of the Definitions 1.1 and 1.2.

**Lemma 1.3:** (a) In an ordered set $\{M, \leq\}$ an antireflexive ordering is defined by

$$\bigwedge_{a,b \in M} a < b :\Leftrightarrow (a \leq b \wedge a \neq b).$$

(b) In an antireflexively ordered set $\{M, <\}$ an order relation is defined by

$$\bigwedge_{a,b \in M} a \leq b :\Leftrightarrow (a < b \vee a = b).$$ ■

The proof of this lemma, being straightforward, is omitted. We note that throughout this book we shall use the sign $<$ in an ordered set $\{M, \leq\}$ and the sign $\leq$ in an antireflexively ordered set $\{N, <\}$ only in the sense of Lemma 1.3. Unfortunately this is not always the case in the literature.

We list a few well-known examples for ordered sets:

1. $\{\boldsymbol{R}, \leq\}$ is a linearly ordered set.[†]

2. If $M$ is a set and $\boldsymbol{P}M$ denotes the power set of $M$, which is defined as the set of all subsets of $M$, then with inclusion as an order relation $\{\boldsymbol{P}M, \subseteq\}$ is an ordered set, and we have the properties

(O1) $\bigwedge_{A \in \boldsymbol{P}M} A \subseteq A,$

(O2) $\bigwedge_{A,B,C \in \boldsymbol{P}M} (A \subseteq B \wedge B \subseteq C \Rightarrow A \subseteq C),$

(O3) $\bigwedge_{A,B \in \boldsymbol{P}M} (A \subseteq B \wedge B \subseteq A \Rightarrow A = B).$

3. If $\{M, \leq\}$ is an ordered set and $M^n$ denotes the product set and

$$x := (x_i) := \begin{pmatrix} x_1 \\ x_2 \\ \vdots \\ x_n \end{pmatrix}, \qquad y := (y_i) := \begin{pmatrix} y_1 \\ y_2 \\ \vdots \\ y_n \end{pmatrix} \in M^n,$$

then by the definition

$$x \leq y :\Leftrightarrow \bigwedge_{i=1(1)n} x_i \leq y_i,$$

$\{M^n, \leq\}$ becomes an ordered set.

In an ordered set $\{M, \leq\}$ an element $a$ is called the *lower neighbor* of an element $b$ if $a < b$ and no other element of $M$ lies between them; i.e., there exists no element $c \in M$ such that $a < c < b$. The concept of lower neighbor can be used in order to draw a figure of any finite ordered set. We just have to assign every element of $M$ to a point of the plane and place $a$ lower than $b$ whenever $a < b$. Then we connect every point to each of its lower neighbors by a straight line segment. The resulting figure is called the *order diagram* of the ordered set $\{M, \leq\}$.

Figure 6 shows the order diagrams of ordered sets of 9 and 16 elements. The set of Fig. 6b consists of two subsets, the respective elements of which are incomparable. An element $x$ of an ordered set $\{M, \leq\}$ is called a *maximal* (resp. *minimal*) *element* if

$$\bigwedge_{a \in M} \neg(x < a) \qquad \left(\text{resp.} \quad \bigwedge_{a \in M} \neg(a < x)\right).$$

[†] $\boldsymbol{R}$ denotes the set of real numbers.

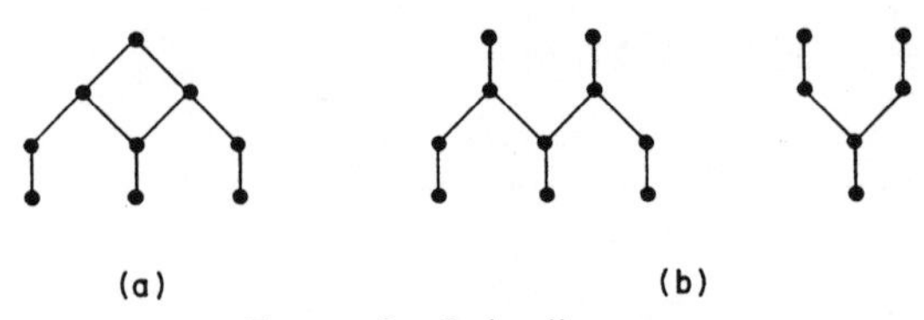


FIGURE 6. Order diagrams.

Every finite ordered set has at least one maximal and one minimal element. In order to see this, let $M = \{a_1, a_2, \ldots, a_n\}$ and set $x_1 = a_1$. Set $x_2 = a_2$ if $a_1 < a_2$, but set $x_2 = x_1$ otherwise. In general, set $x_k = a_k$ if $x_{k-1} < a_k$ and $x_k = x_{k-1}$ otherwise. Then $x_n$ is a maximal element.

An ordered set may have more than one maximal or minimal element. See Fig. 6. An infinite set, however, need have neither a maximal nor a minimal element.

Now we can show that every finite, nonvoid ordered set can be represented by an order diagram. If $M = \{a\}$, this is clear. Let us assume now that the statement is true for sets with $n - 1$ elements, and let $M$ be a set with $n$ elements. Since $M$ consists of a finite number of elements, it has at least one maximal element. Call $T \subseteq M$ the subset of $M$ containing $n - 1$ elements, which is obtained by removing one such maximal element. $T$ has an order diagram. Now we adjoin the element $a$ to the order diagram of $T$ and connect it with all its lower neighbors. The resulting figure is an order diagram of $M$.

Properties of infinite ordered sets may also be illustrated by order diagrams.

In the remainder of this section we define and discuss various constructs associated with ordered sets.

**Definition 1.4:** An element of an ordered set $\{M, \leq\}$ is called the *least element* $o(M)$ (resp. the *greatest element* $i(M)$) if

$$\bigwedge_{a \in M} o(M) \leq a \qquad \left(\text{resp.} \quad \bigwedge_{a \in M} a \leq i(M)\right). \quad \blacksquare$$

That both $o(M)$ and $i(M)$ are well defined is the assertion of the following theorem.

**Theorem 1.5:** An ordered set has at most one least and at most one greatest element.

*Proof*: Suppose that $a$ and $b$ are least elements of $M$. Then $a \leq b \wedge b \leq a \Rightarrow a = b$ by (O3).

The concepts of least and greatest elements and minimal and maximal elements are of course different. They can be illustrated by order diagrams.

See, for instance, Fig. 6. A least element is always minimal and a greatest element always maximal. The converse is not true. See Fig. 6.

The power set $\{\boldsymbol{P}M, \subseteq\}$ of a set $M$ is an important example that contains a least and a greatest element. In particular, $o(\boldsymbol{P}M) = \varnothing$ and $i(\boldsymbol{P}M) = M$.

**Definition 1.6:** Let $\{M, \leq\}$ be a nonvoid ordered set and $T \subseteq M$. An element $a \in M$ is called a *lower* (resp. an *upper*) *bound* of $T$ in $M$ if

$$\bigwedge_{b \in T} a \leq b \qquad \left(\text{resp.} \quad \bigwedge_{b \in T} b \leq a\right).$$

We denote the set of all lower (resp. upper) bounds of $T$ in $M$ by

$$L(T) := \left\{a \in M \mid \bigwedge_{b \in T} a \leq b\right\} \qquad \left(\text{resp.} \quad U(T) := \left\{a \in M \mid \bigwedge_{b \in T} b \leq a\right\}\right).$$

If $T = \{t\}$, we simply write $L(t)$ (resp. $U(t)$) instead of $L(\{t\})$ (resp. $U(\{t\})$). $T$ is called *bounded* in $M$ if $L(T) \neq \varnothing$ and $U(T) \neq \varnothing$. The greatest lower bound (resp. the least upper bound), if it exists, is called the *infimum* (resp. the *supremum*) of $T$. Symbolically

$$\inf T := i(L(T)) \qquad (\text{resp.} \quad \sup T := o(U(T)). \quad \blacksquare$$

Not every subset of an ordered set has an infimum or a supremum nor even lower (resp. upper) bounds. For an illustration see, for instance, the ordered sets drawn in Fig. 6.

According to Theorem 1.5, the infimum (resp. the supremum) of a subset $T \subseteq M$, if it exists, is uniquely determined.

The case $T = \varnothing$ is not excluded in Definition 1.6. Since in the case of the empty set the definition of bounds requires nothing,

$$\bigwedge_{a \in M} \bigwedge_{b \in \varnothing} (a \leq b \wedge b \leq a),$$

every element of $M$ is an upper and a lower bound of the empty subset $\varnothing \subseteq M$.

If an ordered set $\{M, \leq\}$ has a least element $o(M)$ and a greatest element $i(M)$, we have

$$\inf M = o(M), \qquad \sup M = i(M)$$

and

$$\inf \varnothing = i(M), \qquad \sup \varnothing = o(M).$$

If $\varnothing \neq T \subseteq M$, it is always true that $\inf T \leq \sup T$.

If a subset $T$ of an ordered set $\{M, \leq\}$ has an infimum (resp. a supremum) $t$, where in particular $t \in T$, then $t$ is also least and therefore also a minimal (resp. greatest) element and therefore a maximal element of $T$. Conversely, if a subset $T$ of an ordered set has a least (resp. greatest) element $t$, then $t$ is also the infimum (resp. supremum) of $T$ in $M$.

Now let $\{M, \leq\}$ be an ordered set and $T_1$, $T_2$ nonvoid subsets of $M$ each of which is assumed to have an infimum and a supremum in $M$. Then the following properties hold:

(a) $T_1 \subseteq T_2 \Rightarrow \inf T_2 \leq \inf T_1 \wedge \sup T_1 \leq \sup T_2$,

(b) $\bigwedge_{t_1 \in T_1} \bigwedge_{t_2 \in T_2} (t_1 \leq t_2 \Rightarrow t_1 \leq \sup T_1 \leq \inf T_2 \leq t_2)$.

The proofs of these statements, which we forego displaying, consist of a direct use of the preceding definitions and properties.

## 2. COMPLETE LATTICES AND COMPLETE SUBNETS

We begin our discussion of lattices with the following definition.

**Definition 1.7:** Let $\{M, \leq\}$ be an ordered set. Then

(a) $M$ is called a *lattice* if for any two elements $a$, $b \in M$, the $\inf\{a, b\}$ and the $\sup\{a, b\}$ exist;

(b) $M$ is called *conditionally complete* if for every nonempty, bounded subset $T \subseteq M$, the $\inf T$ and the $\sup T$ exist;

(c) $M$ is called *completely ordered* or a *complete lattice* if

(O5) every subset $T \subseteq M$ has an infimum and a supremum. ■

Every finite subset $T = \{a_1, a_2, \ldots, a_n\}$ of a lattice has an infimum and a supremum. We prove this statement by induction. By definition any subset of two elements has an infimum and a supremum. Let us assume now that the assertion is correct for subsets of $n - 1$ elements. Then $\inf\{a_1, a_2, \ldots, a_{n-1}\}$ exists. We consider the element $d := \inf\{\inf\{a_1, a_2, \ldots, a_{n-1}\}, a_n\}$. Then $d$ is obviously a lower bound of $T$. Now let $c$ be any lower bound of $T$. Then $c \leq \inf\{a_1, a_2, \ldots, a_{n-1}\}$ and $c \leq a_n$ and therefore $c \leq d$, i.e., $d$ is the greatest lower bound of $T$, completing the induction. Moreover, we have also observed that $\inf\{a_1, a_2, \ldots, a_n\} = \inf\{\{a_1, a_2, \ldots, a_{n-1}\}, a_n\}$, i.e., that the inf (resp. sup) is associative.

Every finite lattice, therefore, has a least and a greatest element. Every finite lattice, furthermore, is complete (i.e., is a complete lattice) since besides the empty subset, infima and suprema are only to be considered for finite subsets, and we did already observe that the infimum and supremum of the empty subset equal the greatest and the least element, respectively.

Since every lattice is an ordered set, every finite lattice can be represented by an order diagram. The order diagram of a lattice is distinguished by the fact that any two elements are downwardly and upwardly connected with other elements of the set.

Every completely ordered set of course is a lattice. Every complete lattice is conditionally complete. Since the definitions of the infimum and the supremum in an ordered set are completely dual, the following *duality principle* is valid throughout lattice theory:

*If in any lattice theoretic statement or theorem the operations* $\leq$, *inf, sup are replaced by* $\geq$, *sup, inf, respectively, there results a valid lattice theoretic statement or theorem.*

In the definition of a complete lattice, the case $T = M$ is included. Therefore, $\inf M$ and $\sup M$ exist. Since they are elements of $M$, $\inf M$ is the least element and $\sup M$ the greatest element of $M$. Every complete lattice, therefore, has a least element $o(M)$ and a greatest element $i(M)$.

To continue our development, it is convenient to have the following theorem available.

**Theorem 1.8:** Let $\{M, \leq\}$ be an ordered set with the property that every subset has an infimum (or a supremum). Then $\{M, \leq\}$ is a complete lattice.

*Proof*[†]: We are given that every subset $T \subseteq M$ has an infimum, and we show that it has a supremum also. Let $U(T)$ be the set of upper bounds of $T$. (We may suppose that $U(T) \neq \varnothing$ since $\inf \varnothing$ exists by hypothesis and $\inf \varnothing = i(M)$). By assumption the element $u_0 = \inf U(T)$ exists. We show that $u_0 = \sup T$. We have

$$\bigwedge_{t \in T} \bigwedge_{u \in U(T)} t \leq u \Rightarrow \bigwedge_{t \in T} t \leq u_0 = \inf U(T),$$

i.e., $u_0$ is an upper bound of $T$. Now if $u$ is any upper bound of $T$, then $u \in U(T)$ and consequently $u_0 = \inf U(T) \leq u$. $u_0$, therefore, is the least upper bound of $T$, i.e., $u_0 = \sup T = \inf U(T)$. ■

According to Theorem 1.8, in order to verify that an ordered set is a complete lattice, it is sufficient to show that every subset has a greatest lower bound *or* has a least upper bound.

The concepts of a conditionally completely ordered set and a complete lattice are closely related. If a conditionally completely ordered set has a least and a greatest element, it is a complete lattice. The following theorem shows that every conditionally completely ordered set can be completed.

[†] We prove this theorem only for the case that an infimum always exists. The proof in the case of the supremum is dual. In the proofs of many following theorems, we omit the dual case without comment.

**Theorem 1.9:** Let $\{M, \leq\}$ be a conditionally completely ordered set. If we adjoin a least element $o$ and a greatest element $i$ to $M$, then $\{M \cup \{o\} \cup \{i\}, \leq\}$ is a complete lattice.

*Proof*: We show that every subset $T \subseteq M \cup \{o\} \cup \{i\}$ has an infimum. Then by Theorem 1.8 $\{M \cup \{o\} \cup \{i\}, \leq\}$ is a complete lattice.

If $o \in T$, then $o = \inf T$. If $o \notin T$, we consider the set $T' := T \backslash \{i\}$. Then $T' \subseteq M$. If $T' = \varnothing$, then $T = \varnothing$ or $T = \{i\}$. In both cases $i = \inf T$. We assume now $T' \neq \varnothing$. If $T'$ has no lower bound in $M$, then $o = \inf T' = \inf T$. If $T'$ has a lower bound in $M$, we show that $\inf T' = \inf T$. If $T'$ has an upper bound in $M$, then $\inf T'$ exists by hypothesis, and $\inf T' = \inf T$. If $T'$ has no upper bound in $M$, we take any element $c \in T'$ and consider the set $\inf\{\{t, c\} \mid t \in T'\}$. It is bounded in $M$ and therefore has an infimum $j := \inf\{\inf\{t, c\} \mid t \in T'\}$. $j$ is a lower bound of $T'$, and if $l$ is any lower bound of $T'$, we have

$$\bigwedge_{t \in T'} l \leq t \Rightarrow l \leq \{\inf\{t, c\} \mid t \in T'\} \Rightarrow l \leq j,$$

i.e., $j$ is the greatest lower bound of $T'$, and therefore $j = \inf T' = \inf T$. ∎

By Theorem 1.9, every ordered set that is conditionally complete can be made into a complete lattice by adjoining a least and a greatest element. This is a well-known procedure in the case of real numbers $\boldsymbol{R}$. $\{\boldsymbol{R}, \leq\}$ is a conditionally completely ordered set. In real analysis it is shown that every bounded subset of real numbers has an infimum and a supremum. If we adjoin $-\infty$ and $\infty$ to form $\{\boldsymbol{R} \cup \{-\infty\} \cup \{\infty\}, \leq\}$, we obtain a complete lattice with the least element $-\infty$ and the greatest element $\infty$. Similarly, we obtain a complete lattice by adjoining the endpoints to an open interval of real numbers.

The following two theorems provide additional examples of complete lattices.

**Theorem 1.10:** Let $\{M_i, \leq_i\}$, $i = 1(1)n$, be complete lattices. The product set $M = M_1 \times M_2 \times \cdots \times M_n$ is defined as the set of all $n$-tuples $a = (a_1, a_2, \ldots, a_n)$ with $a_i \in M_i$, $i = 1(1)n$. Let $b = (b_1, b_2, \ldots, b_n) \in M$. If we define a relation $\leq$ in $M$ by

$$a \leq b :\Leftrightarrow \bigwedge_{i=1(1)n} a_i \leq_i b_i,$$

then $\{M, \leq\}$ is a complete lattice.

*Proof*: It is clear that $\{M, \leq\}$ is an ordered set. Let $T \subseteq M$ and consider the set of projections $T_j := \{t_j \mid t \in T\}$, where $t = (t_1, t_2, \ldots, t_n)$. Then

$$\bigwedge_{i=1(1)n} \bigwedge_{t_i \in T_i} \inf T_i \leq_i t_i \Rightarrow \bigwedge_{t \in T} (\inf T_1, \inf T_2, \ldots, \inf T_n) \leq t.$$

This means that $(\inf T_i) := (\inf T_1, \inf T_2, \ldots, \inf T_n)$ is a lower bound of $T$. We must show that it is the greatest lower bound. Let $c := (c_i) := (c_1, c_2, \ldots, c_n)$ be any lower bound of $T$. Then

$$\bigwedge_{i=1(1)n} \bigwedge_{t_i \in T_i} c_i \leq_i t_i \Rightarrow \bigwedge_{i=1(1)n} c_i \leq_i \inf T_i \Rightarrow (c_i) \leq (\inf T_i),$$

i.e., $(\inf T_i)$ is the greatest lower bound of $T$. This means $\inf T = (\inf T_i)$. By duality we get $\sup T = (\sup T_i)$. ■

**Theorem 1.11:** The power set of a set $M$ with inclusion $\subseteq$ as an order relation, $\{\boldsymbol{P}M, \subseteq\}$, is a complete lattice. For each subset of $\boldsymbol{P}M$, the infimum is the intersection of the elements of the subset, and the supremum is the union.

*Proof*: $\{\boldsymbol{P}M, \subseteq\}$ is an ordered set with the least element $o(\boldsymbol{P}M) = \varnothing$ and the greatest element $i(\boldsymbol{P}M) = M$. Let $\boldsymbol{A} \subseteq \boldsymbol{P}M$ and consider the sets of all lower bounds and upper bounds of $\boldsymbol{A}$:

$$L(\boldsymbol{A}) := \left\{ B \in \boldsymbol{P}M \;\middle|\; \bigwedge_{A \in \boldsymbol{A}} B \subseteq A \right\},$$

$$U(\boldsymbol{A}) := \left\{ B \in \boldsymbol{P}M \;\middle|\; \bigwedge_{A \in \boldsymbol{A}} A \subseteq B \right\}.$$

Then $\inf \boldsymbol{A}$ and $\sup \boldsymbol{A}$ are, if they exist, defined by

$$\inf \boldsymbol{A} := i(L(\boldsymbol{A})) \qquad \text{and} \qquad \sup \boldsymbol{A} := o(U(\boldsymbol{A})).$$

Now let $X, Y \in \boldsymbol{P}M$ have the following properties:

$$X := \left\{ a \in M \;\middle|\; \bigwedge_{A \in \boldsymbol{A}} a \in A \right\} \qquad \text{(intersection)},$$

$$Y := \left\{ a \in M \;\middle|\; \bigvee_{A \in \boldsymbol{A}} a \in A \right\} \qquad \text{(union)}.$$

We show that $X = \inf \boldsymbol{A}$ and $Y = \sup \boldsymbol{A}$. By the definition of a subset and the definition of $X$, we get

$$\bigwedge_{A \in \boldsymbol{A}} X \subseteq A,$$

i.e., $X$ is lower bound of $\boldsymbol{A}$, $X \in L(\boldsymbol{A})$. We still have to show that $X$ is the greatest lower bound. Let $C \in \boldsymbol{P}M$ be any other lower bound of $\boldsymbol{A}$. Then

$$\bigwedge_{A \in \boldsymbol{A}} C \subseteq A,$$

or by the definition of a subset,

$$\bigwedge_{c \in C} \bigwedge_{A \in \boldsymbol{A}} c \in A.$$

Therefore, $C \subseteq X$, i.e., $X$ is the greatest lower bound, $X = \inf \boldsymbol{A}$. By duality we obtain $\sup \boldsymbol{A} = Y$. ■

As an illustration, Fig. 7 shows the order diagram of the power set $\boldsymbol{P}M$ of the finite set $M := \{a, b, c\}$.

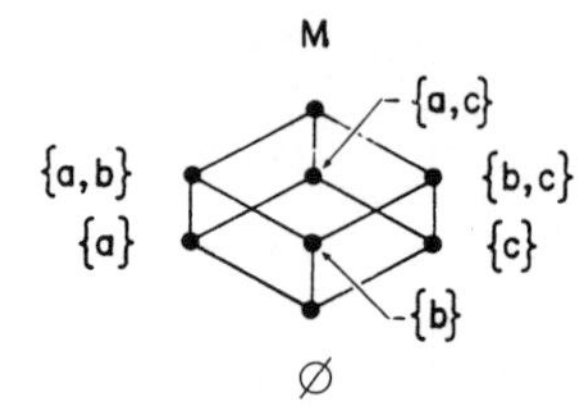


FIGURE 7. Order diagram of the power set of the set $M = \{a, b, c\}$.

Now let $\{M, \leq\}$ be a lattice and $T \subseteq M$. Then $\{T, \leq\}$ is an ordered set. $(T, \leq\}$ may also be a lattice.

Let us consider the example represented in Fig. 8.

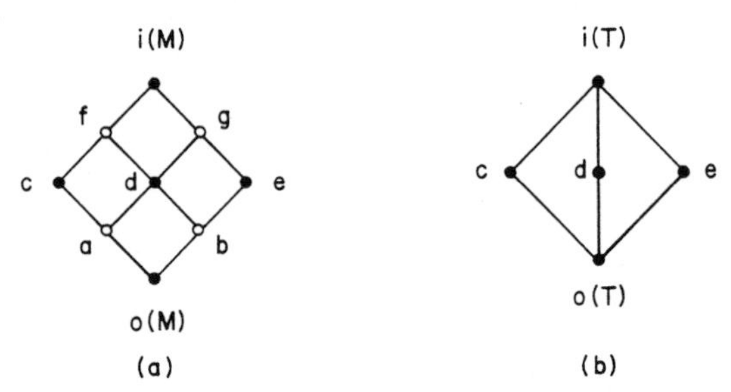


FIGURE 8. Example for the concept of subnet

The ordered set $\{M, \leq\}$ of the nine points in Fig. 8a obviously represents a lattice. The subset $T \subseteq M$ of the solid points in Fig. 8a with respect to the same order relation also represents a lattice, Fig. 8b. In general, however, the infimum and supremum taken in $\{T, \leq\}$ are different from those taken in $\{M, \leq\}$. For example, in $\{M, \leq\}$, $\sup\{c, d\} = f$, while in $\{T, \leq\}$, $\sup\{c, d\} = i(T)$. This leads us to the following definition.

**Definition 1.12:** Let $\{M, \leq\}$ be an ordered set and $T \subseteq M$. If $\{T, \leq\}$ is a lattice, it is called a *subnet* of $\{M, \leq\}$. A subnet is called an *inf-subnet* (resp. a *sup-subnet*) if

$$\bigwedge_{a,b \in T} \inf_M\{a, b\} = \inf_T\{a, b\} \qquad \left(\text{resp.} \quad \bigwedge_{a,b \in T} \sup_M\{a, b\} = \sup_T\{a, b\}\right),$$

where the subscripts $M$ and $T$ indicate the set in which the infimum (resp. supremum) is taken. A subnet of a lattice is called a *sublattice* if it is both an inf-subnet and a sup-subnet. ■

Similar properties hold in complete lattices. Therefore, we give the following definition.

**Definition 1.13:** Let $\{M, \leq\}$ be a complete lattice and $T \subseteq M$. If $\{T, \leq\}$ is also a complete lattice, it is called a *complete subnet* of $\{M, \leq\}$. A complete subnet is called a *complete inf-subnet* (resp. a *complete sup-subnet*) if

$$\bigwedge_{A \subseteq T} \inf_M A = \inf_T A \qquad \left(\text{resp.} \quad \bigwedge_{A \subseteq T} \sup_M A = \sup_T A\right).^{\dagger}$$

A complete subnet is called a *complete sublattice* if it is both a complete inf-subnet and a complete sup-subnet. ■

This definition leads directly to the following theorem and remark.

**Theorem 1.14:** Let $\{M, \leq\}$ be a complete lattice and $\{T, \leq\}$ a complete subnet. Then

$$\bigwedge_{A \subseteq T} \inf_T A \leq \inf_M A \wedge \sup_M A \leq \sup_T A. \quad ■$$

**Remark 1.15:** In a complete lattice $\{M, \leq\}$, the infimum and supremum also exist for the empty set $\varnothing$. Therefore we have

(a) The greatest element of $M$ is equal to the greatest element of every complete inf-subnet, i.e., $i(M) = \inf \varnothing = i(T)$.

(b) The least element of $M$ is equal to the least element of every complete sup-subnet, i.e., $o(M) = \sup \varnothing = o(T)$.

(c) In a complete sublattice $T$ of a complete lattice $M$, the least and the greatest elements of $M$ and $T$ are identical:

$$o(M) = o(T) = \sup \varnothing \wedge i(M) = i(T) = \inf \varnothing. \quad ■$$

In the definition of a complete inf-subnet and a complete sup-subnet, it is presumed that $T$ is a complete lattice. The following theorem enables us to eliminate this requirement.

**Theorem 1.16:** Let $\{M, \leq\}$ be a complete lattice and $T \subseteq M$. $\{T, \leq\}$ is a complete inf-subnet (resp. a complete sup-subnet) of $\{M, \leq\}$ if and only if

$$\bigwedge_{A \subseteq T} \inf_M A \in T \qquad \left(\text{resp.} \quad \bigwedge_{A \subseteq T} \sup_M A \in T\right).$$

† In the cases of equality the subscripts $M$ and $T$ can be omitted.

*Proof*: If $\{T, \leq\}$ is a complete inf-subnet, we have $\inf_M A = \inf_T A$. Since $\inf_T A \in T$, this demonstrates the necessity of the hypothesis. We now demonstrate the sufficiency. Let $A \subseteq T$ and $I := \inf_M A \in T$. Then for all $a \in A, I \leq a$, i.e., $I$ is lower bound of $A$ in $T$. Moreover,

$$\bigwedge_{x \in M} \left( \bigwedge_{a \in A} x \leq a \Rightarrow x \leq I \right) \Rightarrow \bigwedge_{x \in T} \left( \bigwedge_{a \in A} x \leq a \Rightarrow x \leq I \right),$$

i.e., $I$ is the greatest lower bound of $A$ in $\{T, \leq\}$. This implies that $\inf_M A = \inf_T A$. Since every subset $A \subseteq T$ has an infimum in $T$, Theorem 1.8 implies that $\{T, \leq\}$ is a complete lattice. ■

The following corollary is a direct consequence of Theorem 1.16.

**Corollary 1.17:** Let $\{M, \leq\}$ be a complete lattice and $T \subseteq M$. $\{T, \leq\}$ is a complete sublattice of $\{M, \leq\}$ if and only if

$$\bigwedge_{A \subseteq T} (\inf_M A \in T \wedge \sup_M A \in T). \quad ■$$

We illustrate these concepts with two simple examples.

1. Let $\{M, \leq\}$ be an ordered set, $a \in M$ and $T := \{t \in M | t \leq a\}$ (resp. $T := \{t \in M | a \leq t\}$). Then we have

(a) If $\{M, \leq\}$ is a lattice, then $\{T, \leq\}$ is a sublattice.
(b) If $\{M, \leq\}$ is a complete lattice and $\bar{T} = T \cup \{i(M)\}$, then $\{\bar{T}, \leq\}$ is a complete sublattice.

*Proof*: (a) Let $x, y \in T$. Then $x \leq a \wedge y \leq a$ and therefore $\inf_M\{x, y\} \leq \sup_M\{x, y\} \leq a$, i.e., $\inf_M\{x, y\}, \sup_M\{x, y\} \in T$, which proves the assertion by Corollary 1.17.

(b) Let $\varnothing \neq S \subseteq T$. Then for all $s \in S, s \leq a$, and therefore $\inf_M S \leq \sup_M S \leq a$, i.e., $\inf_M S, \sup_M S \in T$. But $\varnothing \subseteq T$ also. Then in order to apply Corollary 1.17, $\inf_M \varnothing = i(M)$ and $\sup_M \varnothing = o(M)$ must be elements of $\bar{T}$. Since $o(M) \leq a$, it is automatically an element of $T$ and therefore of $\bar{T}$. That $i(M) \in \bar{T}$, however, has been assumed explicitly.

2. Let $\{M, \leq\}$ be an ordered set and $T := \{t \in M | t \in [a, b]\}$, where $[a, b]$ denotes an interval in $M$. Then we have

(a) If $\{M, \leq\}$ is a lattice, then $\{T, \leq\}$ is a sublattice.
(b) If $\{M, \leq\}$ is a complete lattice, then $\{T \cup \{o(M)\} \cup \{i(M)\}, \leq\}$ is a complete sublattice.

The proofs of these statements are analogous to the proofs for example 1.

## 3. SCREENS AND ROUNDINGS

We are now going to give an abstract characterization of the essential properties of the sets and subsets displayed in Fig. 1, which are essential for our description of computer arithmetic. To motivate the next definition, let us consider two simple examples from Fig. 1. We recall that all sets listed in Fig. 1 are ordered with respect to certain order relations.

Consider the set $\boldsymbol{I}V_2\boldsymbol{R}$ of interval vectors of dimension 2. The elements of this set are intervals of two-dimensional real vectors. Such an element describes a rectangle in the $x,y$ plane with sides parallel to the axis. Such interval vectors are special elements of the power set $\boldsymbol{P}V_2\boldsymbol{R}$ of real 2-vectors, which is defined as the set of all subsets of real 2-vectors. The following relations hold between these two sets:

1. For each element $a \in \boldsymbol{R}$, there exist upper bounds in $S$ with respect $\boldsymbol{I}V_2\boldsymbol{R}$ with respect to the other relation $\subseteq$. See Fig. 9.

2. For all $a \in \boldsymbol{R}$, the set of upper bounds in the subset $S$ has a least element. least element. See Fig. 9, where this least element is called $c$.

We shall see that these two properties describe the relationship between any set in Fig. 1 and its subsets as listed on its right.

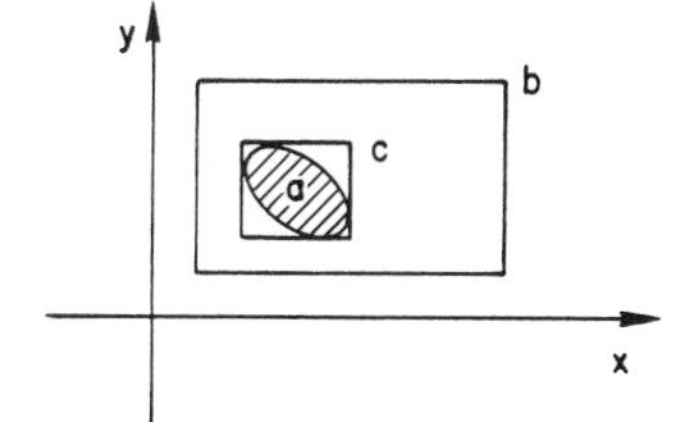


FIGURE 9. Illustration of the concept of a screen $a \in \boldsymbol{P}V_2\boldsymbol{R}$; $b, c, \in \boldsymbol{I}V_2\boldsymbol{R}$; $a \subseteq b \wedge a \subseteq c \wedge \bigwedge_{d \in \boldsymbol{I}V_2\boldsymbol{R}}(a \subseteq d \Rightarrow c \subseteq d)$.

In particular, consider the first row of Fig. 1. For $\boldsymbol{R}$ and the subset $D$ taken to be floating-point numbers, we obtain the above two properties once again:

1. For each element $a \in \boldsymbol{R}$, there exist upper bounds in $S$ with respect to the order relation $\leq$. See Fig. 10.

2. For all $a \in \boldsymbol{R}$, the set of upper bounds in the subset $S$ has a least element. See Fig. 10.

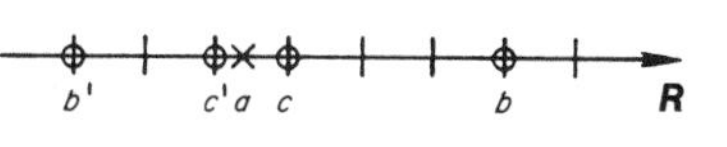


FIGURE 10. Illustration of the concept of a screen. $a \in \boldsymbol{R}$; $b, c \in S$; $a \leq b \wedge a \leq c \wedge \bigwedge_{d \in S}(a \leq d \Rightarrow c \leq d)$.

In the case of this example corresponding properties also hold for the set of lower bounds.

These properties motivate the concept of a screen, which is formalized in the following definition.

**Definition 1.18:** Let $\{M, \leq\}$ be an ordered set. For each element $a \in M$, let $L(a)$ (resp. $U(a)$) be the set of lower (resp. upper) bounds of $a$. A subset $S \subseteq M$ is called a *screen* of $M$ if it fulfills the properties

(S1) $\bigwedge_{a \in M} L(a) \cap S \neq \varnothing \wedge \bigwedge_{a \in M} U(a) \cap S \neq \varnothing$,

(S2) $\bigwedge_{a \in M} \bigvee_{x \in L(a) \cap S} \bigwedge_{b \in L(a) \cap S} b \leq x \wedge \bigwedge_{a \in M} \bigvee_{y \in U(a) \cap S} \bigwedge_{b \in U(a) \cap S} y \leq b$,

i.e., the set $L(a) \cap S$ has a greatest element $x = i(L(a) \cap S)$, and $U(a) \cap S$ has a least element $y = o(U(a) \cap S)$.

If only the left-hand-side properties of (S1) and (S2) hold, then $S$ is called a *lower semiscreen*, and if only the right-hand-side properties hold, $S$ is called an *upper semiscreen*.† Usually we shall write $\{S, \leq\}$ to denote the screen or semiscreen in order to emphasize the ordering. ■

Since in a screen of an ordered set the elements $i(L(a) \cap S)$ and $o(U(a) \cap S)$ always exist, we can define mappings $\varphi : M \to S$ and $\psi : M \to S$ by

(R) $\bigwedge_{a \in M} \varphi a := i(L(a) \cap S) \wedge \bigwedge_{a \in M} \psi a := o(U(a) \cap S)$.

These mappings have properties described in the following lemma.

**Lemma 1.19:** Let $\{M, \leq\}$ be an ordered set and $\{S, \leq\}$ a screen of $M$. The mappings $\varphi$ and $\psi$ defined by (R) have the following properties:

(R1) $\bigwedge_{a \in S} \varphi a = a$, $\quad \bigwedge_{a \in S} \psi a = a$,

(R2) $\bigwedge_{a,b \in M} (a \leq b \Rightarrow \varphi a \leq \varphi b)$, $\quad \bigwedge_{a,b \in M} (a \leq b \Rightarrow \psi a \leq \psi b)$,

(R3) $\bigwedge_{a \in M} \varphi a \leq a$, $\quad \bigwedge_{a \in M} a \leq \psi a$.

If $S$ is a lower (resp. an upper) screen of $M$, then only the function $\varphi$ (resp. $\psi$) can be defined. Then the properties (R1), (R2), (R3) can be demonstrated only for $\varphi$ (resp. $\psi$).

† Screens and upper semiscreens play a central role in the description of computer arithmetic. For the sake of conciseness of expression, we shall, therefore, often speak of an upper screen instead of an upper semiscreen.

*Proof*:

(R1): If $a \in S$, then $a \in U(a) \cap S$, and therefore $\psi a = o(U(a) \cap S) = a$.

(R2): $a \leq b \Rightarrow U(b) \cap S \subseteq U(a) \cap S$.

$$\Rightarrow \psi a = o(U(a) \cap S) \leq o(U(b) \cap S) = \psi b.$$

(R3): $\psi a := o(U(a) \cap S) \in U(a) \cap S$. Therefore $a \leq \psi a$.

The proof of the properties for $\varphi$ is dual. ■

The following theorem expresses a relationship between screens and subnets.

**Theorem 1.20:** Let $\{M, \leq\}$ be a complete lattice. A subset $S \subseteq M$ is a lower screen (resp. an upper screen; resp. a screen) of $M$ if and only if it is a complete sup-subnet (resp. a complete inf-subnet; resp. a complete sublattice).

*Proof*: We prove this theorem only for the case of an upper screen. The rest of the proof is analogous.

(a) First we show that (S1), (S2) $\Rightarrow \{S, \leq\}$ is a complete inf-subnet. Choosing $a = i(M)$ and using (S1) ($U(a) \cap S \neq \emptyset$), we have that $i(M) = i(S)$.

Because of (S2), we can define the mapping $\psi : M \to S$ (which is the subject of Lemma 1.19) with the property $\psi a = o(U(a) \cap S)$. Let $B \subseteq S$. Since $B \subseteq M$ as well, the hypothesis implies that there exists the element $x := \inf_M B$, and we have

$$\bigwedge_{b \in B} (x \leq b \underset{(R2)}{\Rightarrow} \psi x \leq \psi b).$$

On the other hand, $B \subseteq S \Rightarrow \psi b = b$ because of (R1), i.e., $\psi x$ is lower bound of $B$. Therefore we have $\psi x \leq \inf_M B = x$. But by (R3), $x \leq \psi x$. From both inequalities we get by (O3) $x = \psi x$. Since $\psi : M \to S$, then $x = \inf_M B \in S$. Therefore by Theorem 1.16, $\{S, \leq\}$ is a complete inf-subnet of $\{M, \leq\}$, and for all $B \subseteq S$, $\inf_M B = \inf_S B$.

(b) Now we show that (S1) and (S2) hold in any complete inf-subnet $S \subseteq M$. In Remark 1.15 we observed that in a complete inf-subnet $i(M) = i(S) = \inf \emptyset$. Therefore (S1) holds.

Further, for all $A \subseteq S$ we have by hypothesis that $\inf_S A = \inf_M A$. Therefore for all $a \in M$,

$$\inf(U(a) \cap S) \in S. \tag{1}$$

Moreover, for all $b \in U(a) \cap S$, $a \leq b$, and therefore also

$$a \leq \inf(U(a) \cap S). \tag{2}$$

By (1) and (2) we get $\inf(U(a) \cap S) \in U(a) \cap S$, i.e., $\inf(U(a) \cap S)$ is the least element of $U(a) \cap S$ since it is an element of this set itself. We have, therefore, $\inf(U(a) \cap S) = o(U(a) \cap S) \in U(a) \cap S$, which completes the proof of the theorem. ■

In an upper screen of a complete lattice, (S2) requires that the set of upper bounds of an element $a \in M$ has a least element, which itself is an upper bound of $a$. The example given in Fig. 11 shows that the corresponding property for the lower bounds of $a$ is not necessarily valid.

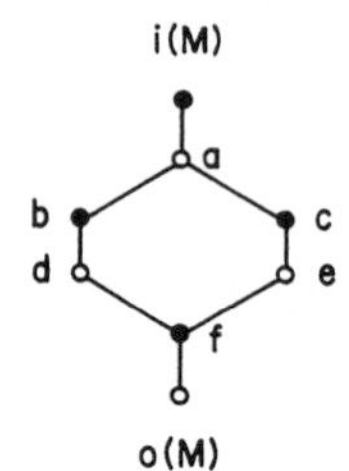


FIGURE 11. Illustration of an upper screen.

Let $\{M, \leq\}$ be the lattice represented by Fig. 11 and $\{S, \leq\}$ the subnet of solid points in the figure, i.e., $S = \{i, b, c, f\}$. $\{S, \leq\}$ is obviously an upper screen of $\{M, \leq\}$ since $i(M) = i(S)$ and for all $A \subseteq S$, $\inf_M A = \inf_S A$.

$\{S, \leq\}$, however, is not a lower screen of $M$. For instance, $o(M) \neq o(S) = f$ and $\sup_M\{b, c\} = a \neq \sup_S\{b, c\} = i$. By Theorem 1.20, for all $x \in M$, the set of upper bounds $U(x) \cap S$ has a least element $\inf(U(x) \cap S)$ with the property $x \leq \inf(U(x) \cap S)$. On the contrary, in general the set $L(x) \cap S$ has no greatest element nor $\sup_S(L(x) \cap S) \leq x \vee \sup_S(L(x) \cap S) \in L(x) \cap S$. In the example given in Fig. 11, we have for $x = a$, for instance,

$$L(a) \cap S = \{b, c, f\} \wedge a < \sup_S(L(a) \cap S) = i(M) \notin L(a) \cap S.$$

This example shows furthermore that on an upper screen the least elements $o(M)$ and $o(S)$ are not necessarily the same.

Now let $S$ be a complete inf-subnet (resp. a complete sup-subnet) of a complete lattice $M$. For all $A \subseteq S$, $\inf_M A = \inf_S A$ (resp. $\sup_M A = \sup_S A$), while the suprema (resp. the infima) differ. See Theorem 1.14. The difference is eliminated by employing the function $\varphi$ (resp. $\psi$) introduced in (R). This property is the subject of the following theorem.

**Theorem 1.21:** Let $\{M, \leq\}$ be a complete lattice and $\{S, \leq\}$ a lower (resp. an upper) screen (or a screen) of $M$ and $\varphi: M \to S$ (and $\psi: M \to S$) defined by

$$\text{(R)} \quad \bigwedge_{a \in M} \varphi a := \sup(L(a) \cap S) \qquad \left(\text{resp.} \quad \bigwedge_{a \in M} \psi a := \inf(U(a) \cap S\right).$$

Then

$$\bigwedge_{A \subseteq S} \inf_S A = \varphi(\inf_M A) \qquad \left(\text{resp.} \quad \bigwedge_{A \subseteq S} \sup_S A = \psi(\sup_M A)\right).$$

*Proof*:

$$\bigwedge_{a \in A \subseteq S} a \leq \sup_M A \underset{(R2)}{\Rightarrow} \bigwedge_{a \in A} \psi a \underset{(R1)}{=} a \leq \psi(\sup_M A) \in S.$$

Therefore

$$\sup_S A \leq \psi(\sup_M A). \tag{3}$$

By Theorem 1.14, however, in a complete subnet we have generally

$$\bigwedge_{A \subseteq S} \sup_M A \leq \sup_S A \underset{(R1),(R2)}{\Rightarrow} \psi(\sup_M A) \leq \sup_S A. \tag{4}$$

Applying (O3) to (3) and (4) proves the theorem. ■

In order to illustrate the definition and theorems given in this chapter, let us consider a few examples.

**Examples:**

1. Let $\boldsymbol{C}$ be the set of complex numbers, and for $r$ a positive real number, let $Z := \{\zeta = \xi + i\eta \in \boldsymbol{C} \,|\, |\xi| \leq r \wedge |\eta| \leq r\}$. Let $M := \boldsymbol{P}Z$, the power set of $Z$. Then $\{M, \subseteq\}$ is a complete lattice. Further, let $S \subseteq M$ be the set of all rectangles of $Z$ with sides parallel to the axes. Also let $\varnothing \in S$. We show that $\{S, \subseteq\}$ is an upper screen of $\{M, \subseteq\}$.

In order to see this, by Theorem 1.16 we have only to show that for every subset $A \subseteq S$, the intersection, which is the infimum in $\{M, \subseteq\}$, is an element of $S$. If $A = \varnothing$, we have $\inf_M \varnothing = i(M) = i(S) = Z$. If $\varnothing \neq A \subseteq S$, it is easy to see by Fig. 12a that the intersection of the elements of $A$ is again an element of $S$. The propertires (S1) and (S2) are illustrated in Figs. 12b and c.

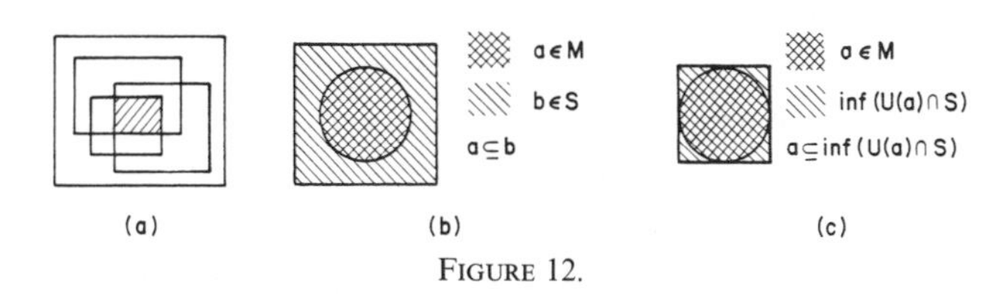


FIGURE 12.

Further, we have $o(M) = o(S) = \varnothing$. Also the condition (S1) for a lower screen holds. See Fig. 13a. The necessary and sufficient criterion for a lower screen given in Theorem 1.16, however, is not valid. For example, in Fig. 13b we have $\sup_M\{a, b\} = a \cup b \subset \sup_S\{a, b\} = c$. Therefore (S2) is not fulfilled either. In Fig. 13c for instance, $\sup_S(L(a) \cap S) = \inf_S(U(a) \cap S) \supseteq a$. Figure 13b further illustrates Theorem 1.21. We have $\sup_S\{a, b\} = \psi(\sup_M\{a, b\})$.

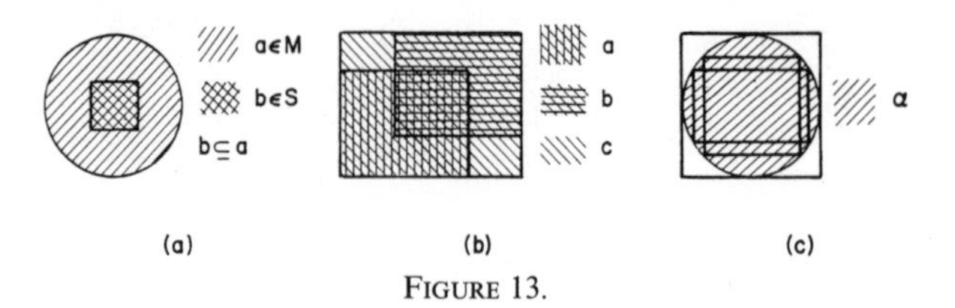


FIGURE 13.

2. Let $\boldsymbol{R}$ be the set of real numbers, $X := \{x \in \boldsymbol{R} \mid |x| \le s\}$, and $M := \boldsymbol{P}X$, the power set of $X$. Then $\{M, \subseteq\}$ is a complete lattice. For all $a_1, a_2 \in X$ with $a_1 \le a_2$, let $\boldsymbol{I}X := \{[a_1, a_2] \mid a_1, a_2 \in X\}$ be the set of intervals over $X$. Also let $\varnothing \in \boldsymbol{I}X$. Then $\boldsymbol{I}X \subset \boldsymbol{P}X$ and $\{\boldsymbol{I}X, \subseteq\}$ is an upper screen of $\{\boldsymbol{P}X, \subseteq\}$. This example just represents the reduction of example 1 to dimension one. All properties can be proved similarly.

3. Again let $X := \{x \in \boldsymbol{R} \mid |x| \le s\}$. Then $\{X, \le\}$ is a linearly ordered, complete lattice. If $S$ is a finite subset of $X$ with the property $-s \in S$ and $s \in S$, then $\{S, \le\}$ is a complete lattice. Then it is easy to see by Theorem 1.16 or by Definition 1.18 and Theorem 1.20 that $\{S, \le\}$ is a screen of $\{X, \le\}$.

4. As in example 2, let $\boldsymbol{I}X$ again denote the set of intervals over $X$ augmented by the empty set and $S$ the subset of $X$ defined in example 3. Now let $\boldsymbol{I}S$ denote the set of intervals over $X$ with endpoints in $S$ and augmented by the empty set. In example 2 we saw that $\{\boldsymbol{I}X, \subseteq\}$ is an upper screen of $\{\boldsymbol{P}X, \subseteq\}$. Now we show that $\{\boldsymbol{I}S, \subseteq\}$ is a screen of $\{\boldsymbol{I}X, \subseteq\}$. $\{\boldsymbol{I}S, \subseteq\}$ is a complete lattice with the property $o(\boldsymbol{I}S) = o(\boldsymbol{I}X) = \varnothing$ and $i(\boldsymbol{I}S) = i(\boldsymbol{I}X) = X$. The infimum (resp. the supremum) in $\boldsymbol{I}S$ is the intersection (resp. the convex hull) as in $\boldsymbol{I}X$. Therefore, for every subset of intervals in $\boldsymbol{I}S$ the infimum and supremum is the same as in $\boldsymbol{I}X$.

We shall see later that the concepts of a screen and of an upper screen describe all essential properties of the sets listed in Fig. 1. Using these concepts, we now specify the notion of rounding.

**Definition 1.22:** Let $M$ be a set and $S \subseteq M$. A mapping $\square : M \to S$ is called a *rounding* if it has the property

(R1) $\bigwedge_{a \in S} \square a = a$ (projection).

A rounding of a complete lattice into a lower (resp. an upper) screen (resp. a screen) is called *monotone* if

(R2) $\bigwedge_{a,b \in M} (a \le b \Rightarrow \square a \le \square b)$ (monotone).

It is called *downwardly* (resp. *upwardly*) *directed* if

(R3) $\bigwedge_{a \in M} \square a \le a$ $\left(\text{resp. } \bigwedge_{a \in M} a \le \square a\right)$ (directed). ■

A general property of monotone mappings, applied to roundings, is described in the following lemma.

**Lemma 1.23:** Let $\{M, \leq\}$ be a complete lattice and $\{S, \leq\}$ a lower (resp. an upper) screen (resp. a screen) of $M$ and $\square: M \to S$ a monotone rounding. Then for all $a \in S$, the set $\square^{-1}a \subseteq M$ is convex.

*Proof*: Let $a_1, a_2 \in \square^{-1}a \subseteq M$. Let $b \in M$ and have the property $a_1 \leq b \leq a_2$. Then by (R2), we get $\square a_1 = \square a \leq \square b \leq \square a_2 = \square a$, and by (O3), $\square b = \square a$, i.e., $b \in \square^{-1}a$. ■

In Definition 1.22 three properties of roundings are enumerated, and we may ask if there do indeed exist roundings with all three of these properties. A characterization of such roundings, which supplies an affirmative answer as well, is the subject of the following theorem.

**Theorem 1.24:** Let $\{M, \leq\}$ be a complete lattice and $\{S, \leq\}$ a lower screen (resp. an upper screen). A mapping $\square: M \to S$ is a monotone downwardly directed (resp. a monotone upwardly directed) rounding if and only if it has the property

$$\text{(R)} \quad \bigwedge_{a \in M} \square a = \sup(L(a) \cap S) \qquad \left(\text{resp.} \quad \bigwedge_{a \in M} \square a = \inf(U(a) \cap S)\right).$$

*Proof*: (a) It is shown in Lemma 1.19 that the mapping defined by (R) has the properties (R1), (R2), and (R3).

(b) We still have to show that a mapping of a complete lattice into a lower or upper screen which has the properties (R1), (R2), and (R3) also fulfills (R). By (R3), $\square a \in L(a) \cap S$. Since $\{S, \leq\}$ is a lower screen, it is a complete sup-subnet of $\{M, \leq\}$. Then there exists the element $\sup(L(a) \cap S) \in S \subseteq M$. Let $b$ be any element of $L(a) \cap S$. Then $b \leq a$ and by (R1) and (R2), $\square b = b \leq \square a$, i.e., $\square a$ is the greatest element of $L(a) \cap S$. Since the greatest element of a set is always its supremum, $\square a = \sup(L(a) \cap S)$. ■

The following remark characterizes the uniqueness property expressed in this theorem.

**Remark:** Since the infimum and the supremum of a subset of a complete lattice are unique, there exists only one monotone downwardly directed rounding (resp. only one monotone upwardly directed rounding) of a complete lattice into a lower (resp. an upper) screen. We shall therefore often use the special symbols $\triangledown$ (resp. $\triangle$) to denote these mappings. They have the property

$$\text{(R)} \quad \bigwedge_{a \in M} \triangledown a = \sup(L(a) \cap S) \qquad \left(\text{resp.} \quad \bigwedge_{a \in M} \triangle a = \inf(U(a) \cap S)\right).$$

These observations are collected into the statement of the following corollary.

**Corollary 1.25:** If $\{M, \leq\}$ is a complete lattice and $\{S, \leq\}$ a screen, then there exists exactly one monotone downwardly directed rounding $\bigtriangledown: M \to S$ and exactly one monotone upwardly directed rounding $\bigtriangleup: M \to S$. These roundings can be defined by property (R), cf. Theorem 1.24. ■

The two montone directed roundings $\bigtriangledown$ and $\bigtriangleup$ are key elements in the set of all roundings. The composition $\bigtriangleup_2 \bigtriangleup_1$ of two such roundings (i.e., a mapping from $M$ into a screen $D \subset M$ followed by a mapping of $D$ into a screen $S \subset D$) is itself a monotone upwardly directed rounding. In a linearly ordered complete lattice every monotone rounding into a screen can be expressed in terms of the two monotone directed roundings $\bigtriangledown$ and $\bigtriangleup$. These properties are made precise by the following two theorems.

**Theorem 1.26:** Let $\{M, \leq\}$ be a complete lattice and let $\{S, \leq\}$ and $\{D, \leq\}$ both be lower (resp. upper) screens of $\{M, \leq\}$ with the property $S \subset D \subseteq M$. Further, let $\bigtriangledown: M \to S$, $\bigtriangledown_1: M \to D$, $\bigtriangledown_2: D \to S$ (resp. $\bigtriangleup: M \to S$, $\bigtriangleup_1: M \to D$, $\bigtriangleup_2: D \to S$) be the associated monotone downwardly (resp. upwardly) directed roundings. Then

$$\bigwedge_{a \in M} \bigtriangledown a = \bigtriangledown_2(\bigtriangledown_1 a) \qquad \left(\text{resp.} \quad \bigwedge_{a \in M} \bigtriangleup a = \bigtriangleup_2(\bigtriangleup_1 a)\right).$$

*Proof*: $S \subset D \Rightarrow L(a) \cap S \subseteq L(a) \cap D \Rightarrow \bigtriangledown a = \sup(L(a) \cap S) < \bigtriangledown_1 a = \sup(L(a) \cap D)$. If we apply the mapping $\bigtriangledown_2$ to this inequality, we obtain

$$\bigtriangledown_2(\bigtriangledown a) \underset{(R1)}{=} \bigtriangledown a \underset{(R2)}{\leq} \bigtriangledown_2(\bigtriangledown_1 a). \tag{5}$$

By (R3) we have $\bigtriangledown_2(\bigtriangledown_1 a) \leq a$, and therefore

$$\bigtriangledown(\bigtriangledown_2(\bigtriangledown_1 a)) \underset{(R1)}{=} \bigtriangledown_2(\bigtriangledown_1 a) \underset{(R2)}{\leq} \bigtriangledown a. \tag{6}$$

From (5) and (6) we get $\bigtriangledown a = \bigtriangledown_2(\bigtriangledown_1 a)$ by (O3). ■

In practical applications of this theorem, $S$ may be a floating-point system and $D$ the set of floating-point numbers of double length on the same computer. We show, however, by means of a simple example that the statement of Theorem 1.26 generally does not hold for monotone but undirected roundings of a complete lattice into a screen.

**Example:** Let $M := [0, 8] \subset \mathbf{R}, S := \{0, 4, 8\}$, and $D := \{0, 1, 2, 3, 4, 5, 6, 7, 8\}$. Let the roundings $\square: M \to S$, $\square_1: M \to D$, and $\square_2: D \to S$ be defined as mappings to the nearest element of the screen, where we additionally assume that the midpoint of two neighboring screenpoints will be mapped onto the lower neighbor in all cases. Then we get for $a = 2.4 \in M$, $\Box_1 a = 2$ and $\Box_2(\Box_1 a) = 0 \neq \Box a = 4$.

In linearly ordered complete lattices we find that in addition to the monotone directed roundings, all monotone roundings play an important role. We discuss a few of their properties, beginning with the following lemma.

**Lemma 1.27:** Let $\{M, \leq\}$ be a linearly ordered complete lattice and $\{S, \leq\}$ be a screen of $\{M, \leq\}$, and let $\bigtriangledown : M \to S$ and $\triangle : M \to S$ be the monotone directed roundings. Then for all $a \in M$, there exists no element $b \in S$ with the property $\bigtriangledown a < b < \triangle a$.

*Proof*: We have $\bigtriangledown a = i(L(a) \cap S)$ and $\triangle a = o(U(a) \cap S)$. Suppose that the conclusion of the lemma is false. Then using (O4), we find that for any $a \in M$, there exists an element $b \in S$ such that $i(L(a) \cap S) < b \leq a \vee a \leq b < o(U(a) \cap S)$. The latter is a contradiction of the definition of the greatest or of the least element. ■

We use this lemma to obtain the following theorem, which characterizes monotone roundings in linearly ordered sets.

**Theorem 1.28:** Let $\{M, \leq\}$ be a linearly ordered complete lattice, $\{S, \leq\}$ a screen of $\{M, \leq\}$, and $\bigtriangledown : M \to S$ resp. $\triangle : M \to S$ the monotone downwardly resp. upwardly directed rounding. For each $a \in M$ let $I := [\bigtriangledown a, \triangle a]$ and let $I_1$ and $I_2$ with $I_1 < I_2$ be subsets† of $M$ which partition $I := I_1 \cup I_2$. Then a mapping $\Box : M \to S$ is a monotone rounding if and only if

$$\bigwedge_{a \in S \subseteq M} \Box a = a$$

and

$$\bigwedge_{a \in M \setminus S} \Box a = \begin{cases} \bigtriangledown a & \text{for all} \quad a \in I_1 \\ \triangle a & \text{for all} \quad a \in I_2. \end{cases}$$

*Proof*: (a) It is clear that every such mapping is a monotone rounding.

(b) We still have to show that every monotone rounding has the properties stated in the theorem. By Theorem 1.24 we have for all $a \in M$:

$$\bigtriangledown a = \sup(L(a) \cap S) \leq a \leq \triangle a = \inf(U(a) \cap S).$$

Applying the mapping $\Box$ to this inequality, we get

$$\Box(\bigtriangledown a) \underset{(R2)}{=} \bigtriangledown a \underset{(R2)}{\leq} \Box a \underset{(R2)}{\leq} \Box(\triangle a) \underset{(R1)}{=} \triangle a.$$

Then Lemma 1.27 implies that $\Box a = (\bigtriangledown a$ or $\triangle a)$. Now let $I_1 := \{x \in I \mid \Box x = \bigtriangledown a\}$ and let $I_2 := \{x \in I \mid \Box x = \triangle a\}$. Since for all $x \in I$, $\bigtriangledown a \leq x \leq \triangle a$, then (R1) and (R2) yield $\bigtriangledown a \leq \Box x \leq \triangle a$, while Lemma 1.27

† If $U$, $V$ are subsets of $\{M, \leq\}$, $U < V$ means $u < v$ for all $u \in U$ and all $v \in V$.

implies that $I_1 \cup I_2 = I$. We still have to show that $I_1 < I_2$. Let us assume that $i_2 \leq i_1$ ((O4)) for any pair $\{i_1, i_2\}$ with $i_1 \in I_1$ and $i_2 \in I_2$. Then (R2) gives $\Box i_2 \leq \Box i_1$, which is a contradiction to our assumption $\Box i_1 = \bigtriangledown a < \triangle a = \Box i_2$ $(a \notin S)$. ■

Theorem 1.28 asserts that every monotone rounding of a linearly ordered complete lattice into a screen can be expressed by the monotone directed roundings $\bigtriangledown$ and $\triangle$. Different monotone roundings are distinguished from each other in an interval $I := [a_1, a_2]$ between neighboring screenpoints $a_1$ and $a_2$ only by the way $I$ is split into the associated subsets $I_1 < I_2$. The monotone directed roundings are extreme cases with respect to this splitting:

$$\bigtriangledown(I_1 = I\backslash\{a_2\}, I_2 = \{a_2\}) \qquad \text{and} \qquad \triangle(I_1 = \{a_1\}, I_2 = I\backslash\{a_1\}).$$

Finally we show by a simple example that in general Lemma 1.27 and Theorem 1.28 are not valid in the case of a complete but not linearly ordered lattice. Let $\{M, \leq\}$ be the complete lattice that appears in Fig. 14.

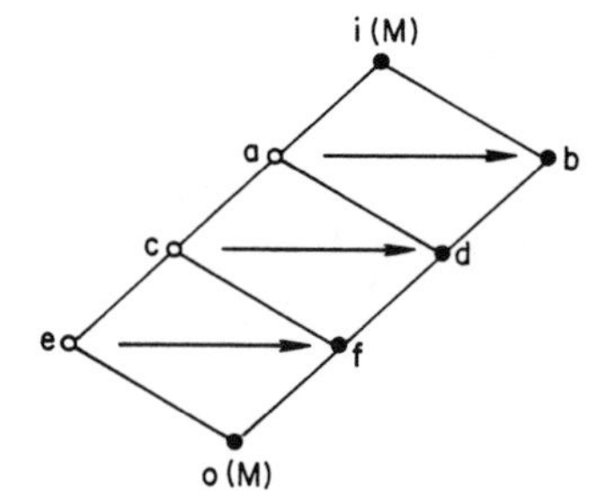


FIGURE 14.

The subset $\{S, \leq\}$ consisting of the solid points in Fig. 14 obviously is a screen of $\{M, \leq\}$. We define a mapping $\Box : M \to S$ by the following properties:

1. All screenpoints are fixed points of the mapping.
2. $\Box a = b$, $\Box c = d$, $\Box e = f$. See Fig. 14.

Then $\Box$ is a monotone rounding. However, neither Lemma 1.27 nor Theorem 1.28 hold. For instance, we obtain $\sup(L(c) \cap S) = f < \Box c = d < \inf(U(c) \cap S) = i(M)$, and consequently $\bigtriangledown c = f < \Box c = d < b < \triangle c = i(M)$.

## 4. ARITHMETIC OPERATIONS AND ROUNDINGS

If $M_1$, $M_2$, and $M$ are nonempty sets, then a mapping $* : M_1 \times M_2 \to M$ of the product set $M_1 \times M_2$ into $M$ is called an *operation*. The image $*(a, b)$ of the operands $a \in M_1$ and $b \in M_2$ is usually written in the dyadic form

$a * b$. If, in particular, $M_1 = M_2 = M$, one speaks of an *inner* operation in contrast to an *outer* operation, where $*: M_1 \times M \to M$. In the latter case $M_1$ is called the operator set. Examples of an inner operation are the addition and multiplication of real numbers and for an outer operation the multiplication of a vector by a scalar or the matrix–vector multiplication. A nonempty set with operations defined for its elements is called an algebraic structure.

The simplest algebraic structure is a set $M$ with one inner operation $*$. It is called a groupoid $\{M, *\}$. An element $e \in M$ is called a left neutral (resp. a right neutral) element if

$$\bigwedge_{a \in M} e * a = a \qquad \left(\text{resp.} \quad \bigwedge_{a \in M} a * e = a\right).$$

$e$ is called a neutral element if it is a left as well as a right neutral element. A neutral element is always unique since the presumption of two such elements $e$ and $e'$ yields $e = e * e' = e'$.

$\{\boldsymbol{R}, +\}$ (resp. $\{\boldsymbol{R}, \cdot\}$) are examples of groupoids with the neutral element 0 (resp. 1). The groupoids $\{\boldsymbol{R}, -\}$ (resp. $\{\boldsymbol{R}, /\}$) have only right neutral elements 0 (resp. 1).

On computers, subtraction is not always the inverse operation for addition, nor division for multiplication. All operations have a more independent nature. It is, however, essential that all operations occurring in Fig. 1 have a right neutral element. See for instance Theorem 1.30 and the example following it.

An operation $*: M_1 \times M_2 \to M$ or $a * b$ is called *ordered* if for subsets $N_1 \subseteq M_1$ and $N_2 \subseteq M_2$ and every fixed $(a', b') \in N_1 \times N_2$ the mapping $a * b'$ as well as the mapping $a' * b$ is monotone. A nonempty set with ordered operations is called an ordered algebraic structure.

$\{\boldsymbol{R}, +, \leq\}$ and $\{\boldsymbol{R}^n, +, \leq\}$ are examples of ordered groups. $\{\boldsymbol{R}, +, \cdot, \leq\}$ is an ordered ring.

In particular, we shall call the triple $\{M, *, \leq\}$ an ordered groupoid if $\{M, *\}$ is a groupoid, $\{M, \leq\}$ is an ordered set, and the following condition (OA) holds.

$$\text{(OA)} \qquad \bigwedge_{a,b,c \in M} (a \leq b \Rightarrow a * c \leq b * c \wedge c * a \leq c * b).$$

(OA) is referred to as the compatibility property between the algebraic and the order structure. (OA) is equivalent to

$$\text{(OA')} \qquad \bigwedge_{a,b,c,d \in M} (a \leq b \wedge c \leq d \Rightarrow a * c \leq b * d).$$

In order to get (OA′) from (OA), one has to apply (OA) twice while (OA′) $\Rightarrow$ (OA) by taking $c = d$.

A key objective of this treatise concerns the question of how operations that are defined on a certain set $M$ can be approximated on a screen or on an upper screen $S$. We shall see later that natural mapping properties such as isomorphism or homomorphism for characterizing this approximation cannot be achieved. Nevertheless, we may develop reasonable and simple compatibility properties between the operations in $S$ and in $M$. Such properties are characterized by the following definition.

**Definition 1.29:** Let $\{M, \leq\}$ be a complete lattice, $\{M, *\}$ a groupoid, and $\{S, \leq\}$ a lower (resp. an upper) screen of $\{M, \leq\}$. A groupoid $\{S, \boxast\}$ is called a *screen groupoid* if

(RG1) $\bigwedge_{a,b \in S} (a * b \in S \Rightarrow a \boxast b = a * b)$.

A screen groupoid is called *monotone* if

(RG2) $\bigwedge_{a,b,c,d \in S} (a * b \leq c * d \Rightarrow a \boxast b \leq c \boxast d)$.

A screen groupoid is called a *lower* (resp. an *upper*) screen groupoid if

(RG3) $\bigwedge_{a,b \in S} a \boxast b \leq a * b \qquad \left(\text{resp.} \quad \bigwedge_{a,b \in S} a * b \leq a \boxast b\right)$. ■

In Definition 1.29 three properties of screen groupoids are enumerated. We may ask if there do indeed exist groupoids with all three of these properties and how such groupoids can be produced. The following theorem supplies an answer to this question.

**Theorem 1.30:** Let $\{M, \leq\}$ be a complete lattice, $\{M, *\}$ a groupoid, $\{S, \leq\}$ a lower (resp. an upper) screen (resp. a screen) of $\{M, \leq\}$. Let $\square : M \to S$ be a mapping and let an operation $\boxast$ in $S$ be defined by

(RG) $\bigwedge_{a,b \in S} a \boxast b := \square(a * b)$.

1. If $\square$ has the property (R$i$), $i = 1, 2, 3$, defined for roundings, then $\{S, \boxast\}$ has the property (RG$i$), $i = 1, 2, 3$, respectively.

2. If the groupoid $\{M, *\}$ has a right neutral element $e$ and $e \in S$, then $\{S, \boxast\}$ is a monotone lower screen groupoid (resp. a monotone upper screen groupoid) if an only if $\boxast$ is defined by (RG), wherein the mapping $\square$ has the properties (R1), (R2), and (R3).

*Proof*: We omit the proof of property 1 since it is straightforward.

2. (a) The direction (RG), (R1), (R2), (R3) $\Rightarrow$ (RG1), (RG2), (RG3) of the proof is given by 1.

(b) We give the proof of the opposite direction in the case of a lower screen. Let

$$x := \bigtriangledown(a * b).$$

Then by (R3) we obtain

$$x = x * e = \bigtriangledown(a * b) \leq a * b. \tag{7}$$

Since $\{S, \boxast\}$ is a screen groupoid and $e \in S$, then $e$ is also a right neutral element in $\{S, \boxast\}$ because

$$\bigwedge_{a \in S} a * e \in S \underset{(\mathrm{RG1})}{\Rightarrow} \bigwedge_{a \in S} (a \boxast e = a * e = a).$$

Applying (RG2) to (7), therefore, we get

$$x \boxast e = x = \bigtriangledown(a * b) \leq a \boxast b. \tag{8}$$

(RG3) yields $a \boxast b \leq a * b$. If we apply the rounding $\bigtriangledown$ to this inequality, we get by (R1) and (R2)

$$\bigtriangledown(a \boxast b) = a \boxast b \leq \bigtriangledown(a * b). \tag{9}$$

From (8), (9), and (O3) we obtain $a \boxast b = \bigtriangledown(a * b)$. ■

The hypotheses of Theorem 1.30 under which the monotone directed roundings $\bigtriangledown$ and $\bigtriangleup$ are unique also show that there exists exactly one monotone lower screen groupoid (resp. exactly one monotone upper screen groupoid) on a lower (resp. an upper) screen. Because of this fact, we shall use the special signs $\mathbin{\bigtriangledown\!\!\!\!\ast}$ (resp. $\mathbin{\bigtriangleup\!\!\!\!\ast}$) for the associated operations.

Theorem 1.30 implies that $\mathbin{\bigtriangledown\!\!\!\!\ast}$ (resp. $\mathbin{\bigtriangleup\!\!\!\!\ast}$) can also be defined by the property

$$(\mathrm{RG}) \quad \bigwedge_{a,b \in S} a \mathbin{\bigtriangledown\!\!\!\!\ast} b = \bigtriangledown(a * b) \qquad \left(\text{resp.} \quad \bigwedge_{a,b \in S} a \mathbin{\bigtriangleup\!\!\!\!\ast} b = \bigtriangleup(a * b)\right).$$

This formula can now be used in place of Definition 1.29 to construct the result of a computation in a lower (resp. an upper) screen groupoid. We illustrate this by an example.

**Example:** Let $Z := \{\zeta := \xi + i\eta \in \mathbf{C} \,|\, |\xi| \leq s \wedge |\eta| \leq s \wedge s > 1\}$, $M := \mathbf{P}Z$ and $S \subset M$ the set of all rectangles of $Z$ with sides parallel to the axes. We already know from example 1 in Section 1.3 that $\{S, \subseteq\}$ is an upper screen of $\{M, \subseteq\}$. Now we define a multiplication in $M$ by

$$\bigwedge_{a,b \in M} a \cdot b := \{\alpha \cdot \beta \,|\, \alpha, \beta \in Z \wedge \alpha \in a \wedge \beta \in b\}.$$

In order not to leave the set $Z$ while executing this product, we replace the real and/or imaginary part of the product $\alpha \cdot \beta$ by $s$ whenever the former and/or the latter do in fact exceed $s$. Then $\{M, \cdot\}$ is a groupoid with the neutral element $\{1\}$. Now set $s = 5$ and consider two special elements $a, b \in M$. See Fig. 15a.

$$a\colon \quad \{\zeta = \xi + i\eta \,|\, \xi \in [1,2] \wedge \eta \in [0,1]\},$$
$$b\colon \quad \{e^{i\pi/4}\} = \{\tfrac{1}{2}\sqrt{2}(1+i)\}.$$

Here $a$, $b$ are also elements of $S \subset M$. Figure 15b shows the product $a \cdot b$ as well as the construction of the product $a \mathbin{\triangle\!\!\!\!\!\cdot}\, b$ in the upper screen groupoid using the formula $a \mathbin{\triangle\!\!\!\!\!\cdot}\, b = \triangle(a \cdot b)$.

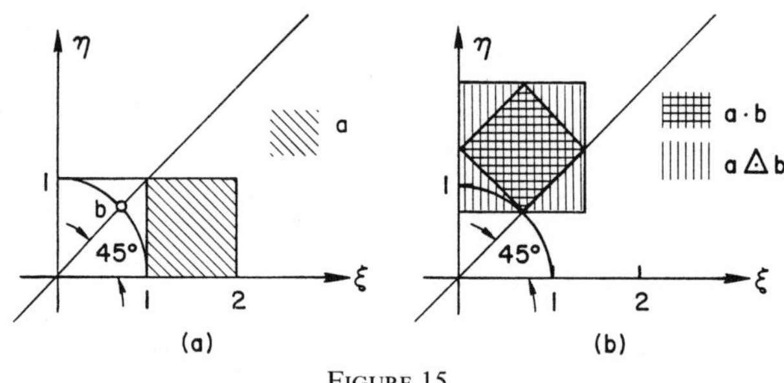


FIGURE 15.

When the groupoids are ordered, additional properties can be derived for them. If $\{M, *, \leq\}$ is an ordered groupoid, then every monotone screen groupoid $\{S, \boxed{*}, \leq\}$ is also an ordered groupoid which follows directly from (OA′) and (RG2). Additional properties are given in the following theorem.

**Theorem 1.31:** Let $\{M, *, \leq\}$ be a completely ordered groupoid with a right neutral element $e \in M$ and $\{S, \mathbin{\nabla\!\!\!\!\!\ast}, \leq\}$ the monotone lower screen groupoid (resp. $\{S, \mathbin{\triangle\!\!\!\!\!\ast}, \leq\}$ the monotone upper screen groupoid) with $e \in S$. Then for all $a, b \in M$ the following inequalities hold:

$$(\nabla a) \mathbin{\nabla\!\!\!\!\!\ast} (\nabla b) \leq \nabla(a * b) \leq a * b \qquad (\text{resp.} \quad a * b \leq \triangle(a * b) \leq (\triangle a) \mathbin{\triangle\!\!\!\!\!\ast} (\triangle b)).$$

*Proof*: Using Theorem 1.30 and (R3) we have for all $a, b \in S$ that

$$a \mathbin{\nabla\!\!\!\!\!\ast} b = \nabla(a * b) \leq a * b \qquad (\text{resp.} \quad a * b \leq \triangle(a * b) = a \mathbin{\triangle\!\!\!\!\!\ast} b).$$

Moreover, (R3) implies $\nabla a \leq a$ and $\nabla b \leq b$. Then using (OA′) and (RG2) we obtain $(\nabla a) \mathbin{\nabla\!\!\!\!\!\ast} (\nabla b) \leq a \mathbin{\nabla\!\!\!\!\!\ast} b$. ■

The inequalities $(\nabla a) \mathbin{\nabla\!\!\!\!\!\ast} (\nabla b) \leq \nabla(a * b)$ (resp. $\triangle(a * b) \leq (\triangle a) \mathbin{\triangle\!\!\!\!\!\ast} (\triangle b)$) of this theorem assert in general that the monotone lower (resp. upper) screen groupoid is not a homomorphic image of the ordered groupoid $\{M, *, \leq\}$. In concrete cases such as in interval arithmetic, it is easy to show by simple examples that the equality sign is not valid in general.

The outer inequality of Theorem 1.31, viz.,

$$(\triangledown a) \mathbin{\triangledown\!\!\!\!\!\ast} (\triangledown b) \le a * b \qquad (\text{resp.} \quad a * b \le (\vartriangle a) \mathbin{\vartriangle\!\!\!\!\!\ast} (\vartriangle b)),$$

asserts that the computation on the screen is always on one side of the computation in the groupoid $\{M, *, \le\}$. This property remains valid even in expressions containing many operations.

Moreover, if several different operations are defined in $M$ and if they are all approximated in $S$ by lower (resp. upper) screen groupoids, then the previous assertion is also valid for expressions containing several different operations. In particular, this is the case in interval arithmetic, where the calculations are done on an upper screen with inclusion as an order relation. See Chapter 4.

Properties of comparison of computations in different lower (resp. upper) screen groupoids are described in the following theorem.

**Theorem 1.32:** Let $\{M, *, \le\}$ be a completely ordered groupoid with the right neutral element $e$. Let $S$ and $D$ be lower (resp. upper) screens of $\{M, \le\}$ with the property $e \in S \subseteq D \subseteq M$ and $\triangledown : M \to S, \triangledown_1 : M \to D, \triangledown_2 : D \to S$ the monotone downwardly (resp. $\vartriangle : M \to S, \vartriangle_1 : M \to D, \vartriangle_2 : D \to S$ upwardly) directed roundings and $\{S, \mathbin{\triangledown\!\!\!\!\!\ast}, \le\}$, $\{D, \mathbin{\triangledown\!\!\!\!\!\ast}_1, \le\}$, $\{S, \mathbin{\triangledown\!\!\!\!\!\ast}_2, \le\}$ (resp. $\{S, \mathbin{\vartriangle\!\!\!\!\!\ast}, \le\}$, $\{D, \mathbin{\vartriangle\!\!\!\!\!\ast}_1, \le\}$, $\{S, \mathbin{\vartriangle\!\!\!\!\!\ast}_2, \le\}$) the corresponding monotone lower (resp. upper) screen groupoids. Then

1. $\{S, \mathbin{\triangledown\!\!\!\!\!\ast}, \le) = \{S, \mathbin{\triangledown\!\!\!\!\!\ast}_2, \le\} \qquad (\text{resp.} \quad \{S, \mathbin{\vartriangle\!\!\!\!\!\ast}, \le\} = \{S, \mathbin{\vartriangle\!\!\!\!\!\ast}_2, \le\})$.
2. $\bigwedge_{a,b \in M} (\triangledown a) \mathbin{\triangledown\!\!\!\!\!\ast} (\triangledown b) \le (\triangledown_1 a) \mathbin{\triangledown\!\!\!\!\!\ast}_1 (\triangledown_1 b) \le a * b$

$$(\text{resp.} \quad \bigwedge_{a,b \in M} (a * b \le (\vartriangle_1 a) \mathbin{\vartriangle\!\!\!\!\!\ast}_1 (\vartriangle_1 b) \le (\vartriangle a) \mathbin{\vartriangle\!\!\!\!\!\ast} (\vartriangle b)),$$

i.e., operations in a finer monotone lower (resp. upper) screen groupoid always lead to nondegraded results.

*Proof*:

1. By Theorem 1.30 we have for all $a, b \in T : a \mathbin{\triangledown\!\!\!\!\!\ast}_2 b = \triangledown_2(a \mathbin{\triangledown\!\!\!\!\!\ast}_1 b)$, and therefore by Theorem 1.26 $a \mathbin{\triangledown\!\!\!\!\!\ast} b = \triangledown(a * b) = \triangledown_2(\triangledown_1(a * b)) = \triangledown_2(a \mathbin{\triangledown\!\!\!\!\!\ast}_1 b) = a \mathbin{\triangledown\!\!\!\!\!\ast}_2 b$.

2. $$\begin{aligned}(\triangledown a) \mathbin{\triangledown\!\!\!\!\!\ast} (\triangledown b) &= \triangledown((\triangledown a) * (\triangledown b)) = \triangledown_2(\triangledown_1((\triangledown a) * (\triangledown b)) \\ &= \triangledown_2((\triangledown a) \mathbin{\triangledown\!\!\!\!\!\ast}_1 (\triangledown b)) \underset{(\mathrm{R3})}{\le} (\triangledown a) \mathbin{\triangledown\!\!\!\!\!\ast}_1 (\triangledown b) \\ &= (\triangledown_2(\triangledown_1 a)) \mathbin{\triangledown\!\!\!\!\!\ast}_1 (\triangledown_2(\triangledown_1 b)) \underset{(\mathrm{OA}'),(\mathrm{RG2})}{\le} (\triangledown_1 a) \mathbin{\triangledown\!\!\!\!\!\ast}_1 \triangledown_1 b). \quad \blacksquare\end{aligned}$$

Some of the concepts and results of this section will also be needed in the case of outer operations. In order to have them available, we state the following Definition 1.33 and Theorem 1.34.

**Definition 1.33:** Let $\{M, \leq\}$ be a complete lattice and $\{S, \leq\}$ a lower (resp. an upper) screen of $\{M, \leq\}$. Further, let $N$ be an operator set of $M$ with an outer operation $*: N \times M \to M$. Let $T \subseteq N$. An operation $\boxast$: $T \times S \to S$ is called an *outer screen operation* if

(RG1) $\bigwedge_{\alpha \in T} \bigwedge_{a \in S} (\alpha * a \in S \Rightarrow \alpha \boxast a = \alpha * a).$

An outer screen operation is called *monotone* if

(RG2) $\bigwedge_{\alpha,\beta \in T} \bigwedge_{a,b \in S} (\alpha * a \leq \beta * b \Rightarrow \alpha \boxast a \leq \beta \boxast b).$

An outer screen operation is called *lower* (resp. *upper*) screen operation if

(RG3) $\bigwedge_{\alpha \in T} \bigwedge_{a \in S} \alpha \boxast a \leq \alpha * a \quad \left(\text{resp. } \bigwedge_{\alpha \in T} \bigwedge_{a \in S} \alpha * a \leq \alpha \boxast a\right).$ ■

**Theorem 1.34:** Let $\{M, \leq\}$ be a complete lattice, $\{S, \leq\}$ a lower (resp. an upper) screen (resp. a screen) of $\{M, \leq\}$, and $\square: M \to S$ a mapping. Further, let $N$ be an operator set of $M$ with an operation $*: N \times M \to M$. Let $T \subseteq N$, and let an operation $\boxast: T \times S \to S$ be defined by

(RG) $\bigwedge_{\alpha \in T} \bigwedge_{a \in S} \alpha \boxast a := \square(\alpha * a).$

1. If $\square$ has the property R$i$, $i = 1, 2, 3$, defined for rounding, then the operation $\boxast$ has the property (RG$i$), $i = 1, 2, 3$, respectively.
2. If there exists an identity operator $\varepsilon \in N$ with the property

$$\bigwedge_{a \in M} \varepsilon * a = a$$

and if $\varepsilon \in T \subseteq N$, then an operation $\boxast: T \times S \to S$ is a monotone lower screen operation (resp. a monotone upper screen operation) if and only if $\boxast$ is defined by (RG), wherein the mapping $\square$ has the properties (R1), (R2), and (R3). ■

We omit the proof of this theorem since it is completely analogous to that of Theorem 1.30. Since the monotone directed roundings are unique, there exists exactly one monotone lower (resp. upper) screen operation, which we therefore denote by $\triangledown\!\!\!*$ (resp. $\triangle\!\!\!*$). By Theorem 1.34 these operations can also be defined by the property

(RG) $\bigwedge_{\alpha \in T} \bigwedge_{a \in S} \alpha \triangledown\!\!\!* a := \triangledown(\alpha * a) \quad \left(\text{resp. } \bigwedge_{\alpha \in T} \bigwedge_{a \in S} \alpha \triangle\!\!\!* a := \triangle(\alpha * a)\right).$

# Chapter 2 / RINGOIDS AND VECTOIDS

**Summary:** In this chapter we shall develop the concepts of a ringoid and a vectoid as well as those of ordered or weakly ordered ringoids and vectoids. The development will be self-contained and, in particular, independent of Chapter 1. Chapters 3 and 4 will bring the contents of the first two chapters together.

This chapter begins with the definition of a ringoid and the derivation of its most important properties. Then we show that the power set of a ringoid is a ringoid, that the matrices over an ordered or weakly ordered ringoid form an ordered or weakly ordered ringoid, and that the complexification of a weakly ordered ringoid also leads to a weakly ordered ringoid. Then we define the concept of a vectoid and derive its properties. The power set of a vectoid turns out to be a vectoid. The $n$-tuples over a ringoid $R$ form a vectoid over $R$, as do the matrices over $R$. Finally, we show that ringoids and vectoids, the ordering of which is only conditionally complete, can always be completed without loss of algebraic or order properties.

## 1. RINGOIDS

We begin with the following definition of a ringoid.

**Definition 2.1:** A nonempty set $R$ in which an addition $(+)$ and a multiplication $(\cdot)$[†] are defined is called a *ringoid* if the following properties hold:

[†] We often write $ab$ instead of $a \cdot b$. We adopt the convention that multiplication and division are to be executed before addition and subtraction. However, the usual priorities dictated by brackets are assumed. Several operations of the same priority are to be executed from left to right, i.e., $a_1 * a_2 * \cdots * a_{n-1} * a_n := (a_1 * a_2 * \cdots * a_{n-1}) * a_n$, $n \geq 3$.

(D1) $\bigwedge_{a,b \in R} a + b = b + a.$

(D2) $\bigvee_{o \in R} \bigwedge_{a \in R} a + o = a.$

(D3) $\bigvee_{e \in R \setminus \{o\}} \bigwedge_{a \in R} a \cdot e = e \cdot a = a.$

(D4) $\bigwedge_{a \in R} a \cdot o = o \cdot a = o.$

(D5) There exists an element $x \in R$ such that

(a) $x + e = o,$
(b) $x \cdot x = e,$
(c) $\bigwedge_{a,b \in R} x(ab) = (xa)b = a(xb),$
(d) $\bigwedge_{a,b \in R} x(a + b) = (xa) + (xb).$

(D6) $x$ is unique.

A ringoid $R$ is called *division ringoid* if a division $/: R \times (R \setminus N) \to R$ (with respect to which properties (D7)–(D9) hold) is defined. Here $N \subseteq R$ is some subset of $R$ which contains $o$.

(D7) $\bigwedge_{a \in R} a/e = a.$

(D8) $\bigwedge_{a \in R \setminus N} o/a = o.$

(D9) $\bigwedge_{a \in R} \bigwedge_{b \in R \setminus N} x(a/b) = (xa)/b = a/(xb).$

A ringoid is called *weakly ordered* if $\{R, \leq\}$ is an ordered set and

(OD1) $\bigwedge_{a,b,c \in R} (a \leq b \Rightarrow a + c \leq b + c),$

(OD2) $\bigwedge_{a,b \in R} (a \leq b \Rightarrow xb \leq xa).$

A weakly ordered ringoid is called *ordered* if

(OD3) $\bigwedge_{a,b,c \in R} (o \leq a \leq b \wedge c \geq o \Rightarrow a \cdot c \leq b \cdot c \wedge c \cdot a \leq c \cdot b).$

A division ringoid is called *weakly ordered* if it is a weakly ordered ringoid. A division ringoid is called *ordered* if

(OD4) $\bigwedge_{a,b,c \in R} (o < a \leq b \wedge c > o \Rightarrow o \leq a/c \leq b/c \wedge c/a \geq c/b \geq o).$

Expressions such as linearly ordered ringoid, completely ordered ringoid, and completely and weakly ordered ringoid will be used in this book when the ordering in the ringoid is linear or complete, as the case may be (similarly for division ringoids). A ringoid is called *inclusion-isotonally ordered* with respect to an order relation $\{R, \subseteq\}$ if for all operations $* \in \{+, \cdot\}$,

(OD5) $\bigwedge_{a,b,c,d \in R} (a \subseteq b \wedge c \subseteq d \Rightarrow a * c \subseteq b * d)$.

A division ringoid is called *inclusion-isotonally ordered* if

(OD6) $\bigwedge_{a,b \in R} \bigwedge_{c,d \in R \setminus N} (a \subseteq b \wedge c \subseteq d \Rightarrow a/c \subseteq b/d)$. ∎

A ringoid (as well as a division ringoid) is just a set of elements in which three special elements $o$, $e$, and $x$ exist and for which the rules concerning the operations $+, \cdot, /$ are given by Definition 2.1. (D7) and (D9) imply that $e \notin N$ and that for all $a \notin N$, $xa \notin N$ as well. A ringoid may be weakly ordered or ordered with respect to one order relation as well as inclusion-isotonally ordered with respect to another order relation. The third special element $x$ in a ringoid has many properties that in the real or complex number field in particular distinguish the element $-1$. This motivates the following definition.

**Definition 2.2:** In a ringoid $\{R, +, \cdot\}$ we define a minus operator by

$$\bigwedge_{a \in R} -a := x \cdot a \tag{1}$$

and a subtraction by

$$\bigwedge_{a,b \in R} a - b := a + (-b). \quad ∎ \tag{2}$$

Setting $a = e$ in (1) and using (D3) gives

$$x = -e. \tag{3}$$

Using (3), several of the rules of Definition 2.1 can be written in a simpler form:

(D5) There exists an element $-e \in R$ such that

(a) $(-e) + e = o$,
(b) $(-e) \cdot (-e) = e$,
(c) $\bigwedge_{a,b \in R} -(ab) = (-a)b = a(-b)$,
(d) $\bigwedge_{a,b \in R} -(a + b) = (-a) + (-b) = -a - b$,

(D9) $\bigwedge_{a \in R} \bigwedge_{b \in R \setminus N} -(a/b) = (-a)/b = a/(-b)$,

(OD2) $\bigwedge_{a,b \in R} (a \leq b \Rightarrow -b \leq -a)$.

Every ring with unit element is also a ringoid. Thus a ringoid represents a certain generalization of a ring with unit element. In order to see this, we recall that a ring is an additive group with an associative multiplication with respect to which a neutral element exists. The two distributive laws $a(b + c) = ab + ac$ and $(a + b)c = ac + bc$ are also valid. From these properties the rules (D1), (D2), (D3), (D5a), (D5c), (D5d), and (D6) of a ringoid follow immediately. The proof of (D4) and (D5b) runs as follows:

(D4): $a \cdot o = a \cdot ((-e) + e) = -a + a = o = ((-e) + e) \cdot a = o \cdot a,$

(D5b): $(-e) \cdot o = (-e)((-e) + e) = (-e)(-e) + (-e) = o$

$$\Rightarrow (-e)(-e) = e.$$

Here the last implication follows from the uniqueness of the additive inverse of $-e$.

In a ringoid many familiar properties of a ring with unit element still hold. Some among others are summarized in the following theorem.

**Theorem 2.2:** Let $\{R, +, \cdot\}$ be a ringoid with the neutral elements $o$ and $e$. Then for all $a, b, c, d \in \boldsymbol{R}$ the following properties hold:

(a) $o$ and $e$ are unique, and $-e \neq o$.
(b) $o - a = -a$.
(c) $-a = (-e) \cdot a = a \cdot (-e)$.
(d) $-(-a) = a$.
(e) $-(a - b) = -a + b = b - a$.
(f) $(-a)(-b) = a \cdot b$.
(g) $-e$ is the unique solution of the equation $(-e) \cdot z = e$.
(h) $a \cdot e = o$ implies $a = o$ and $-a = o$ implies $a = o$.
(i) $a - z = a \Rightarrow z = o$, i.e., $o$ is the only right neutral element of subtraction.

In a division ringoid $\{R, N, +, \cdot, /\}$ the following additional properties hold:

(j) $(-a)/(-b) = a/b$.
(k) $(-e)/(-e) = e$.
(l) If for all $a$, $a/a = e$, then $e$ is the only right neutral element of division.

In a weakly ordered ringoid $\{R, +, \cdot, \leq\}$ we have

(m) $a \leq b \wedge c \leq d \Rightarrow a + c \leq b + d$,
(n) $a < b \Rightarrow -b < -a$.

In an ordered ringoid we have moreover

(o) $o \leq a \leq b \wedge o \leq c \leq d \Rightarrow o \leq ac \leq bd \wedge o \leq ca \leq db$,
(p) $a \leq b \leq o \wedge c \leq d \leq o \Rightarrow o \leq bd \leq ac \wedge o \leq db \leq ca$,
(q) $a \leq b \leq o \wedge o \leq c \leq d \Rightarrow ad \leq bc \leq o \wedge da \leq cb \leq o$.

In a linearly ordered ringoid $\{R, +, \cdot, \leq\}$ we have further

(r) $a \leq b \wedge c \geq o \Rightarrow ac \leq bc \wedge ca \leq cb$,
(s) $a \leq b \wedge c \leq o \Rightarrow ac \geq bc \wedge ca \geq cb$,
(t) $a + (-a) = o$, i.e., $-a$ is an additive inverse of $a$,
(u) $e > o$ and $-e < o$.

In an ordered division ringoid $\{R, N, +, \cdot, /, \leq\}$ we have

(v) $a > o \wedge b > o \Rightarrow a/b \geq o$,
(w) $a < o \wedge b > o \Rightarrow a/b \leq o \wedge b/a \leq o$,
(x) $a < o \wedge b < o \Rightarrow a/b \geq o$.

In a completely and weakly ordered ringoid we have

(y) $$\bigwedge_{\emptyset \neq A \in PR} (\inf A = -\sup(-A) \wedge \sup A = -\inf(-A)).$$

In an inclusion-isotonally ordered ringoid $\{R, +, \cdot, \subseteq\}$, (OD5) is valid with subtraction

(z) $$\bigwedge_{a,b,c,d \in R} (a \subseteq b \wedge c \subseteq d \Rightarrow a - c \subseteq b - d).$$

*Proof*: The proof of most of the properties is straightforward. Thus we give a concise sketch of each such proof. This procedure will be typical in this chapter.

(a): A neutral element is unique. $(-e)(-e) = e \Rightarrow_{(D4)} -e \neq o$.
(b): $o - a = o + (-a) = -a$.
(c): (D5c) $\Rightarrow_{b=e} -a = (-e)a = a(-e)$.
(d): $-(-a) = (-e)((-e)a) =_{(D5c)} ((-e)(-e))a =_{(D5b)} a$.
(e): $-(a - b) = (-e)(a + (-e)b) =_{(D5d)} (-e)a + (-e)((-e)b) =_{(d)} -a + b =_{(D1)} b - a$.
(f): $(-a)(-b) = ((-e)a)((-e)b) =_{(D5d)} ((-e)((-e)a))b =_{(d)} ab$.
(g): Assume that $a$ solves $(-e)z = e$. Then $-e = (-e) \cdot e = (-e)((e) \cdot a) =_{(d)} a$.
(h): $e \cdot a = o \Rightarrow a = o$ by (D3). $-a = (-e)a = o \Rightarrow_{(D4)} (-e)((-e)a) = o \Rightarrow_{(d)} a = o$.
(i): $a - o = a + (-o) =_{(D4)} a + o =_{(D2)} a$, i.e., $o$ is a right neutral element of subtraction. Assume $o'$ is another one. Then $\bigwedge_{a \in R} a - o' = a + (-o') = a \Rightarrow_{(a)} -o' = o \Rightarrow_{(h)} o' = o$.
(j): $(-a)/(-b) = ((-e)a)/((-e)b) =_{(D9)} ((-e)((-e)a))/b =_{(d)} a/b$.
(k): $(-e)/(-e) =_{(j)} e/e =_{(D7)} e$.
(l): Assume that $e$ and $e'$ *are* right neutral elements. Then $e = e'/e' = e'$.
(m): $a \leq b \wedge c \leq d \Rightarrow a + c \leq b + c \leq b + d$.
(n): $a < b :\Leftrightarrow a \leq b \wedge a \neq b \Rightarrow_{(OD2)} -b \leq -a \wedge -b \neq -a \Leftrightarrow: -b < -a$.

(o): $o \le a \le b \wedge o \le c \le d \Rightarrow_{(OD3)} o \le ac \le bc \le bd$.

(p): $a \le b \le o \wedge c \le d \le o \Rightarrow o \le -b \le -a \wedge o \le -d \le -c \Rightarrow_{(o)} o \le (-b)(-d) \le (-a)(-c) \Rightarrow_{(b)} o \le bd \le ac$.

(q): $a \le b \le o \wedge o \le c \le d \Rightarrow o \le -b \le -a \wedge o \le c \le d \Rightarrow_{(o)} o \le (-b)c \le (-a)d \Rightarrow ad \le bc \le o$.

(r): 1. $o \le a \le b \wedge c \ge o \Rightarrow_{(OD3)} o \le ac \le bc$.
2. $a \le b \le o \wedge c \ge o \Rightarrow_{(q)} ac \le bc$.
3. $a \le o \le b \wedge c \ge o \Rightarrow ac \le_{(q)} o \le_{(o)} bc$.

(s): $a \le b \wedge c \le o \Rightarrow a \le b \wedge o \le -c \Rightarrow_{(r)} a(-c) \le b(-c) \Rightarrow_{(OD2)} bc \le ac$.

(t): $b := a + (-a) \Rightarrow -b = (-e)(a + (-a)) =_{(D5d),(D2)} a + (-a) = b$. $(O4) \Rightarrow b \ge o \vee b \le o \Rightarrow_{(OD2)} -b = b \le o \vee -b = b \ge o \Rightarrow_{(O3)} b = o$.

(u): $(D3) \Rightarrow e \ne o \Rightarrow_{(O4)} e > o \vee e < o$. Assume $e < o \Rightarrow_{(p)} e \cdot e = e \ge o$, i.e., $e > o$, which is a contradiction $\Rightarrow e > o \Rightarrow_{(n)} -e < o$.

(v): (OD4).

(w): $a < o \wedge b > o \Rightarrow_{(n)} -a > o \wedge b > o \Rightarrow_{(v)} (-a)/b \ge o \Rightarrow_{(OD2)} (-a)/(-b) =_{(j)} a/b \le o$.

(x): $a < o \wedge b < o \Rightarrow_{(n)} -a > o \wedge -b > o \Rightarrow_{(v)} (-a)/(-b) = a/b \ge o$.

(y): 1. $\bigwedge_{a \in A}(\inf A \le a \Rightarrow_{(OD2)} -a \le -\inf A \Rightarrow \sup(\{-e\} \cdot A) \le -\inf A \Rightarrow_{(OD2)} \inf A \le -\sup(-A)$.
2. $\bigwedge_{a \in A}(-a \le \sup(\{-e\} \cdot A) \Rightarrow_{(OD2)} -\sup(-A) \le a \Rightarrow -\sup(-A) \le \inf A$.
1. $\wedge$ 2. $\Rightarrow_{(O3)} \inf A = -\sup(-A)$.

(z): Employing (OD5) with multiplication, we obtain $\bigwedge_{c,d \in R}(c \subseteq d \Rightarrow (-e) \cdot c \subseteq (-e) \cdot d \Rightarrow -c \subseteq -d)$, $a \subseteq b \wedge c \subseteq d \Rightarrow a \subseteq b \wedge -c \subseteq -d \Rightarrow a - c = a + (-c) \subseteq b + (-d) = b - d)$. ■

Since Theorem 2.2 is quite ramified, we offer the following comments, by way of summary. Theorem 2.2 shows that in a ringoid or division ringoid the subtraction operator has the same properties as does subtraction in the real or complex number field. Compared to other algebraic structures, a ringoid is distinguished by the fact that the existence of inverse elements is not assumed. Nevertheless, subtraction is not an independent operation. It can be defined by means of the operations of multiplication and addition.

In a weakly ordered (resp. an ordered) ringoid for all elements that are comparable with $o$ with respect to $\le$ and $\ge$, the same rules for inequalities hold as in the cases of complex (resp. real) numbers. Since in a linearly ordered ringoid, all elements are comparable to $o$ with respect to $\le$ and $\ge$, then for all elements the same rules for inequalities hold as for inequalities among real numbers.

It is evident but important to note for further applications that $\{\boldsymbol{R}, +, \cdot\}$ and $\{\boldsymbol{C}, +, \cdot\}$ are ringoids, while $\{\boldsymbol{R}, \{o\}, +, \cdot, /\}$ and $\{\boldsymbol{C}, \{o\}, +, \cdot, /\}$ are division ringoids. $\{\boldsymbol{R}, +, \cdot, \leq\}$ is a linearly ordered ringoid. However, if the order relation is defined componentwise, $\{\boldsymbol{C}, +, \cdot, \leq\}$ is only a weakly ordered ringoid. The properties (OD1) and (OD2) are easily verified. We show that (OD3) is not valid by means of a simple example. Let

$$0 \leq \alpha := e^{i3\pi/8} \leq \beta := 2 \cdot e^{i3\pi/8} \wedge 0 \leq \gamma := e^{i\pi/4}.$$

Then

$$\alpha \cdot \gamma = e^{i5\pi/8} \qquad \text{and} \qquad \beta \cdot \gamma = 2 \cdot e^{i5\pi/8}.$$

Then $Re(\beta \cdot \gamma) \leq Re(\alpha \cdot \gamma)$.

The following theorem plays a key role in subsequent chapters when we study the spaces listed in Fig. 1.

**Theorem 2.3:** In a linearly ordered ringoid $\{R, +, \cdot, \leq\}$ property (D6) is a consequence of the remaining properties (D1), (D2), (D3), (D4), (D5), (O1), (O2), (O3), (O4), (OD1), (OD2), and (OD3).

*Proof*: Setting $b = e$ in (D5c) implies

$$\bigwedge_{a \in R} x \cdot a = a \cdot x. \tag{1}$$

Suppose that two elements $x, y \in R$ fulfill the properties (D5) and (OD2). Then, without loss of generality, we can assume by (O4) that $x \leq y$. Then

$$x \leq y \underset{\text{(OD2)}}{\Rightarrow} xy \leq xx \underset{\text{(D5b)}}{=} e \wedge yy \underset{\text{(D5b)}}{=} e \leq yx. \tag{2}$$

By (1) $xy = yx$, and therefore by (2) and (O3) $xy = e$. If we multiply this equation by $x$, we obtain $y = x$. ■

Now let $M$ be a nonempty set with elements $a, b, c, \ldots$, and let $\boldsymbol{P}M$ be the power set of $M$. For any operation $*$ defined for the elements in $M$, we define a corresponding operation $*$ in $\boldsymbol{P}M$ by

$$\bigwedge_{A,B \in \boldsymbol{P}M} A * B := \{a * b \mid a \in A \wedge b \in B\}. \tag{3}$$

If $A$ or $B$ is the empty set, then the product is defined to be the empty set. The latter case is formally included in (3).

Let $\{\boldsymbol{P}M\}$ denote the subset of $\boldsymbol{P}M$ consisting of the elements $\{a\}$, $\{b\}$, $\{c\}, \ldots$. Then the mapping $\varphi: \{\boldsymbol{P}M\} \to M$ defined by $\bigwedge_{a \in M} \{a\} \leftrightarrow a$ is obviously an isomorphism.[†]

[†] Let $M$ and $N$ be algebraic structures and let a one-to-one correspondence exist between the operations $*$ in $M$ and $\circledast$ in $N$. A one-to-one mapping $\varphi: M \to N$ is called an isomorphism if $\bigwedge_{a,b \in M} \varphi a \circledast \varphi b = \varphi(a * b)$.

Let $\{M, *\}$ be a group with the neutral element $e$. The power set $\{\boldsymbol{P}M, *\}$ is not a group. In order to see this, we consider the equation $A * X = \{e\}$ in $\boldsymbol{P}M$:

$$A * X = \{a * x \,|\, a \in A \wedge x \in X\} = \{e\}.$$

The only solutions of this equation occur for $A = \{a\}$ and $X = \{a^{-1}\}$ with $a \in M$.

From this observation we conclude that the power set of a ring (resp. a field; resp. a vector space; resp. a matrix algebra) is not a ring (resp. a field; resp. a vector space; resp. a matrix algebra). However, the following theorem shows that ringoid properties are preserved upon passage to the power set.

**Theorem 2.4:** If $\{R, +, .\}$ is a ringoid with the neutral elements $o$ and $e$, then $\{\boldsymbol{P}R, +, \cdot\}$ is also a ringoid with the neutral elements $\{o\}$ and $\{e\}$. If $\{R, N, +, \cdot, /\}$ is a division ringoid, then $\{\boldsymbol{P}R, N^*, +, \cdot, /\}$ is also a division ringoid, where $N^* := \{A \in \boldsymbol{P}R \,|\, A \cap N \neq \emptyset\}$. Furthermore, $\{\boldsymbol{P}R, +, \cdot, \subseteq\}$ (resp. $\{\boldsymbol{P}R, N^*, +, \cdot, /, \subseteq\}$) is inclusion-isotonally ordered.

*Proof*: The properties (D1), (D2), (D3), (D4), (D7), (D8), and (OD5) (resp. (OD6)) follow directly from the definition of the operations in the power set and the corresponding properties in $R$. The properties (D5) and (D9) in $\boldsymbol{P}M$ are easily verified for $X = \{-e\}$. There remains to prove (D6).

(D6): We have to show that there exists only one element in $\boldsymbol{P}M$ that has the properties (D5a)–(D5d). Let $X$ be any element with these properties. Then we have, in particular

(D5b) $X \cdot X = \{x \cdot \tilde{x} \,|\, x, \tilde{x} \in X\} = \{e\}$

and

(D5c) $\bigwedge_{A,B \in \boldsymbol{P}M} X(AB) = (XA)B = A(XB).$

From (D5b) we obtain directly that

$$\bigwedge_{x,\tilde{x} \in X} xx = x\tilde{x} = \tilde{x}x = \tilde{x}\tilde{x} = e.$$

With this we obtain from (D5c) with $A = \{x\}$ and $B = \{\tilde{x}\}$ for any $x, \tilde{x} \in X$:

$$X \cdot \{e\} = \{e\} \cdot \{\tilde{x}\} = \{x\} \cdot \{e\}.$$

This means that $X = \{x\}$ contains only one element $x \in R$. $X = \{x\}$ still has to satisfy (D5a)–(D5d) in $\boldsymbol{P}M$ for all $A, B \in \boldsymbol{P}M$, and in particular for $A = \{a\}$ and $B = \{b\}$ with $a, b \in M$. Because of the isomorphism $\{\boldsymbol{P}M\} \leftrightarrow M$, the properties (D5a)–(D5d) for $A = \{a\}$ and $B = \{b\}$ are equivalent to (D5a)–

(D5d) in $R$. Since $R$ is a ringoid, there exists only one such element $x = -e$. Therefore $X = \{-e\}$ is the only element in $PR$ that satisfies (D5). ∎

Using Theorem 2.4, it is clear that $\{\boldsymbol{PR}, N, +, \cdot, /, \subseteq\}$ with $N := \{A \in \boldsymbol{PR} \mid 0 \in A\}$ and $\{\boldsymbol{PC}, N', +, \cdot, /, \subseteq\}$ with $N' := \{A \in \boldsymbol{PC} \mid 0 \in A\}$ are inclusion-isotonally ordered division ringoids with the neutral elements $\{0\}$ and $\{1\}$.

Now we consider matrices over a given ringoid $R$. Let $P := \{1, 2, \ldots, m\}$ and $Q := \{1, 2, \ldots, n\}$. Then a matrix is defined as a mapping $A: P \times Q \to R$. We denote the set of all such matrices by $M_{mn}R$. If $m = n$, we simply write $M_nR$. Then

$$M_{mn}R = \left\{ \begin{pmatrix} a_{11} & a_{12} & \cdots & a_{1n} \\ a_{21} & a_{22} & \cdots & a_{2n} \\ \vdots & \vdots & \cdots & \vdots \\ a_{m1} & a_{m2} & \cdots & a_{mn} \end{pmatrix} \middle| \, a_{ij} \in R, i = 1(1)m, j = 1(1)n \right\}.$$

Whenever possible we shall use the more concise notation $A = (a_{ij})$, $B = (b_{ij}) \in M_{mn}R$. Equality of and addition and multiplication for matrices are defined by

$$(a_{ij}) = (b_{ij}) :\Leftrightarrow \bigwedge_{i=1(1)m} \bigwedge_{j=1(1)n} a_{ij} = b_{ij},$$

$$\bigwedge_{(a_{ij}),(b_{ij}) \in M_{mn}R} (a_{ij}) + (b_{ij}) := (a_{ij} + b_{ij}),$$

$$\bigwedge_{(a_{ij}) \in M_{mn}R} \bigwedge_{(b_{ij}) \in M_{np}R} (a_{ij}) \cdot (b_{ij}) := \left( \sum_{j=1}^{n} a_{ij} b_{jk} \right) \in M_{mp}R.$$

If $\{R, \leq\}$ is an ordered set, an order relation $\leq$ is defined in $M_{mn}R$ by

$$(a_{ij}) \leq (b_{ij}) :\Leftrightarrow \bigwedge_{i=1(1)m} \bigwedge_{j=1(1)n} a_{ij} \leq b_{ij}.$$

In the case $m = n$ we denote the set of diagonal matrices[†] with a constant diagonal element by $\{M_nR\}$. Then the mapping $\varphi: \{M_nR\} \to R$ defined by

$$\bigwedge_{a \in M} \begin{pmatrix} a & o & \cdots & o \\ o & a & \cdots & o \\ \vdots & \vdots & & \vdots \\ o & o & \cdots & a \end{pmatrix} \to a$$

is obviously an isomorphism.[‡] The following theorem establishes ringoid properties for $\{M_mR\}$.

[†] A matrix is called a diagonal matrix if $a_{ij} = 0$ for all $i \neq j$. If the diagonal entries also are zero, we call the matrix the zero matrix; if the diagonal entries are $e$, we call the matrix the unit matrix.

[‡] See the footnote on p. 47.

**Theorem 2.5:** If $\{R, +, \cdot\}$ is a ringoid with the neutral elements $o$ and $e$, then $\{M_nR, +, \cdot, \leq\}$ is also a ringoid, and its neutral elements are

$$O = \begin{pmatrix} o & o & \cdots & o \\ o & o & \cdots & o \\ \vdots & \vdots & & \vdots \\ o & o & \cdots & o \end{pmatrix}, \qquad E = \begin{pmatrix} e & o & \cdots & o \\ o & e & \cdots & o \\ \vdots & \vdots & & \vdots \\ o & o & \cdots & e \end{pmatrix},$$

and

$$-E = \begin{pmatrix} -e & o & \cdots & o \\ o & -e & \cdots & o \\ \vdots & \vdots & & \vdots \\ o & o & \cdots & -e \end{pmatrix}.$$

If, moreover, $\{R, +, \cdot, \leq\}$ is weakly ordered (resp. ordered), then $\{M_nR, +, \cdot, \leq\}$ is a weakly ordered (resp. an ordered) ringoid.

*Proof*: The properties (D1), (D2), (D3), (D4), (OD1), and (OD2) follow immediately from the definition of the operations in $M_nR$ and the corresponding properties in $R$. The properties (D5) are easily verified for $X = -E$. There remains to demonstrate (D6) and (OD3).

(D6): Let $X = (x_{ij})$ be any element of $M_nR$ with properties (D5). Then we have in particular that

(D5a): $X + E = O$.

This means that $x_{ij} = 0$ for all $i \neq j$. The diagonal matrix $X$ must be shown to satisfy the properties (D5a)–(D5d) in $M_nR$. From (D5a) and (D5b) we obtain

$$x_{ii} + e = 0 \qquad \text{for all} \quad i = 1(1)n,$$
$$x_{ii} \cdot x_{ii} = e \qquad \text{for all} \quad i = 1(1)n.$$

From (D5c) and (D5d) we obtain for diagonal matrices $A, B \in M_nR$ with constant diagonal elements $a, b \in R$:

$$\bigwedge_{a,b \in R} x_{ii}(ab) = (x_{ii}a)b = a(x_{ii}b) \qquad \text{for all} \quad i = 1(1)n,$$

$$\bigwedge_{a,b \in R} x_{ii}(a + b) = (x_{ii}a) + (x_{ii}b) \qquad \text{for all} \quad i = 1(1)n.$$

These four relations imply that all diagonal elements $x_{ii}$ of $X$ have the properties (D5a)–(D5d) in $R$. Since $R$ is a ringoid, $x_{ii} = -e$ for all $i = 1(1)n$ and $X = -E$.

(OD3): $O \le A \le B \wedge C \ge O$

$$\Rightarrow o \le a_{ij} \le b_{ij} \wedge c_{ij} \ge o \qquad \text{for all} \quad i = 1(1)n$$

$$\underset{(\mathrm{OD3})_R}{\Rightarrow} o \le a_{ir}c_{rj} \le b_{ir}c_{rj} \qquad \text{for all} \quad i, j, r = 1(1)n$$

$$\underset{\text{Theorem 2.2(m)}}{\Rightarrow} \sum_{r=1}^{n} a_{ir}c_{rj} \le \sum_{r=1}^{n} b_{ir}c_{rj} \quad \text{for all} \quad i, j = 1(1)n$$

$$\Rightarrow AC \le BC. \blacksquare$$

Theorem 2.5 shows that $\{M_n\boldsymbol{R}, +, \cdot, \le\}$ is an ordered ringoid and that $\{M_n\boldsymbol{C}, +, \cdot, \le\}$ is a weakly ordered ringoid. Furthermore, from Theorem 2.4 we conclude that $\{\boldsymbol{PM_nR}, +, \cdot, \subseteq\}$ and $\{\boldsymbol{PM_nC}, +, \cdot, \subseteq\}$ are inclusion-isotonally ordered ringoids.

Now we consider the process of complexification of a given ringoid (resp. division ringoid) $R$. We denote the set of all pairs over $R$ by $CR := \{(a, b) | a, b \in R\}$. Equality, addition, multiplication, and division of elements $\alpha = (a_1, a_2)$, $\beta = (b_1, b_2) \in CR$ are defined by

$$(a_1, a_2) = (b_1, b_2) :\Leftrightarrow a_1 = b_1 \wedge a_2 = b_2,$$

$$\bigwedge_{\alpha,\beta \in CR} \alpha + \beta := (a_1 + b_1, a_2 + b_2),$$

$$\bigwedge_{\alpha,\beta \in CR} \alpha \cdot \beta := (a_1 b_1 - a_2 b_2, a_1 b_2 + a_2 b_1),$$

$$\bigwedge_{\alpha \in CR} \bigwedge_{\beta \in CR \setminus N} \alpha/\beta := ((a_1 b_1 + a_2 b_2)/b, (a_2 b_1 - a_1 b_2)/b) \text{ with } b := b_1 b_1 + b_2 b_2$$

and

$$\bar{N} := \{\gamma = (c_1, c_2) \in CR | c_1 c_1 + c_2 c_2 \in N\}.$$

If $\{R, \le\}$ is an ordered set, an order relation $\le$ is defined in $CR$ by

$$(a_1, a_2) \le (b_1, b_2) :\Leftrightarrow a_1 \le b_1 \wedge a_2 \le b_2.$$

Here $a_1$ is called the real part of $(a_1, a_2)$, while $a_2$ is called its imaginary part. If we denote the set of all elements of $CR$ with vanishing imaginary part by $\{CR\}$, the mapping $\varphi: \{CR\} \to R$ defined by $\bigwedge_{a \in R}(a, o) \leftrightarrow a$ is obviously an isomorphism.

The following theorem shows that ringoid properties are preserved under complexification.

**Theorem 2.6:** If $\{R, +, \cdot\}$ is a ringoid with the neutral elements $o$ and $e$, then $\{CR, +, \cdot\}$ is also a ringoid with the neutral elements $\omega = (o, o)$ and $\varepsilon = (e, o)$. Moreover, $-\varepsilon = (-e, o)$. If $\{R, N, +, \cdot, /\}$ is a division ringoid, then $\{C, \bar{N}, +, \cdot, /\}$ is also a division ringoid. If $\{R, +, \cdot, \le\}$ is a weakly ordered ringoid, $\{CR, +, \cdot, \le\}$ is likewise a weakly ordered ringoid.

*Proof*: The properties (D1), (D2), (D3), (D4), (D7), (D8), (OD1) and (OD2) follow directly from the definition of the operations and the corresponding properties in $R$. The properties (D5) and (D9) can be shown to be valid for $\xi = (-e, o) \in \boldsymbol{C}R$. (D5a) and (D5b) are easily verified for this $\xi$. (D5c), (D5d), and (D9) can also be proved in a straightforward manner by using the property

$$\xi\alpha = (-e, o)(a_1, a_2) = (-a_1, -a_2).$$

The only remaining property is (D6).

(D6): Let us assume that $\eta = (x, y)$ is any element of $\boldsymbol{C}R$ that satisfies (D5). $\eta$ must satisfy (D5a), which means

$$\eta + \varepsilon = \omega \Leftrightarrow x + e = o \wedge y = o,$$

i.e., $\eta$ is of the form $\eta = (x, o)$. Since $\eta$ satisfies (D5a)–(D5d) for all $\alpha, \beta \in \boldsymbol{C}R$, it does so in particular for $\alpha = (a, o)$ and $\beta = (b, o)$ with $a, b \in R$. Because of the isomorphism $\{\boldsymbol{C}R\} \leftrightarrow R$, the properties (D5a)–(D5d) in $\boldsymbol{C}R$ for elements $(a, o)$ and $(b, o)$ are equivalent to (D5a)–(D5d) in $R$. Since $R$ is a ringoid, there exists only one such element $x = -e$. Therefore $\xi = (-e, o)$ is the only element in $\boldsymbol{C}R$ that satisfies (D5). ■

## 2. VECTOIDS

We begin our discussion of vectoids with the following definition.

**Definition 2.7:** Let $\{R, +, \cdot\}$ be a ringoid with elements $a, b, c, \ldots$, and with neutral elements $o$ and $e$. Let $\{V, +\}$ be a groupoid with elements $\boldsymbol{a}, \boldsymbol{b}, \boldsymbol{c}, \ldots$, and the following properties:

(V1) $\bigwedge_{\boldsymbol{a},\boldsymbol{b} \in V} \boldsymbol{a} + \boldsymbol{b} = \boldsymbol{b} + \boldsymbol{a}$,

(V2) $\bigvee_{\boldsymbol{o} \in V} \bigwedge_{\boldsymbol{a} \in V} \boldsymbol{a} + \boldsymbol{o} = \boldsymbol{a}$.

$V$ is called an *R-vectoid* if there is an outer multiplication $\cdot : R \times V \rightarrow V$ defined, which with the abbreviation

$$\bigwedge_{\boldsymbol{a} \in V} -\boldsymbol{a} := (-e) \cdot \boldsymbol{a}$$

has the following properties:

(VD1) $\bigwedge_{a \in R} \bigwedge_{\boldsymbol{a} \in V} (a \cdot \boldsymbol{o} = \boldsymbol{o} \wedge o \cdot \boldsymbol{a} = \boldsymbol{o})$,

(VD2) $\bigwedge_{\boldsymbol{a} \in V} e \cdot \boldsymbol{a} = \boldsymbol{a}$,

(VD3) $\bigwedge_{a \in R} \bigwedge_{\boldsymbol{a} \in V} -(a \cdot \boldsymbol{a}) = (-a) \cdot \boldsymbol{a} = a \cdot (-\boldsymbol{a})$,

(VD4) $\bigwedge_{\boldsymbol{a},\boldsymbol{b} \in V} -(\boldsymbol{a} + \boldsymbol{b}) = (-\boldsymbol{a}) + (-\boldsymbol{b})$.

An $R$-vectoid is called *multiplicative* if a multiplication in $V$ with the following properties is defined:

(V3) $\bigvee_{\boldsymbol{e} \in V \setminus \{\boldsymbol{o}\}} \bigwedge_{\boldsymbol{a} \in V} \boldsymbol{a} \cdot \boldsymbol{e} = \boldsymbol{e} \cdot \boldsymbol{a} = \boldsymbol{a}$,

(V4) $\bigwedge_{\boldsymbol{a} \in V} \boldsymbol{a} \cdot \boldsymbol{o} = \boldsymbol{o} \cdot \boldsymbol{a} = \boldsymbol{o}$,

(V5) $-\boldsymbol{e} + \boldsymbol{e} = \boldsymbol{o}$,

(VD5) $\bigwedge_{\boldsymbol{a},\boldsymbol{b} \in V} -(\boldsymbol{a}\boldsymbol{b}) = (-\boldsymbol{a})\boldsymbol{b} = \boldsymbol{a}(-\boldsymbol{b})$.

Now let $\{R, +, \cdot, \leq\}$ be a weakly ordered ringoid. Then an $R$-vectoid is called *weakly ordered* if $\{V, \leq\}$ is an ordered set[†] which has the following two properties:

(OV1) $\bigwedge_{\boldsymbol{a},\boldsymbol{b},\boldsymbol{c} \in V} (\boldsymbol{a} \leq \boldsymbol{b} \Rightarrow \boldsymbol{a} + \boldsymbol{c} \leq \boldsymbol{b} + \boldsymbol{c})$,

(OV2) $\bigwedge_{\boldsymbol{a},\boldsymbol{b} \in V} (\boldsymbol{a} \leq \boldsymbol{b} \Rightarrow -\boldsymbol{b} \leq -\boldsymbol{a})$.

A weakly ordered $R$-vectoid is called *ordered* if

(OV3) $\bigwedge_{a,b \in R} \bigwedge_{\boldsymbol{a},\boldsymbol{b} \in V} (o \leq a \leq b \wedge \boldsymbol{o} \leq \boldsymbol{a} \Rightarrow a \cdot \boldsymbol{a} \leq b \cdot \boldsymbol{a}$
$\wedge\, o \leq a \wedge \boldsymbol{o} \leq \boldsymbol{a} \leq \boldsymbol{b} \Rightarrow a \cdot \boldsymbol{a} \leq a \cdot \boldsymbol{b})$.

A multiplicative vectoid is called *weakly ordered* if it is a weakly ordered vectoid. A multiplicative vectoid is called *ordered* if it is an ordered vectoid and

(OV4) $\bigwedge_{\boldsymbol{a},\boldsymbol{b},\boldsymbol{c} \in V} (\boldsymbol{o} \leq \boldsymbol{a} \leq \boldsymbol{b} \wedge \boldsymbol{o} \leq \boldsymbol{c} \Rightarrow \boldsymbol{a} \cdot \boldsymbol{c} \leq \boldsymbol{b} \cdot \boldsymbol{c} \wedge \boldsymbol{c} \cdot \boldsymbol{a} \leq \boldsymbol{c} \cdot \boldsymbol{b})$.

An $R$-vectoid that may be multiplicative is called *inclusion-isotonally ordered* with respect to order relations $\{R, \subseteq\}$ and $\{V, \subseteq\}$ if

(OV5) (a) $\bigwedge_{a,b \in R} \bigwedge_{\boldsymbol{c},\boldsymbol{d} \in V} (a \subseteq b \wedge \boldsymbol{c} \subseteq \boldsymbol{d} \Rightarrow a \cdot \boldsymbol{c} \subseteq b \cdot \boldsymbol{d})$,

(b) $\bigwedge_{\boldsymbol{a},\boldsymbol{b},\boldsymbol{c},\boldsymbol{d} \in V} (\boldsymbol{a} \subseteq \boldsymbol{b} \wedge \boldsymbol{c} \subseteq \boldsymbol{d} \Rightarrow \boldsymbol{a} * \boldsymbol{c} \subseteq \boldsymbol{b} * \boldsymbol{d})$, $\quad * \in \{+, \cdot\}$. ■

[†] Since it is always clear by the context which order relation is meant, we denote the order relation in $V$ and $R$ by the same sign $\leq$.

A vectoid (resp. a multiplicative vectoid) is just a set of elements in which one (resp. two) special element $\boldsymbol{o}$ (and $\boldsymbol{e}$) exists and which obeys rules concerning inner and outer operations as enumerated in Definition 2.7. A vectoid may be weakly ordered or ordered with respect to one order relation and inclusion-isotonally ordered with respect to another order relation. In the following definition, we use the minus operator of a vectoid in order to define a subtraction.

**Definition 2.8:** In an $R$-vectoid $\{V, R\}$ we define a subtraction by

$$\bigwedge_{\boldsymbol{a},\boldsymbol{b},\in V} \boldsymbol{a} - \boldsymbol{b} := \boldsymbol{a} + (-\boldsymbol{b}). \quad \blacksquare$$

Every vector space is a vectoid. This means that a vectoid represents a certain substructure or a generalization of a vector space. In order to see this, we recall that the latter is an additive group with an outer multiplication with the elements of a field with the following properties:

(1) $a(\boldsymbol{a} + \boldsymbol{b}) = a\boldsymbol{a} + a\boldsymbol{b}$,
(2) $(a + b)\boldsymbol{a} = a\boldsymbol{a} + b\boldsymbol{a}$,
(3) $a(b\boldsymbol{a}) = (ab)\boldsymbol{a}$,
(4) $e\boldsymbol{a} = \boldsymbol{a}$.

These properties imply (V1) and (V2) directly. (VD2) is the same as (4). (VD3) follows from (3) upon setting $a = -e$ (resp. $b = -e$), and (VD4) follows from (1) upon setting $a = -e$. To show (VD1), we argue as follows:

$$o \cdot \boldsymbol{a} = (-e + e) \cdot \boldsymbol{a} \underset{(2)}{=} -\boldsymbol{a} + \boldsymbol{a} = \boldsymbol{o},$$

$$a \cdot \boldsymbol{o} = a\boldsymbol{o} + a\boldsymbol{o} - a\boldsymbol{o} \underset{(1)}{=} a(\boldsymbol{o} + \boldsymbol{o}) - a\boldsymbol{o} = a\boldsymbol{o} - a\boldsymbol{o} = \boldsymbol{o}.$$

A vectoid possesses many of the familiar properties of a vector space. Some of them are summarized by the following theorem.

**Theorem 2.9:** Let $\{V, R\}$ be a vectoid with the neutral element $\boldsymbol{o}$ and with the neutral element $\boldsymbol{e}$ in case that a multiplication exists. Then for all $a, b \in R$, and for all $\boldsymbol{a}, \boldsymbol{b}, \boldsymbol{c}, \boldsymbol{d} \in V$, the following properties hold:

(a) $\boldsymbol{o}$ and $\boldsymbol{e}$ are unique.
(b) $\boldsymbol{o} - \boldsymbol{a} = -\boldsymbol{a}$.
(c) $-(-\boldsymbol{a}) = \boldsymbol{a}$.
(d) $-(\boldsymbol{a} - \boldsymbol{b}) = -\boldsymbol{a} + \boldsymbol{b} = \boldsymbol{b} - \boldsymbol{a}$.
(e) $(-a)(-\boldsymbol{a}) = a\boldsymbol{a}$.
(f) $-\boldsymbol{a} = \boldsymbol{o} \Leftrightarrow \boldsymbol{a} = \boldsymbol{o}$.
(g) $\boldsymbol{a} - \boldsymbol{z} = \boldsymbol{a} \Leftrightarrow \boldsymbol{z} = \boldsymbol{o}$, i.e., $\boldsymbol{o}$ is the only right neutral element of subtraction.

In a multiplicative vectoid, we have further

(h) $-\boldsymbol{a} = (-\boldsymbol{e})\boldsymbol{a} = \boldsymbol{a}(-\boldsymbol{e})$,
(i) $(-\boldsymbol{a})(-\boldsymbol{b}) = \boldsymbol{ab}$,
(j) $-\boldsymbol{e}$ is the unique solution of the equation $(-\boldsymbol{e}) \cdot \boldsymbol{z} = \boldsymbol{e}$.

In a weakly ordered vectoid $\{V, R, \leq\}$, we have

(k) $\boldsymbol{a} \leq \boldsymbol{b} \wedge \boldsymbol{c} \leq \boldsymbol{d} \Rightarrow \boldsymbol{a} + \boldsymbol{c} \leq \boldsymbol{b} + \boldsymbol{d}$,
(l) $\boldsymbol{a} < \boldsymbol{b} \Rightarrow -\boldsymbol{b} < -\boldsymbol{a}$.

In an ordered vectoid, we have additionally

(m) $o \leq a \leq b \wedge \boldsymbol{o} \leq \boldsymbol{c} \leq \boldsymbol{d} \Rightarrow \boldsymbol{o} \leq ac \leq bd$,
(n) $a \leq b \leq o \wedge \boldsymbol{o} \leq \boldsymbol{c} \leq \boldsymbol{d} \Rightarrow ad \leq bc \leq \boldsymbol{o}$,
(o) $a \leq b \leq o \wedge \boldsymbol{c} \leq \boldsymbol{d} \leq \boldsymbol{o} \Rightarrow \boldsymbol{o} \leq bd \leq ac$,
(p) $o \leq a \leq b \wedge \boldsymbol{c} \leq \boldsymbol{d} \leq \boldsymbol{o} \Rightarrow bc \leq ad \leq \boldsymbol{o}$.

In an ordered multiplicative vectoid, we have further

(q) $\boldsymbol{o} \leq \boldsymbol{a} \leq \boldsymbol{b} \wedge \boldsymbol{o} \leq \boldsymbol{c} \leq \boldsymbol{d} \Rightarrow \boldsymbol{o} \leq \boldsymbol{ac} \leq \boldsymbol{bd} \wedge \boldsymbol{o} \leq \boldsymbol{ca} \leq \boldsymbol{db}$,
(r) $\boldsymbol{a} \leq \boldsymbol{b} \leq \boldsymbol{o} \wedge \boldsymbol{o} \leq \boldsymbol{c} \leq \boldsymbol{d} \Rightarrow \boldsymbol{ad} \leq \boldsymbol{bc} \leq \boldsymbol{o} \wedge \boldsymbol{da} \leq \boldsymbol{cb} \leq \boldsymbol{o}$,
(s) $\boldsymbol{a} \leq \boldsymbol{b} \leq \boldsymbol{o} \wedge \boldsymbol{c} \leq \boldsymbol{d} \leq \boldsymbol{o} \Rightarrow \boldsymbol{o} \leq \boldsymbol{bd} \leq \boldsymbol{ac} \wedge \boldsymbol{o} \leq \boldsymbol{db} \leq \boldsymbol{ca}$.

In a completely and weakly ordered vectoid, we have

(t) $\bigwedge_{\boldsymbol{a} \in \boldsymbol{P}V} \inf \boldsymbol{a} = -\sup(-\boldsymbol{a}) \wedge \sup \boldsymbol{a} = -\inf(-\boldsymbol{a})$,

(u) In an inclusion-isotonally ordered vectoid $\{V, R, \subseteq\}$, (OV5b) holds for subtraction:

$$\bigwedge_{\boldsymbol{a},\boldsymbol{b},\boldsymbol{c},\boldsymbol{d} \in V} (\boldsymbol{a} \subseteq \boldsymbol{b} \wedge \boldsymbol{c} \subseteq \boldsymbol{d} \Rightarrow \boldsymbol{a} - \boldsymbol{c} \subseteq \boldsymbol{b} - \boldsymbol{d}).$$

*Proof*: We omit the proof of this theorem since using the corresponding properties of a vectoid, it is completely analogous to that of Theorem 2.2. ■

Theorem 2.9 allows us to assert that in a vectoid or a multiplicative vectoid the same rules for the minus operator hold as in a vector space or a matrix algebra over the real number field. Compared to other algebraic structures, a vectoid is distinguished by the fact that the existence of inverse elements is not assumed. Nevertheless, subtraction is not an independent operation. It can be defined by means of the outer multiplication and addition.

In an ordered vectoid (resp. multiplicative vectoid) for all elements that are comparable with $\boldsymbol{o}$ with respect to $\leq$ and $\geq$, the same rules for inequalities hold as in the vector space $\boldsymbol{R}^n$ over the real numbers (resp. in the algebra of matrices over the real numbers). In a weakly ordered vectoid for all elements that are comparable with $\boldsymbol{o}$ with respect to $\leq$ and $\geq$, the same rules for inequalities hold as in the vector space $\boldsymbol{C}^n$ over the complex numbers.

In a multiplicative vectoid, the inner operations $+$ and $\cdot$ always have the properties (D1)–(D5) of a ringoid. These properties (D1)–(D5a) are identical with (V1)–(V5), while (D5b)–(D5d) can easily be proved:

$$\text{(D5b)}: \quad (-\boldsymbol{e})(-\boldsymbol{e}) \underset{\text{Theorem 2.9(i)}}{=} \boldsymbol{e} \cdot \boldsymbol{e} = \boldsymbol{e}.$$

$$\text{(D5c)}: \quad (-\boldsymbol{e})(\boldsymbol{a} \cdot \boldsymbol{b}) \underset{\text{(h)}}{=} -(\boldsymbol{ab}) \underset{\text{(VD5)}}{=} (-\boldsymbol{a})\boldsymbol{b} = \boldsymbol{a}(-\boldsymbol{b}) \underset{\text{(h)}}{=} ((-\boldsymbol{e})\boldsymbol{a})\boldsymbol{b} = \boldsymbol{a}((-\boldsymbol{e})\boldsymbol{b}).$$

$$\text{(D5d)}: \quad (-\boldsymbol{e})(\boldsymbol{a} + \boldsymbol{b}) \underset{\text{(h)}}{=} -(\boldsymbol{a} + \boldsymbol{b}) \underset{\text{(VD4)}}{=} (-\boldsymbol{a}) + (-\boldsymbol{b}) \underset{\text{(h)}}{=} (-\boldsymbol{e})\boldsymbol{a} + (-\boldsymbol{e})\boldsymbol{b}.$$

It can be shown by examples [11], however, that a multiplicative vectoid is not necessarily a ringoid.

Power sets over vectoids are vectoids as well. This property is made precise in the following theorem.

**Theorem 2.10:** If $\{V, R\}$ is a vectoid with the neutral element $\boldsymbol{o}$, then $\{\boldsymbol{P}V, \boldsymbol{P}R, \subseteq\}$ is an inclusion-isotonally ordered vectoid with the neutral element $\{\boldsymbol{o}\}$. If $\{V, R\}$ is multiplicative and $\boldsymbol{e}$ is the neutral element of multiplication, then $\{\boldsymbol{P}V, \boldsymbol{P}R, \subseteq\}$ is an inclusion-isotonally ordered multiplicative vectoid and $\{\boldsymbol{e}\}$ is its neutral element of multiplication.

*Proof*: The proof is straightforward. All properties can be obtained from the corresponding properties in $\{V, R\}$ and the definition of the operations in the power set. ■

Once again let us consider matrices over a given ringoid $R$. In addition to the inner operations and the order relation defined above in $M_{mn}R$, we now define an outer multiplication by

$$\bigwedge_{a \in R} \bigwedge_{(a_{ij}) \in M_{mn}R} a \cdot (a_{ij}) := (a \cdot a_{ij}).$$

Then the following theorem holds.

**Theorem 2.11:** Let $\{R, +, \cdot\}$ be a ringoid with the neutral elements $o$ and $e$. Then $\{M_{mn}R, R\}$ is a vectoid. The neutral element is the matrix all components of which are $o$. If $\{R, +, \cdot, \leq\}$ is a weakly ordered (resp. an ordered) ringoid, then $\{M_{mn}R, R, \leq\}$ is a weakly ordered (resp. an ordered) vectoid.

*Proof*: We omit the proof of this theorem since it is straightforward. ■

For $m = 1$ we obtain matrices that consist only of one column. Such matrices are called vectors. The product set $V_nR := R \times R \times \cdots \times R = M_{1n}$ represents the set of all vectors with $n$ components. In this case the last theorem specializes into the following form.

**Theorem 2.12:** Let $\{R, +, \cdot\}$ be a ringoid with the neutral elements $o$ and $e$. Then $\{V_nR, R\}$ is a vectoid. The neutral element is the zero vector, i.e., the vector all components of which are $o$. If $\{R, +, \cdot, \leq\}$ is a weakly ordered (resp. an ordered) ringoid, then $\{V_nR, R, \leq\}$ is a weakly ordered (resp. an ordered) vectoid. ■

Another important specialization of the $m \times n$-matrices is the set of $n \times n$-matrices $M_nR$. In this case we obtain the following specialization of Theorem 2.11.

**Theorem 2.13:** Let $\{R, +, \cdot\}$ be a ringoid with the neutral elements $o$ and $e$. Then $\{M_nR, R\}$ is a multiplicative vectoid. The neutral elements are the zero matrix and the unit matrix. If $\{R, +, \cdot, \leq\}$ is a weakly ordered (resp. an ordered) ringoid, then $\{M_nR, R, \leq\}$ is a weakly ordered (resp. an ordered) multiplicative vectoid.

*Proof*: (V1)–(V5) follow immediately from the properties (D1)–(D5a) of the ringoid $M_nR$. Likewise, (OV4) follows from (OD3). The properties (VD1)–(VD4) and (OV1)–(OV3) were already verified in Theorem 2.11. The remaining property (VD5) follows from (D5c) in $M_nR$ and the property (h) of Theorem 2.9. ■

Among the consequences of these theorems we may note again that $\{V_n\mathbf{R}, \mathbf{R}, \leq\}$ is an ordered vectoid, while $\{V_n\mathbf{C}, \mathbf{C}, \leq\}$ is a weakly ordered vectoid, $\{M_n\mathbf{R}, \mathbf{R}, \leq\}$ is an ordered multiplicative vectoid, and $\{M_n\mathbf{C}, \mathbf{C}, \leq\}$ is a weakly ordered multiplicative vectoid. We shall use these theorems later for many other applications.

We have also defined a multiplication of an $m \times n$ matrix by an $n \times p$ matrix. The most important case of this multiplication is that of a matrix by a vector. The result then is a vector, and the structure to be described in the following theorem matches with that of a vectoid.

**Theorem 2.14:** Let $\{R, +, \cdot\}$ be a ringoid, $\{M_nR, +, \cdot\}$ the ringoid of $n \times n$ matrices over $R$, and $V_nR$ the set of $n$-tuples over $R$. Then $\{V_nR, M_nR\}$ is a vectoid. The neutral element is the zero vector. If $\{M_nR, +, \cdot, \leq\}$ is a weakly ordered (resp. an ordered) ringoid, then $\{V_nR, M_nR, \leq\}$ is a weakly ordered (resp. an ordered) vectoid.

*Proof*: The properties (V1), (V2), (VD1), (VD2), (VD4), (OV1), and (OV2) follow immediately from the definition of the operations. The properties

(VD3) and (OV3) can be proved in complete analogy to (D5c) and (OD3) in $M_n R$ (Theorem 2.5). ■

As a consequence of the theorem, of course, $\{V_n\mathbf{R}, M_n\mathbf{R}, \leq\}$ is an ordered vectoid, while $\{V_n\mathbf{C}, M_n\mathbf{C}, \leq\}$ is a weakly ordered vectoid.

We conclude this section by pointing out that ringoids and vectoids also occur in sets of mappings. Let $\{R, +, \cdot\}$ be a ringoid with the neutral elements $o$ and $e$ and let $M$ be the set of mappings of a given set $S$ into $R$, i.e., $M := \{x \mid x{:}S \to R\}$. In $M$ we define an equality, addition, and outer and inner multiplication respectively by

$$x = y :\Leftrightarrow \bigwedge_{t \in S} x(t) = y(t),$$

$$\bigwedge_{x,y \in M} (x + y)(t) := x(t) + y(t),$$

$$\bigwedge_{a \in R} \bigwedge_{x \in M} (a \cdot x)(t) := a \cdot x(t),$$

$$\bigwedge_{x,y \in M} (x \cdot y)(t) := x(t) \cdot y(t).$$

If $\{R, \leq\}$ is an ordered set, we define an order relation $\leq$ in $M$ by

$$x \leq y :\Leftrightarrow \bigwedge_{t \in S} x(t) \leq y(t).$$

Then the following theorem, which characterizes sets of mappings in terms of the structures in question, is easily verified.

**Theorem 2.15:** Let $\{R, +, \cdot\}$ be a ringoid with the neutral elements $o$ and $e$. Then $\{M, R\}$ is a multiplicative vectoid. The neutral elements are $o(t) = o$ and $e(t) = e$ for all $t \in S$. If $\{R, +, \cdot, \leq\}$ is a weakly ordered (resp. an ordered) ringoid, then $\{M, R, \leq\}$ is a weakly ordered (resp. an ordered) vectoid. With respect to the inner operations, $\{M, +, \cdot\}$ is a ringoid. ■

## 3. COMPLETELY ORDERED RINGOIDS AND VECTOIDS

By definition, the real numbers are a conditionally complete and a linearly ordered field (see Chapter 5). Derived spaces such as those occurring in the second column of Fig. 1 are also conditionally complete with respect to the order relation $\leq$. If the real numbers are completed in the customary manner, the least element $-\infty$ and the greatest element $+\infty$ fail to satisfy several of the algebraic properties of a field. For example, $a + \infty = b + \infty$ even if $a < b$, so that the cancellation law is not valid.

We have already seen that some properties of the spaces $\boldsymbol{R}, M_n\boldsymbol{R}, V_n\boldsymbol{R}$, $C, M_nC, V_nC$, for instance, are described in terms of the structures of ordered or weakly ordered ringoids and vectoids. If the latter are conditionally complete, they always can be completed so that the least and greatest element also satisfy all the algebraic and order properties. This is expressed by the following two theorems. Moreover, because of this fact, it is not necessary to distinguish between conditionally complete and completely ordered or weakly ordered ringoids and vectoids. Thus we may always assume that the ordering is complete. Formal justification of this assumption is provided by the following theorems.

**Theorem 2.16:** Let $\{R, +, \cdot, \leq\}$ be an ordered (resp. a weakly ordered) ringoid, which is conditionally complete and has the neutral elements $o$ and $e$. If we adjoin a least element $-p$ and a greatest element $+p$ and define the addition and multiplication by the tables of Fig. 16, then $\{R \cup \{-p\} \cup \{+p\}, +, \cdot, \leq\}$ is a completely ordered (resp. weakly ordered) ringoid. Let $\{R, N, +, \cdot, /, \leq\}$ be a division ringoid to which we adjoin $-p$ and $p$.[†] Let divisions involving $-p$ and $p$ be defined as in Fig. 16. Then $\{R \cup \{-p\} \cup \{+p\}, N, +, \cdot, /, \leq\}$ is a completely (resp. weakly) ordered ringoid.

*Proof*:

(D1): The table for addition in Fig. 16 is symmetric with respect to the main diagonal.

(D2): $p + o = p$ and $-p + o = -p$.

(D3): $(-p) \cdot e = -p$ and $(+p) \cdot e = +p$.

(D4): $p \cdot o = o$ and $(-p) \cdot o = o$ and $o \cdot p = o$ and $o \cdot (-p) = o$.

(D5c):
$$\begin{aligned}(-e)(ap) &= (-a)p = a(-p) = (-e)(pa)\\ &= (-p)a = p(-a)\\ &= \begin{cases} -p & \text{for} \quad a \in R \wedge a > 0\\ +p & \text{for} \quad a \in R \wedge a < 0.\end{cases}\end{aligned}$$

$$\begin{aligned}(-e)(a(-p)) &= (-a)(-p) = ap = (-e)((-p)a)\\ &= pa = (-p)(-a)\\ &= \begin{cases} +p & \text{for} \quad a \in R \wedge a > 0\\ -p & \text{for} \quad a \in R \wedge a < 0.\end{cases}\end{aligned}$$

$$(-e)(pp) = -p = (-p)p = p(-p) = (-e)((-p)(-p)).$$
$$(-e)(p(-p)) = (-e)((-p)p) = p = (-p)(-p) = pp.$$

[†] We often simply write $p$ instead of $+p$.

addition $a + b$

| $a \setminus b$ | $b \in R$ | $-p$ | $+p$ |
|---|---|---|---|
| $a \in R$ | $a + b$ | $-p$ | $+p$ |
| $-p$ | $-p$ | $-p$ | $p - p$ |
| $+p$ | $+p$ | $p - p$ | $p + p$ |

with $-(p - p) = p - p$.

multiplication $a \cdot b$

| $a \setminus b$ | $-p < b < o$ | $o < b < p$ | $-p$ | $-e$ | $o$ | $e$ | $+p$ | $b \| o$ |
|---|---|---|---|---|---|---|---|---|
| $-p < a < o$ | $a \cdot b$ | $a \cdot b$ | $+p$ | $-a$ | $o$ | $a$ | $-p$ | $a \cdot b$ |
| $o < a < +p$ | $a \cdot b$ | $a \cdot b$ | $-p$ | $-a$ | $o$ | $a$ | $+p$ | $a \cdot b$ |
| $-p$ | $+p$ | $-p$ | $+p$ | $+p$ | $o$ | $-p$ | $-p$ | $(-p)b$ |
| $-e$ | $-b$ | $-b$ | $+p$ | $e$ | $o$ | $-e$ | $-p$ | $-b$ |
| $o$ | $o$ | $o$ | $o$ | $o$ | $o$ | $o$ | $o$ | $o$ |
| $e$ | $b$ | $b$ | $-p$ | $-e$ | $o$ | $e$ | $+p$ | $b$ |
| $+p$ | $-p$ | $+p$ | $-p$ | $-p$ | $o$ | $+p$ | $+p$ | $(+p)b$ |
| $a \| o$ | $a \cdot b$ | $a \cdot b$ | $a(-p)$ | $-a$ | $o$ | $a$ | $a(+p)$ | $a \cdot b$ |

division $a/b$

| $a \setminus b$ | $-p < b < o$ | $o < b < +p$ | $-p$ | $e$ | $+p$ | $b \| o$ |
|---|---|---|---|---|---|---|
| $-p < a < o$ | $a/b$ | $a/b$ | $o$ | $a$ | $o$ | $a/b$ |
| $o < a < +p$ | $a/b$ | $a/b$ | $o$ | $a$ | $o$ | $a/b$ |
| $-p$ | $+p$ | $-p$ | $(-p)/(-p)$ | $-p$ | $(-p)/(+p)$ | $(-p)/b$ |
| $o$ | $o$ | $o$ | $o$ | $o$ | $o$ | $o$ |
| $+p$ | $-p$ | $+p$ | $(+p)/(-p)$ | $+p$ | $(+p)/(+p)$ | $(+p)/b$ |
| $a \| o$ | $a/b$ | $a/b$ | $a/(-p)$ | $a$ | $a/(+p)$ | $a/b$ |

with $(-p)/(-p) = p/p > o \wedge (-p)/(+p) = p/(-p) < o$.

FIGURE 16. Definition of the operations for the least and the greatest element of a ringoid.

(D5d): For all $a \in R$ we have

$$(-e)(a+p) = -p = (-a) + (-p)$$

and

$$(-e)(a-p) = p = (-a) + p,$$
$$(-e)(p+p) = -p = -p - p,$$
$$(-e)(-p-p) = p = p + p, \quad (-e)(p-p) = -p + p.$$

(D6): $p \cdot p = p > e$ and $(-p) \cdot (-p) = p > e$.

Suppose $p = e$. Then by (D5a), $-e + p = o$. But referring to Fig. 16, we see that $-e + p = p$. Thus $p = o$, which is a contradiction.

(D7): $(-p)/e = -p$ and $p/e = p$.

(D8): $o/(-p) = o$ and $o/p = o$.

(D9): The proof is analogous to the proof of (D5c).

(OD1): $-p \leq a \leq p \wedge -p \neq c \neq p \Rightarrow -p + c = -p \leq a + c \leq p + c = p,$
$-p \leq a \leq p \wedge c = -p \Rightarrow -p - p = -p \leq a - p \leq p - p,$
$-p \leq a \leq p \wedge c = p \Rightarrow -p + p \leq a + p \leq p + p = p.$

(OD2): $-p \leq a \leq p \Rightarrow (-e)(-p) = p \geq -a \geq (-e)p = -p.$

(OD3): $o \leq a \leq p \wedge o \leq c \neq p \Rightarrow ac \leq pc \wedge ca \leq cp,$
$o \leq a \leq p \wedge o \leq c = p \Rightarrow ap \leq pp = p,$
$o \leq a \leq b \neq p \wedge c = p \Rightarrow o \leq ap \leq bp.$

(OD4): $o < a \leq b \neq p \wedge p > o \Rightarrow o \leq a/p = o \leq b/p = o \wedge p/a = p \geq p/b = p \geq o,$
$o < a \leq p \wedge p \neq c > o \Rightarrow o \leq a/c \leq p/c = p \wedge c/a \geq c/p = o \geq o,$
$o < a \leq p \wedge p > o \Rightarrow o \leq a/p \leq p/p \wedge p/a \geq p/p \geq o.$ ∎

Occasionally elements $-p$ and $+p$ are defined on computers. Their implementation should be made with extreme care and the user informed in the case of an occurrence of such an element. The results $0 \cdot (-p) = (-p) \cdot 0 = 0 \cdot p = p \cdot 0 = 0$ may be unexpected.

Whenever a result in Fig. 16 is not explicitly defined, a more precise definition is not necessary in order to satisfy the axioms of an ordered or weakly ordered ringoid. In any specific definition, however, (D5c) and (D9) are to be matched.

We conclude Chapter 2 with the following theorem, which is the vectoidal analog of Theorem 2.16.

addition $\boldsymbol{a} + \boldsymbol{b}$

| $\boldsymbol{a}$ \ $\boldsymbol{b}$ | $\boldsymbol{b} \in V$ | $-\boldsymbol{p}$ | $+\boldsymbol{p}$ |
|---|---|---|---|
| $\boldsymbol{a} \in V$ | $\boldsymbol{a} + \boldsymbol{b}$ | $-\boldsymbol{p}$ | $+\boldsymbol{p}$ |
| $-\boldsymbol{p}$ | $-\boldsymbol{p}$ | $-\boldsymbol{p}$ | $\boldsymbol{p} - \boldsymbol{p}$ |
| $+\boldsymbol{p}$ | $+\boldsymbol{p}$ | $\boldsymbol{p} - \boldsymbol{p}$ | $+\boldsymbol{p}$ |

with $-(\boldsymbol{p} - \boldsymbol{p}) = \boldsymbol{p} - \boldsymbol{p}$

outer multiplication $a \cdot \boldsymbol{a}$

| $a$ \ $\boldsymbol{a}$ | $-\boldsymbol{p} < \boldsymbol{a} < \boldsymbol{o}$ | $\boldsymbol{o} < \boldsymbol{a} < +\boldsymbol{p}$ | $-\boldsymbol{p}$ | $\boldsymbol{o}$ | $+\boldsymbol{p}$ | $\boldsymbol{a} \parallel \boldsymbol{o}$ |
|---|---|---|---|---|---|---|
| $-p < a < o$ | $a\boldsymbol{a}$ | $a\boldsymbol{a}$ | $+\boldsymbol{p}$ | $\boldsymbol{o}$ | $-\boldsymbol{p}$ | $a\boldsymbol{a}$ |
| $o < a < p$ | $a\boldsymbol{a}$ | $a\boldsymbol{a}$ | $-\boldsymbol{p}$ | $\boldsymbol{o}$ | $+\boldsymbol{p}$ | $a\boldsymbol{a}$ |
| $-p$ | $+\boldsymbol{p}$ | $-\boldsymbol{p}$ | $+\boldsymbol{p}$ | $\boldsymbol{o}$ | $-\boldsymbol{p}$ | $(-p)\boldsymbol{a}$ |
| $-e$ | $-\boldsymbol{a}$ | $-\boldsymbol{a}$ | $+\boldsymbol{p}$ | $\boldsymbol{o}$ | $-\boldsymbol{p}$ | $-\boldsymbol{a}$ |
| $o$ | $\boldsymbol{o}$ | $\boldsymbol{o}$ | $\boldsymbol{o}$ | $\boldsymbol{o}$ | $\boldsymbol{o}$ | $\boldsymbol{o}$ |
| $e$ | $\boldsymbol{a}$ | $\boldsymbol{a}$ | $-\boldsymbol{p}$ | $\boldsymbol{o}$ | $+\boldsymbol{p}$ | $\boldsymbol{a}$ |
| $+p$ | $-\boldsymbol{p}$ | $+\boldsymbol{p}$ | $-\boldsymbol{p}$ | $\boldsymbol{o}$ | $+\boldsymbol{p}$ | $(+p)\boldsymbol{a}$ |
| $a \parallel o$ | $a\boldsymbol{a}$ | $a\boldsymbol{a}$ | $a(-\boldsymbol{p})$ | $\boldsymbol{o}$ | $a\boldsymbol{p}$ | $a\boldsymbol{a}$ |

multiplication $\boldsymbol{a} \cdot \boldsymbol{b}$

| $\boldsymbol{a}$ \ $\boldsymbol{b}$ | $-\boldsymbol{p} < \boldsymbol{b} < \boldsymbol{o}$ | $\boldsymbol{o} < \boldsymbol{b} < +\boldsymbol{p}$ | $-\boldsymbol{p}$ | $\boldsymbol{o}$ | $\boldsymbol{e}$ | $+\boldsymbol{p}$ | $\boldsymbol{b} \parallel \boldsymbol{o}$ |
|---|---|---|---|---|---|---|---|
| $-\boldsymbol{p} < \boldsymbol{a} < \boldsymbol{o}$ | $\boldsymbol{ab}$ | $\boldsymbol{ab}$ | $+\boldsymbol{p}$ | $\boldsymbol{o}$ | $\boldsymbol{a}$ | $-\boldsymbol{p}$ | $\boldsymbol{ab}$ |
| $\boldsymbol{o} < \boldsymbol{a} < +\boldsymbol{p}$ | $\boldsymbol{ab}$ | $\boldsymbol{ab}$ | $-\boldsymbol{p}$ | $\boldsymbol{o}$ | $\boldsymbol{a}$ | $+\boldsymbol{p}$ | $\boldsymbol{ab}$ |
| $-\boldsymbol{p}$ | $+\boldsymbol{p}$ | $-\boldsymbol{p}$ | $+\boldsymbol{p}$ | $\boldsymbol{o}$ | $-\boldsymbol{p}$ | $-\boldsymbol{p}$ | $(-\boldsymbol{p})\boldsymbol{b}$ |
| $\boldsymbol{o}$ | $\boldsymbol{o}$ | $\boldsymbol{o}$ | $\boldsymbol{o}$ | $\boldsymbol{o}$ | $\boldsymbol{o}$ | $\boldsymbol{o}$ | $\boldsymbol{o}$ |
| $\boldsymbol{e}$ | $\boldsymbol{b}$ | $\boldsymbol{b}$ | $-\boldsymbol{p}$ | $\boldsymbol{o}$ | $\boldsymbol{e}$ | $+\boldsymbol{p}$ | $\boldsymbol{b}$ |
| $+\boldsymbol{p}$ | $-\boldsymbol{p}$ | $+\boldsymbol{p}$ | $-\boldsymbol{p}$ | $\boldsymbol{o}$ | $+\boldsymbol{p}$ | $+\boldsymbol{p}$ | $(+\boldsymbol{p})\boldsymbol{b}$ |
| $\boldsymbol{a} \parallel \boldsymbol{o}$ | $\boldsymbol{ab}$ | $\boldsymbol{ab}$ | $\boldsymbol{a}(-\boldsymbol{p})$ | $\boldsymbol{o}$ | $\boldsymbol{a}$ | $\boldsymbol{a}(+\boldsymbol{p})$ | $\boldsymbol{ab}$ |

FIGURE 17. Definition of the operations for the least and greatest element of a vectoid.

**Theorem 2.17:** Let $\{R, +, \cdot, \leq\}$ be a completely and weakly ordered ringoid with the neutral elements $o$ and $e$, and let $\{V, R, \leq\}$ be an ordered (resp. weakly ordered) vectoid that is conditionally complete. If we adjoin a least element $-\boldsymbol{p}$ and a greatest element $+\boldsymbol{p}$ and define addition and outer multiplication by the tables of Fig. 17, then $\{V \cup \{-\boldsymbol{p}\} \cup \{+\boldsymbol{p}\}, R, \leq\}$ is a completely ordered (resp. a completely and weakly ordered) vectoid. If $\{V, R, \leq\}$ is multiplicative and if for $-\boldsymbol{p}$ and $+\boldsymbol{p}$ a multiplication is defined as in Fig. 17, then $\{V \cup \{-\boldsymbol{p}\} \cup \{+\boldsymbol{p}\}, R, \leq\}$ is a completely ordered (resp. a completely and weakly ordered) multiplicative vectoid.

*Proof*: The proof of this theorem is analogous to the proof of the previous theorem. ■

Whenever a result in Fig. 17 is not explicitly defined, a more precise definition is not necessary in order to fulfill the axioms of an ordered or weakly ordered vectoid. In any specific definition, however, (VD3) (resp. (VD5)) are to be matched.

# Chapter 3 / DEFINITION OF COMPUTER ARITHMETIC

**Summary:** We have established that the spaces listed in the leftmost element in every row in Fig. 1 are ringoids and vectoids with certain order properties such as being weakly ordered or ordered or inclusion-isotonally ordered. We are now going to show that these structures recur in the subsets on the right-hand side of Fig. 1 provided that the arithmetic and mapping properties are properly defined. In Section 1 of this chapter we give a heuristic approach to the mapping concept of a semimorphism. In Section 2 we derive some fundamental properties of semimorphisms and certain other mappings. In Section 3 we specify the vertical definition of computer arithmetic and show that it produces ordered and weakly ordered ringoids in certain subsets of Fig. 1. In Section 4 we introduce the horizontal definition of computer arithmetic by means of semimorphisms. We show expressly for all rows of Fig. 1 that do not include interval sets that weakly ordered or ordered ringoids and vectoids are invariant with respect to semimorphisms. The horizontal definition will also be used in Chapter 4 in order to derive similar results for the rows of Fig. 1 that correspond to sets of intervals.

## 1. INTRODUCTION

Let $M$ be a set in which certain operations and relations are defined, and let $\bar{M}$ be a set of rules or axioms given for these operations. By way of example, the commutative law of addition might be one such rule. Then we call the pair $\{M, \bar{M}\}$ a structure. We shall also refer to $\{M, \bar{M}\}$ as the structure in the set $M$. The real or complex numbers, vectors, or matrices are well-known structures. We now assume that certain operations in $\{M, \bar{M}\}$ cannot be performed in a finite amount of time. As an example, take the addition of irrational numbers. We are then obliged to replace $\{M, \bar{M}\}$ by a structure $\{S, \bar{S}\}$, the operations of which are easily and speedily performable and which lead to good approximations to the operations in $\{M, \bar{M}\}$. We shall always take $S \subseteq M$.

In all significant applications there exist at least two operations in $M$: an addition $+$ and a multiplication $\cdot$, which have neutral elements or an identity operator in the case of an outer multiplication. Let us denote these elements by $o$ resp. $e$. In addition, we shall always have a minus operator. We assume therefore that the subset $S$ has the property

(S3) $o, e \in S \wedge \bigwedge_{a \in S} -a \in S.$

We further assume that the elements of $M$ are mapped into the subset $S$ by a rounding defined by

(R1) $\bigwedge_{a \in S} \square a = a.$

In attempting to approximate a given structure $\{M, \bar{M}\}$ by a structure $\{S, \bar{S}\}$ with $S \subseteq M$, we are motivated to employ a mapping having useful properties such as an isomorphism or a homomorphism. For the sake of clarity, we give the definition of these concepts for ordered algebraic structures.

**Definition 3.1:** Let $\{M, \bar{M}\}$ and $\{S, \bar{S}\}$ be ordered algebraic structures, and let a one-to-one correspondence exist between the corresponding operations and order relation(s). A mapping $\square : M \to S$ is called a *homomorphism* if it is an algebraic homomorphism, i.e., if

$$\bigwedge_{a,b \in M} (\square a) \boxast (\square b) = \square(a * b) \tag{1}$$

for all corresponding operations $*$ and $\boxast$ and if it is an order homomorphism, i.e., if

$$\bigwedge_{a,b \in M} (a \le b \Rightarrow \square a \le \square b). \tag{2}$$

A homomorphism is called an isomorphism if $\square : M \to S$ is a one-to-one mapping of $M$ onto $S$ and

$$\bigwedge_{a,b \in S} (a \le b \Rightarrow \square^{-1} a \le \square^{-1} b). \quad \blacksquare$$

Thus an isomorphism requires a one-to-one mapping. However, in our applications $S$ is in general a subset of $M$ of different cardinality. There does not exist a one-to-one mapping and certainly no isomorphism between sets of different cardinality.

A homomorphism preserves the structure of a group, a ring, or a vector space. However, we show by a simple example that the approximation of $\{M, \bar{M}\}$ by $\{S, \bar{S}\}$ by means of a homomorphism cannot be realized in a sensible way.

**Example:** Let $S$ be a floating-point system[†] that is defined as a set of numbers of the form $x = m \cdot b^e$. $m$ is called the mantissa, $b$ the base of the system, and $e$ the exponent. Let $b = 10$, let the range of the mantissa be $m = -0.9(0.1)0.9$, and let the exponent $e \in \{-1, 0, 1\}$. If $\square : \boldsymbol{R} \to S$ denotes the rounding to the nearest number of $S$ and $x = 0.34$, $y = 0.54 \in \boldsymbol{R}$, we get

$$\square x = 0.3, \qquad \square y = 0.5,$$
$$(\square x) \boxplus (\square y) = \square((\square x) + (\square y)) = \square(0.8) = 0.8,$$
$$\square(x + y) = \square(0.88) = 0.9, \text{ i.e.,}$$
$$(\square x) \boxplus (\square y) \neq \square(x + y). \quad \blacksquare$$

It seems, however, desirable to stay as close to a homomorphism as possible. We therefore derive some necessary conditions for a homomorphism. If we restrict (1) to elements of $S$, then because of (R1) we obtain

(RG) $\bigwedge_{a,b \in S} a \boxast b = \square(a * b).$

We shall use this formula to define the operation $\boxast$ in $S$ by means of the corresponding operation $*$ in $M$ and the rounding $\square : M \to S$. From (2) we see that the rounding has to be monotone

(R2) $\bigwedge_{a,b \in M} (a \leq b \Rightarrow \square a \leq \square b).$

Now in the case of multiplication in (1), replace $a$ by the negative multiplicative unit $-e$. Then

$$\bigwedge_{b \in M} \square(-b) = \square(-e) \boxdot \square b \underset{(S3),(R1)}{=} (-e) \boxdot \square b$$
$$\underset{(R)}{=} \square(-\square b) \underset{(S3),(R1)}{=} -\square b,$$

i.e.,

(R4) $\bigwedge_{a \in M} \square(-a) = -\square a.$

Thus the rounding has to be antisymmetric.

[†] See Chapter 5 for a general definition of a floating-point system.

Specifically (see Definition 3.2), we shall call a mapping a semimorphism if it has the properties (R1), (R2), (R4), and (RG).

The conditions (R1), (R2), and (R4) do not define the rounding uniquely. We shall see, however, in this chapter and in Chapter 4 that in all cases of Fig. 1, ordered or weakly ordered ringoids and vectoids as well as inclusion-isotonally ordered ringoids and vectoids are invariant with respect to semimorphisms. We shall see further in Chapter 6 that semimorphisms can be implemented on computers by fast algorithms in all cases of Fig. 1. These properties give to the concept of semimorphism a significance that is quite independent of the present derivation of it via the necessary conditions for a homomorphism.

## 2. PRELIMINARIES

In ringoids and vectoids there is a minus operator available. We use it in order to extend the concepts of a screen and of a rounding. We do this in the following definition, where we also formalize the notion of a semimorphism.

**Definition 3.2:** Let $M$ be a ringoid or a vectoid and $o$ the neutral element of addition and $e$ the neutral element of the multiplication if it exists. Further, let $\{M, \leq\}$ be a complete lattice.

(a) A screen $\{S, \leq\}$ (resp. a lower; resp. an upper screen) is called *symmetric* if

(S3) $o, e \in S \wedge \bigwedge_{a \in S} -a \in S.$

(b) A rounding of $M$ into a symmetric screen $S$, $\square : M \to S$, is called *antisymmetric* if

(R4) $\bigwedge_{a \in M} \square(-a) = -\square a$ (antisymmetric).

(c) A mapping of the ringoid or vectoid $M$ into a symmetric lower (resp. upper) screen (resp. a symmetric screen) is called a *semimorphism* if it is a monotone and antisymmetric rounding, i.e., has the properties (R1), (R2), (R4), and if all inner and outer operations in $S$ are defined by

(RG) $\bigwedge_{a,b} a \boxast b := \square(a * b).$ ■

An interdependence of the special roundings $\bigtriangledown$ and $\triangle$ is the subject of the following theorem.

**Theorem 3.3:** If $M$ is a complete and weakly ordered ringoid or vectoid, $\{S, \leq\}$ a symmetric screen, and $\bigtriangledown : M \to S$ resp. $\triangle : M \to S$ the monotone downwardly resp. upwardly directed roundings, then

$$\bigwedge_{a \in M} (\triangledown a = -\triangle(-a) \wedge \triangle a = -\triangledown(-a)).$$

*Proof*: We give the proof only in the case of a ringoid. In the case of a vectoid, the proof can be given analogously by using the corresponding properties of Theorem 2.9.

If $\{M, +, \cdot\}$ is a ringoid, then $\{\boldsymbol{P}M, +, \cdot\}$ is also a ringoid. With

$$L(a) := \{b \in M \mid b \leq a\} \qquad \text{and} \qquad U(a) := \{b \in M \mid a \leq b\},$$

we obtain because of (OD2) that

$$-(L(a) \cap S) := \{b \in S \mid -b \leq a\} = \{b \in S \mid -a \leq b\} = U(-a) \cap S$$

and

$$-(U(a) \cap S) := \{b \in S \mid a \leq -b\} = \{b \in S \mid b \leq -a\} = L(-a) \cap S.$$

Using this and Theorem 2.2(y), we obtain

$$\sup(L(a) \cap S) = -\inf(-(L(a) \cap S)) = -\inf(U(-a) \cap S)$$

and

$$\inf(U(a) \cap S) = -\sup(-(U(a) \cap S)) = -\sup(L(-a) \cap S).$$

This and the general property $\triangledown a = \sup(L(a) \cap S) \wedge \triangle a = \inf(U(a) \cap S)$ prove the theorem. ■

This theorem can easily be illustrated in the case that $M$ is a linearly ordered ringoid and $S$ is a finite subset of $M$. See Fig. 18. It has been proved, independently, for much more general cases such as complexifications and vector or matrix sets.

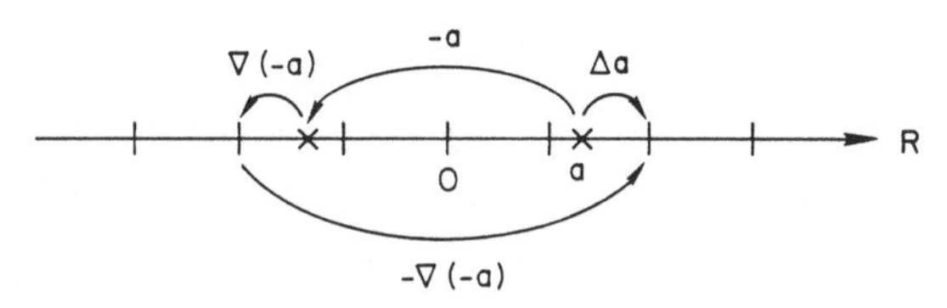


FIGURE 18. Dependence of the directed roundings in a linearly ordered ringoid.

Theorem 3.3 asserts that under its hypothesis the monotone directed roundings are not independent of each other. Thus only one of the monotone directed roundings is necessary when both are to be implemented on computers. The other one can be obtained by sign changes.

If $S \neq M$ and we take an $a \in S$ with $a \notin M$, then $\bigtriangledown a < a < \bigtriangleup a$. A comparison of Definition 3.2 and Theorem 3.3 shows that the monotone directed roundings are not antisymmetric under the hypothesis of the theorem. In contrast to this result, we shall consider in Chapter 4 several monotone upwardly directed roundings that are antisymmetric. In this case the screen and the monotone directed roundings are defined with respect to an order relation that is different from the one that orders the ringoid (resp. vectoid).

For applications in the next sections and Chapter 4, we provide the following two lemmas.

**Lemma 3.4:** Let $\{R, +, \cdot\}$ be a ringoid with the neutral elements $o$ and $e$, $\{R, \leq\}$ a complete lattice, $\{S, \leq\}$ a symmetric screen (resp. a symmetric lower; resp. upper screen) of $\{R, \leq\}$, and $\square: R \to S$ a semimorphism

(a) Then the structure $\{S, \boxplus, \boxdot\}$ has the properties (D1), (D2) for $o$, (D3) for $e$, (D4), (D5) for $-e$, as well as

(RG1) $\bigwedge_{a,b \in S} (a * b \in S \Rightarrow a \circledast b = a * b)$ for $* \in \{+, \cdot\}$,

(RG2) $\bigwedge_{a,b,c,d \in S} (a * b \leq c * d \Rightarrow a \circledast b \leq c \circledast d)$ for $* \in \{+, \cdot\}$

and

$$\bigwedge_{a \in S} -a = (-e) \boxdot a. \tag{3}$$

If $\{S, \boxplus, \boxdot\}$ is a ringoid, then (RG), as well as (RG1) and (RG2), also holds for subtraction, and (3) establishes the identity of the minus operators in $R$ and in $S$.

(b) If the rounding $\square: R \to S$ is upwardly (resp. downwardly) directed, then

(RG3) $\bigwedge_{a,b \in S} a \circledast b \leq a * b$ (resp $\bigwedge_{a,b \in S} a * b \leq a \circledast b$).

If $\{S, \boxplus, \boxdot\}$ is a ringoid, then (RG3) also holds for subtraction.

(c) If $\{R, +, \cdot, \leq\}$ is a weakly ordered (or an ordered) ringoid, then $\{S, \boxplus, \boxdot, \leq\}$ has in addition the properties (OD1) and (OD2) for $-e$ (and (OD3) in the ordered case).

*Proof*: (a): The proof of the properties (D1)–(D5b) is obtained in a straightforward manner by employing (RG) and the corresponding properties of the ringoid $R$. In order to prove (D5c), we prove (3) first:

(3): $\bigwedge_{a \in S} (-e) \boxdot a = \square(-a) \underset{(R4)}{=} -\square a \underset{(R1)}{=} -a \underset{(S3)}{\in} S.$

Then

$$\text{(D5c)}: \quad (-e) \boxdot (a \boxdot b) \underset{(RG)}{=} \square((-e) \cdot \square(ab)) \underset{(R4)}{=} \square(\square(-(ab)))$$
$$\underset{(R1)}{=} \square(-(ab)) \underset{(D5c)_R}{=} \square((-a)b) \underset{(D5c)_R}{=} \square(a(-b))$$
$$\underset{(RG)}{=} (-a) \boxdot b \underset{(RG)}{=} a \boxdot (-b) \underset{(3)}{=} ((-e) \boxdot a) \boxdot b$$
$$\underset{(3)}{=} a \boxdot ((-e) \boxdot b).$$

The proof of (D5d) is similar. (RG1) follows from (RG) and (R1). In case that $S$ is a ringoid, we have for all $a, b \in S$ that $a \boxminus b := a \boxplus (\boxminus b) =_{(3)} a \boxplus (-b) =_{(RG)+} \square(a - b)$, which is (RG1) for subtraction. (RG2) follows from (RG) and (R2).

(b): (RG3) follows immediately by employing (RG) and (R3).

(c): (OD1): $a \le b \underset{(OD1)_R}{\Rightarrow} a + c \le b + c \underset{(R2),(RG)}{\Rightarrow} a \boxplus c \le b \boxplus c.$

(OD2): $a \le b \underset{(OD2)_R}{\Rightarrow} -b \le -a \underset{(3)}{\Rightarrow} (-e) \boxdot b \le (-e) \boxdot a.$

(OD3): $o \le a \le b \wedge c \ge o \underset{(OD3)_R}{\Rightarrow} ac \le bc \underset{(R2),(RG)}{\Rightarrow} a \boxdot c \le b \boxdot c.$ ■

The following Lemma 3.5, which deals with the case of division ringoids, can be proved in an entirely similar manner.

**Lemma 3.5:** Let $\{R, N, +, \cdot, /\}$ be a division ringoid with the neutral elements $o$ and $e$, $\{R, \le\}$ a complete lattice, $\{S, \le\}$ a symmetric screen (resp. a symmetric lower; resp. upper screen) of $\{R, \le\}$, and $\square : R \to S$ a semimorphism.

(a) Then the structure $\{S, S \cap N, \boxplus, \boxdot, \boxslash\}$ has the properties (D1), (D2) for $o$, (D3) for $e$, (D4), (D5), and (D9) for $-e$, (D7), (D8), (RG1), and (RG2) for $* \in \{+, \cdot, /\}$.

(b) If the rounding $\square : R \to S$ is downwardly (resp. upwardly) directed, then (RG3) holds for $* \in \{+, \cdot, /\}$.

(c) If $\{R, N, +, \cdot, /, \le\}$ is an ordered division ringoid, then (OD1)–(OD4) hold in $\{S, S \cap N, \boxplus, \boxdot, \boxslash, \le\}$. ■

The properties (S3), (R1)–(R4), and (RG) were in fact used within the proofs of the last two lemmas. If we change these properties or if they are not strictly implemented on computers (even with respect to quantifiers), we get a different structure or no describable structure at all in the subset $S$.

The last two lemmas assert that if we generate $S$ as hypothesized, we nearly reobtain for $S$ itself the structure of a ringoid (resp. a division ringoid). What is missing is the establishment of the property (D6). A universal proof of this latter property seems to be a more difficult task, and we are obliged to establish it individually for all cases in Fig. 1.

Since neither an isomorphism nor a homomorphism between the basic set and the screen is achievable, we define arithmetic on the screen through semimorphisms. As was done in the case of a groupoid, we characterize the compatibility properties, which we then obtain between the operations in the basic set and in the screen by the following two definitions.

**Definition 3.6:** Let $\{R, +, \cdot\}$ be a ringoid, $\{R, \leq\}$ a complete lattice, and $\{S, \leq\}$ a symmetric screen (resp. a symmetric lower; resp. upper screen). A ringoid $\{S, \boxplus, \boxdot\}$ is called a *screen ringoid* if[†]

$$\text{(RG1)} \quad \bigwedge_{a,b \in S} (a * b \in S \Rightarrow a \circledast b = a * b), \quad * \in \{+, -, \cdot\}$$

and

$$\text{(RG4)} \quad \bigwedge_{a \in S} -a = \boxminus a.$$

A screen ringoid is called *monotone* if

$$\text{(RG2)} \quad \bigwedge_{a,b,c,d \in S} (a * b \leq c * d \Rightarrow a \circledast b \leq c \circledast d), \quad * \in \{+, -, \cdot\}.$$

It is called a *lower* (resp. *upper*) screen ringoid if

$$\text{(RG3)} \quad \bigwedge_{a,b \in S} a \circledast b \leq a * b \qquad \left(\text{resp. } \bigwedge_{a,b \in S} a * b \leq a \circledast b\right), \qquad * \in \{+, -, \cdot\}.$$

If $\{R, N, +, \cdot, /\}$ is a division ringoid, then a division ringoid $\{S, S \cap N, \boxplus, \boxdot, \boxslash\}$ is called a *screen division ringoid* (resp. a *monotone*; resp. *lower* or *upper* screen division ringoid) if (RG1) (resp. (RG2); resp. (RG3)) also holds for division with $b, d \in S \backslash (S \cap N)$. ■

A screen ringoid $\{S, \boxplus, \boxdot\}$ always has the same neutral elements as the ringoid $\{R, +, \cdot\}$ itself since

$$\bigwedge_{a \in S} \left(a + o \in S \wedge a \cdot e \in S \underset{\text{(RG1)}}{\Rightarrow} a \boxplus o = a + o = a \wedge a \boxdot e = a \cdot e = a\right).$$

**Definition 3.7:** Let $\{V, R\}$ be a vectoid, $\{V, \leq\}$ a complete lattice, and $\{S, \leq\}$ a symmetric screen (resp. a symmetric lower; resp. upper screen) of $V$, and $T$ a screen ringoid of $R$. A vectoid $\{S, T\}$ is called a *screen vectoid* if for its operations $\boxplus$ and $\boxdot$

$$\text{(RG1)} \quad \bigwedge_{\boldsymbol{a},\boldsymbol{b} \in S} (\boldsymbol{a} + \boldsymbol{b} \in S \Rightarrow \boldsymbol{a} \boxplus \boldsymbol{b} = \boldsymbol{a} + \boldsymbol{b})$$

$$\wedge \bigwedge_{a \in T} \bigwedge_{\boldsymbol{a} \in S} (a \cdot \boldsymbol{a} \in S \Rightarrow a \boxdot \boldsymbol{a} = a \cdot \boldsymbol{a}).$$

[†] The property (RG$i$) for subtraction can be proved by using (RG4) and (RG$i$) for addition for all $i = 1, 2, 3$.

A screen vectoid is called *monotone* if

$$\text{(RG2)} \quad \bigwedge_{\boldsymbol{a},\boldsymbol{b},\boldsymbol{c},\boldsymbol{d} \in S} (\boldsymbol{a} + \boldsymbol{b} \leq \boldsymbol{c} + \boldsymbol{d} \Rightarrow \boldsymbol{a} \boxplus \boldsymbol{b} \leq \boldsymbol{c} \boxplus \boldsymbol{d})$$

$$\wedge \bigwedge_{a,b \in T} \bigwedge_{\boldsymbol{a},\boldsymbol{b} \in S} (a \cdot \boldsymbol{a} \leq b \cdot \boldsymbol{b} \Rightarrow a \boxdot \boldsymbol{a} \leq b \boxdot \boldsymbol{b}).$$

It is called a *lower* (resp. an *upper*) screen vectoid if

$$\text{(RG3)} \quad \bigwedge_{\boldsymbol{a},\boldsymbol{b} \in S} \boldsymbol{a} \boxplus \boldsymbol{b} \leq \boldsymbol{a} + \boldsymbol{b} \qquad \left(\text{resp.} \quad \bigwedge_{\boldsymbol{a},\boldsymbol{b} \in S} \boldsymbol{a} + \boldsymbol{b} \leq \boldsymbol{a} \boxplus \boldsymbol{b}\right)$$

$$\wedge \bigwedge_{a \in T} \bigwedge_{\boldsymbol{a} \in S} a \boxdot \boldsymbol{a} \leq a \cdot \boldsymbol{a} \qquad \left(\text{resp.} \quad \bigwedge_{a \in T} \bigwedge_{\boldsymbol{a} \in S} a \cdot \boldsymbol{a} \leq a \boxdot \boldsymbol{a}\right).$$

If $\{V, R\}$ is a multiplicative vectoid, a multiplicative vectoid $\{S, T\}$ is called a *multiplicative screen vectoid* (resp. *monotone*; resp. *lower* or *upper* multiplicative screen vectoid) if (RG1) (resp. (RG2); resp. (RG3)) also holds for the inner multiplication. ■

As before, it is easy to see that a screen vectoid $\{S, T\}$ always has the same neutral elements as the vectoid $\{V, R\}$ itself.

Since in Definition 3.7 $S$ is a symmetric screen of $V$, then by using (S3), we have for all $\boldsymbol{a} \in S$ that $-\boldsymbol{a} := (-e) \cdot \boldsymbol{a} \in S$. Because of (RG1), we get therefore $\boxminus \boldsymbol{a} := (-e) \boxdot \boldsymbol{a} = (-e) \cdot \boldsymbol{a} = -\boldsymbol{a}$, i.e., in a screen vectoid we always have

$$\text{(RG4)} \quad \bigwedge_{\boldsymbol{a} \in S} \boxminus \boldsymbol{a} = -\boldsymbol{a}.$$

With this property, (RG1) (resp. (RG2); resp. (RG3)) for subtraction is easily proved by using (RG1) (resp. (RG2); resp. (RG3)) for addition.

In the case of a linearly ordered ringoid we can use Lemma 3.4 and Theorem 2.3 to obtain the following theorem.

**Theorem 3.8:** Let $\{R, +, \cdot, \leq\}$ be a completely and linearly ordered ringoid, $\{S, \leq\}$ a symmetric screen, and $\square : R \to S$ a semimorphism. Then $\{S, \boxplus, \boxdot, \leq\}$ is a completely and linearly ordered monotone screen ringoid. ■

Since the real numbers are a completely and linearly ordered ringoid, Theorem 3.8 asserts, for example, that the floating-point numbers are a linearly ordered monotone screen ringoid if their arithmetic is defined by a semimorphism. This process can be repeated each time passing into a coarser subset and each time obtaining a linearly ordered monotone screen ringoid. Theorem 3.8 enables us to define the arithmetic in the sets $D$ and $S$ in Fig. 1. These in turn can be used to define the arithmetic in the corresponding columns in Fig. 1. This latter process is the vertical definition of computer arithmetic given in Section 3.

## 3. THE VERTICAL DEFINITION OF COMPUTER ARITHMETIC

Let us now assume that $\{R, N, +, \cdot, /, \leq\}$ is a completely and linearly ordered division ringoid. We know already by Theorem 2.6 that the process of complexification leads to a completely and weakly ordered division ringoid $\{\boldsymbol{C}R, \bar{N}, +, \cdot, /, \leq\}$. As for the matrices over $R$ and $\boldsymbol{C}R$, we know by Theorem 2.5 that $\{M_nR, +, \cdot, \leq\}$ is a completely ordered ringoid and that $\{M_n\boldsymbol{C}R, +, \cdot, \leq\}$ is a completely and weakly ordered ringoid. The definition of the operations in $M_nR$, $\boldsymbol{C}R$, and $M_n\boldsymbol{C}R$ by those of $R$ as given preceding Theorems 2.5 and 2.6, we call the vertical definition of the arithmetic.

If now $D$ is a symmetric screen of $R$ and $S$ a symmetric screen of $D$ and if we define the arithmetic in $D$ and $S$ by semimorphisms (stepwise), then by Theorem 3.8 we obtain once more completely and linearly ordered ringoids. Therefore, we can again define an arithmetic and structure in the derived sets $M_nD, \boldsymbol{C}D, M_n\boldsymbol{C}D$ and $M_nS, \boldsymbol{C}S, M_n\boldsymbol{C}S$ by the vertical method and get the same results and properties as in the corresponding sets over $R$. See Fig. 19.

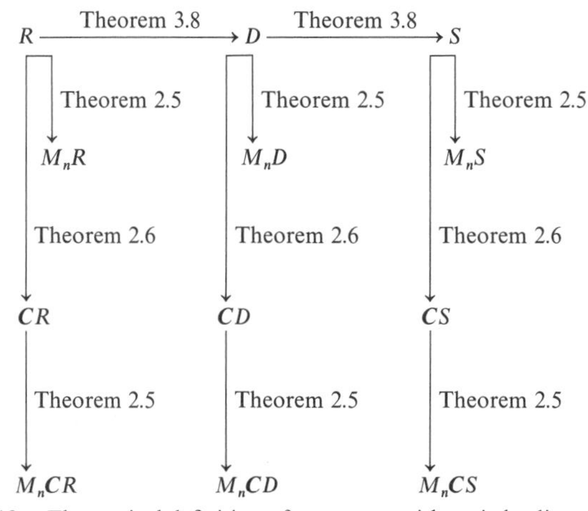


FIGURE 19. The vertical definition of computer arithmetic leading to ringoids.

As an example, $R$ may be the linearly ordered ringoid of real numbers, $S$ the subset of single precision floating-point numbers, and $D$ the set of double precision floating-point numbers of a given computer.

A similar process can be used in order to define vectoids over $R$, $D$, and $S$. Figure 20 lists the resulting spaces and the theorems with which we proved the vectoid properties. The spaces listed immediately under $R$, $D$, and $S$ are ordered vectoids, and those listed under $\boldsymbol{C}R, \boldsymbol{C}D$ and $\boldsymbol{C}S$ are weakly ordered vectoids.

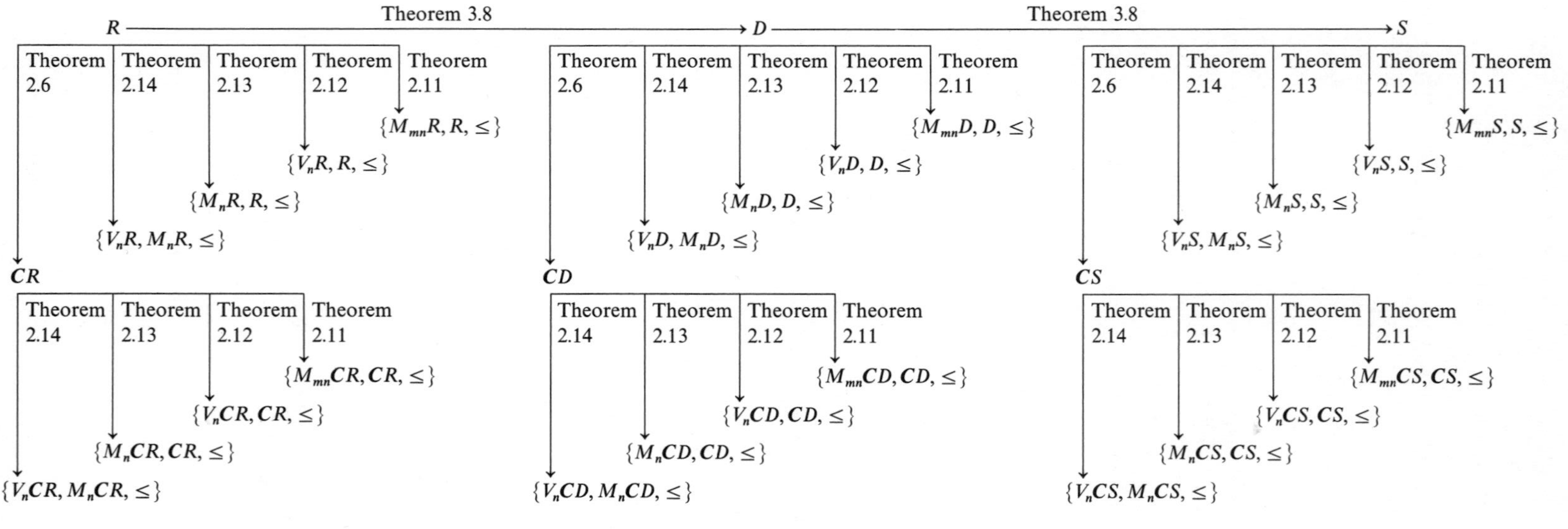


FIGURE 20. The vertical definition of computer arithmetic leading to vectoids.

The vertical method of defining arithmetic is used on most existing computers in order to specify the arithmetic of floating-point vectors, matrices, complexifications, and so on. The operations in $M_nS$, for instance, are defined by the operations in $S$ and the usual formulas for addition and multiplication of real matrices. We saw that the resulting structures can be described in terms of ordered or weakly ordered ringoids and vectoids.

This will also be one of the main results of the horizontal method of defining the arithmetic in the subsets by means of semimorphisms, which we discuss in Section 4. The horizontal method goes further, leading to simple compatibility properties such as (RG1)–(RG4) between the operations in the basic set and the screen. It is the connection between these two classes of operations that is of central interest.

In Chapter 5 we derive error estimates for the operations defined by both methods. It will turn out that the error formulas for the operations of the horizontal method are much simpler and more accurate than those of the vertical method and consequently allow a simpler and more accurate error analysis of numerical methods.

## 4. THE HORIZONTAL DEFINITION OF COMPUTER ARITHMETIC BY SEMIMORPHISMS

The first result of the horizontal definition of computer arithmetic by means of semimorphisms, of course, is contained in Theorem 3.8. In order to define the arithmetic in all the other spaces listed in Figs. 19 and 20 by means of semimorphisms, we first prove the following theorem.

**Theorem 3.9:** Let $\{R, +, \cdot\}$ be a ringoid, $\{R, \leq\}$ a complete lattice, $\{S, \leq\}$ a symmetric screen (resp. a symmetric lower; resp. upper screen), and $\square : R \to S$ a monotone and antisymmetric rounding. Consider the product sets $V_nS := S \times S \times \cdots \times S \subseteq V_nR := R \times R \times \cdots \times R$. Then $\{V_nR, \leq\}$ is a complete lattice if the order relation is defined componentwise and $\{V_nS, \leq\}$ is a symmetric screen (resp. a symmetric lower; resp. upper screen) of $\{V_nR, \leq\}$. Further, the mapping $\square : V_nR \to V_nS$ defined by

$$\bigwedge_{\boldsymbol{a}=(a_i)\in V_nR} \square \boldsymbol{a} := \begin{pmatrix} \square a_1 \\ \square a_2 \\ \vdots \\ \square a_n \end{pmatrix}$$

is a monotone and antisymmetric rounding.

*Proof*: Although the proof is straightforward, we give it in all detail in order to review the concepts dealt with. As a product set of complete lattices, $\{V_nR, \leq\}$ is itself a complete lattice. The subset $\{V_nS, \leq\}$ is an ordered set and as a product set of complete lattices also a complete lattice and therefore a complete subnet of $\{V_nR, \leq\}$. The supremum (resp. infimum) in $V_nS$ equals the vector of suprema (resp. infima) taken in $S$. Since $\{S, \leq\}$ is a lower (resp. an upper) screen of $R$, the latter equals the suprema (resp. infima) taken in $R$. Consequently, the supremum (resp. infimum) taken in $V_nS$ equals the supremum (resp. infimum) taken in $V_nR$. Then by Theorem 1.20, $V_nS$ is a lower (resp. an upper) screen of $\{V_nR, \leq\}$. Since $S$ is a symmetric screen of $R$, then the zero vector is an element of $V_nS$, and for all $a = (a_i) \in V_nS$, $-a = (-a_i) \in V_nS$, i.e., $V_nS$ is a symmetric screen (resp. a symmetric lower; resp. upper screen) of $\{V_nR, \leq\}$. The mapping $\square: V_nR \to V_nS$ is defined by rounding componentwise. This directly implies that the properties (R1), (R2), and (R4) of the rounding $\square: R \to S$ also hold for the rounding $\square: V_nR \to V_nS$. ■

A matrix of $M_{mn}R$ may be viewed as an element of the $m \times n$-dimensional product set. Therefore, Theorem 3.9 remains valid if $V_nR$ is replaced by $M_{mn}R$ or $M_nR$ and $V_nS$ by $M_{mn}S$ or $M_nS$. The rounding for matrices has to be defined by rounding the components. The following two theorems define the arithmetic in the sets listed in Fig. 19 by means of semimorphisms.

**Theorem 3.10:** Let $\{R, +, \cdot, \leq\}$ be a ringoid, $\{R, \leq\}$ a complete lattice, $\{S, \leq\}$ a symmetric screen of $\{R, \leq\}$, and $\{S, \boxplus, \boxdot, \leq\}$ a ringoid defined by a semimorphism $\square: R \to S$. Further, let $\{M_nR, +, \cdot, \leq\}$ be the ringoid of $n \times n$ matrices over $R$. Operations in $M_nS$ are specified by the semimorphism $\square: M_nR \to M_nS$ defined by

$$\bigwedge_{A=(a_{ij}) \in M_nR} \square A := (\square a_{ij})$$

and

$$\text{(RG)} \quad \bigwedge_{A,B \in M_nS} A \circledast B := \square(A * B), \qquad \text{for} \quad * \in \{+, \cdot\}.$$

(a) Then $\{M_nS, \boxplus, \boxdot\}$ is a monotone screen ringoid of $M_nR$.

(b) If the rounding $\square: R \to S$ is upwardly (resp. downwardly) directed, then $\{M_nS, \boxplus, \boxdot\}$ is a lower (resp. an upper) screen ringoid.

(c) If $\{R, +, \cdot, \leq\}$ is a weakly ordered (resp. an ordered) ringoid, then $\{M_nS, \boxplus, \boxdot, \leq\}$ is a weakly ordered (resp. an ordered) screen ringoid of $M_nR$.

*Proof*: (a): By Theorem 3.9 $\{M_nS, \leq\}$ is a symmetric screen of $\{M_nR, \leq\}$, and $\square: M_nR \to M_nS$ is a monotone and antisymmetric rounding. Now, Lemma 3.4 implies the validity of the properties (D1)–(D5), (RG1), (RG2), and (RG4). There remains only to prove (D6).

(D6): Let $O$ and $E$ denote the neutral elements of $\{M_nR, +, \cdot, \leq\}$. By Lemma 3.4 we know that the element $-E$ satisfies (D5). We show that it is unique with respect to this property.

Let $X = x_{ij}$ be any element of $M_nS$ that satisfies (D5). Then by (D5a),

$$X \boxplus E = O, \tag{4}$$

and by the definition of the addition and the rounding, we get

$$x_{ij} \boxplus o = o \underset{(D2)_s}{\Rightarrow} x_{ij} = o \qquad \text{for all} \quad i \neq j,$$

i.e., $X$ is a diagonal matrix.

Specializing to the diagonal in (4), we get

$$x_{ii} \boxplus e = o \qquad \text{for all} \quad i = 1(1)n, \tag{5}$$

and by (D5b) for $X$:

$$x_{ii} \boxdot x_{ii} = e, \qquad \text{for all} \quad i = 1(1)n. \tag{6}$$

From (D5c) and (D5d) we get for two diagonal matrices with the constant diagonal entries $a$ and $b$

$$\bigwedge_{a,b \in S} x_{ii} \boxdot (a \boxdot b) = (x_{ii} \boxdot a) \boxdot b = a \boxdot (x_{ii} \boxdot b) \tag{7}$$

and

$$\bigwedge_{a,b \in S} x_{ii} \boxdot (a \boxplus b) = x_{ii} \boxdot a \boxplus x_{ii} \boxdot b. \tag{8}$$

Relations (5)–(8) imply that $x_{ii}$ has to satisfy the axiom (D5) in $S$. Since $\{S, \boxplus, \boxdot\}$ is a screen ringoid of $\{R, +, \cdot\}$, $x_{ii} = -e$ for all $i = 1(1)n$, and therefore $X = -E \in M_nS$. (b) and (c) are simple consequences of Lemma 3.4(b) and (c). ■

**Theorem 3.11:** Let $\{R, +, \cdot\}$ be a ringoid, $\{R, \leq\}$ a complete lattice, $\{S, \leq\}$ a symmetric screen of $\{R, \leq\}$, and $\{S, \boxplus, \boxdot\}$ a ringoid defined by a semimorphism $\square: R \to S$. Further, let $\{CR, +, \cdot\}$ be the ringoid of pairs $(a, b)$ over $R$, and let operations in $CS$ be specified by the semimorphism $\square: CR \to CS$ defined by

$$\bigwedge_{\alpha = (a_1, a_2) \in CR} \square\alpha := (\square a_1, \square a_2)$$

and

$$\text{(RG)} \qquad \bigwedge_{\alpha, \beta \in CS} \alpha \circledast \beta := \square(\alpha * \beta), \quad * \in \{+, \cdot\}.$$

(a) Then $\{CS, \boxplus, \boxdot\}$ is a monotone screen ringoid of $CR$.

(b) If $\square: R \to S$ is upwardly (resp. downwardly) directed, then $\{CS, \boxplus, \boxdot\}$ is a lower (resp. upper) screen ringoid of $CR$.

(c) If $\{R, +, \cdot, \leq\}$ is a weakly ordered ringoid, then $\{CS, \boxplus, \boxdot, \leq\}$ is a weakly ordered screen ringoid of $CR$.

(d) Now let $\{R, N, +, \cdot, /\}$ be a division ringoid, $\{CR, \bar{N}, +, \cdot, /\}$, with $\bar{N} = \{(c_1, c_2) \in CR \mid c_1 c_1 + c_2 c_2 \in N\}$, its complexification, and $\{S, N \cap S, \boxplus, \boxdot, \boxslash\}$ the division ringoid defined by the semimorphism $\square: R \to S$. If a division by an element $\beta \in CS$ is defined by the semimorphism (RG), with $\beta \notin \bar{N} \cap CS$, then $\{CS, \bar{N} \cap CS, \boxplus, \boxdot, \boxslash\}$ is a monotone screen division ringoid of $CR$.

*Proof*: (a): By Theorem 3.9 $\{CS, \leq\}$ is a symmetric screen of $\{CR, \leq\}$, and $\square: CR \to CS$ is a monotone and antisymmetric rounding. Lemma 3.4 provides the properties (D1)–(D5), (RG1), (RG2), and (RG4). There remains to prove (D6).

(D6): Let $o$ and $e$ be the neutral elements of $\{R, +, \cdot, \leq\}$. Then $\omega = (o, o)$ and $\varepsilon = (e, o)$ are the neutral elements of $\{CR, +, \cdot, \leq\}$. By Lemma 3.4 we know that the element $(-e, o)$ satisfies (D5) in $CS$. We show that it is unique with respect to this property.

Let $\eta = (x, y)$ be any element of $CS$ that satisfies (D5). Then by (D5a),

$$\eta \boxplus (e, o) = (o, o),$$

and by the definition of addition and the rounding, we get

$$y \boxplus o = o \underset{(D2)_S}{\Rightarrow} y = o,$$

i.e., $\eta$ is of the form $\eta = (x, o)$.

Let us now denote by $\{CS\}$ the set of all elements of $CS$ with vanishing imaginary part (second component). Then the mapping $\varphi: \{CS\} \to S$ defined by

$$\bigwedge_{a \in S} (a, o) \to a$$

is a ringoid isomorphism.

The element $\eta = (x, o)$ has to satisfy (D5) for all $\alpha, \beta \in CS$, and in particular for $\alpha = (a, o)$ and $\beta = (b, o)$ with $a, b \in S$. Because of the isomorphism $\{CS\} \leftrightarrow S$, the properties (D5) in $CS$ for the elements $(a, o)$ and $(b, o)$ are equivalent to (D5) in $S$. Since $\{S, \boxplus, \boxminus\}$ is a screen ringoid, there exists only one such element $x = -e$. Therefore, $\eta = (-e, o)$ is the only element in $CS$ that satisfies (D5).

(b) and (c) are simple consequences of Lemma 3.4(b) and (c).

(d): This follows from Lemma 3.5. ■

The operations defined in $M_n S$ and $CS$ by Theorems 3.10 and 3.11 are quite different from those defined in the same sets by the vertical method in Theorems 2.5 and 2.6.

Theorems 3.10 and 3.11 already specify the definition of the arithmetic by semimorphisms for several of the sets displayed under $D$ in Fig. 19.

As an example, let $R$ denote the linearly ordered set of real numbers, $S$ the subset of single precision floating-point numbers, and $D$ the set of double precision floating-point numbers of a given computer. Then the arithmetic and structure are defined in $M_nD$ by Theorem 3.10, in $CD$ by Theorem 3.11, and in $M_nCD$ once again by Theorem 3.10. See Fig. 21.

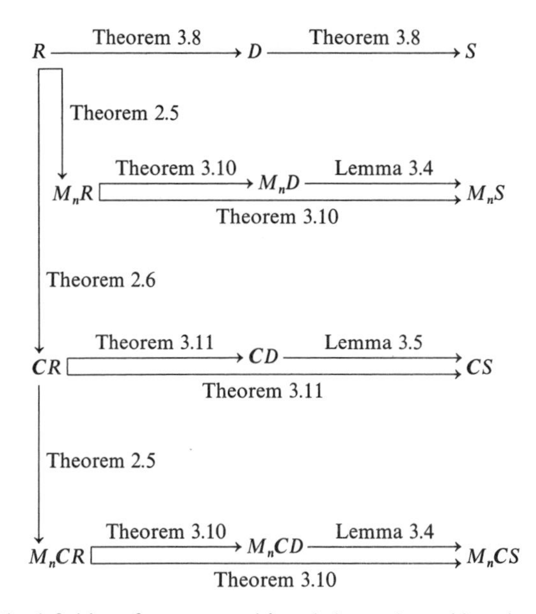


FIGURE 21. The definition of computer arithmetic by semimorphisms leading to ringoids.

These theorems, however, cannot be used directly to define the arithmetic and structure in $M_nS$, $CS$, and $M_nCS$. The reason for this is that in Theorems 3.10 and 3.11 the arithmetic in the spaces $M_nR$, $CR$, and $M_nCR$ is defined by the vertical method. The arithmetic and structure among the spaces $D$, $M_nD$, $CD$, and $M_nCD$, however, are no longer connected by a vertical definition.

Nevertheless, the following consideration leads to the desired results. If $S$ is a symmetric screen of $D$, then so is $M_nS$ of $M_nD$. See Fig. 21.

For a monotone and antisymmetric rounding $\lozenge : D \to S$, we consider the semimorphism $\lozenge : M_nD \to M_nS$ defined by

$$\bigwedge_{A=(a_{ij})\in M_nD} \lozenge A := (\lozenge a_{ij})$$

and

$$\text{(RG)} \quad \bigwedge_{A,B \in M_nS} A \mathbin{\diamond\!\!\!\!\ast} B := \Diamond(A \boxast B), \quad \text{for} \quad * \in \{+, \cdot\}.$$

Then apart from (D6), $\{M_nS, \diamondplus, \diamonddot, \leq\}$ has all desired properties as we know from Lemma 3.4.

In order to see that $\{M_nS, \diamondplus, \diamonddot, \leq\}$ is a ringoid, we consider the composition of the mappings $\Diamond \Box : R \to S$. This composition is also a monotone and antisymmetric rounding. By Theorem 3.10, therefore, the semimorphism $\Diamond \Box : M_nR \to M_nS$ with the property

$$\text{(RG)} \quad \bigwedge_{A,B \in M_nS} A \mathbin{\diamond\!\!\!\!\ast} B := \Diamond(\Box(A * B)), \qquad \text{for} \quad * \in \{+, \cdot\}$$

leads to a ringoid in $M_nS$. This ringoid is identical to the one defined by the semimorphism $\Diamond : M_nD \to M_nS$. $\{M_nS, \diamondplus, \diamonddot, \leq\}$, therefore, is an ordered (resp. weakly ordered) monotone screen ringoid of $M_nD$.

Using the composition of the mappings $\Box : \boldsymbol{C}R \to \boldsymbol{C}D$ and $\Diamond : \boldsymbol{C}D \to \boldsymbol{C}S$, it can be proved analogously by Lemma 3.5 and Theorem 3.11 that $\{\boldsymbol{C}S, \bar{N} \cap \boldsymbol{C}S, \diamondplus, \diamonddot, \diamondslash, \leq\}$ is a weakly ordered monotone screen division ringoid of $\boldsymbol{C}D$.

Similarly, Lemma 3.4 and Theorem 3.10 lead to the result that $\{M_n\boldsymbol{C}S, \diamondplus, \diamonddot, \leq\}$ is a weakly ordered monotone screen ringoid of $\{M_n\boldsymbol{C}D, \boxplus, \boxdot, \leq\}$. See Fig. 21.

In Fig. 21 the chain $R \supseteq D \supseteq S$ can also be extended rightward by means of coarser symmetric screens. Again it is clear by Lemmas 3.4 and 3.5 and Theorems 3.10 and 3.11 that we obtain ringoids in all lines of Fig. 21. In this sense the ringoid structures in all lines of Fig. 21 are invariant with respect to semimorphisms.

We now state and prove the following two theorems that enable us to define the arithmetic in the sets listed in Fig. 20 by means of semimorphisms.

**Theorem 3.12:** Let $\{VR\}$ be a vectoid, $\{V, \leq\}$ a complete lattice, $\{S, \leq\}$ a symmetric screen (resp. a symmetric lower; resp. upper screen) of $\{V, \leq\}$, and $\{T, \oplus, \odot\}$ a screen ringoid of $\{R, +, \cdot\}$. Further, let $\Box : V \to S$ be a semimorphism, i.e., a monotone and antisymmetric rounding with which operations are defined by

$$\text{(RG)} \quad \bigwedge_{\boldsymbol{a},\boldsymbol{b} \in S} \boldsymbol{a} \boxplus \boldsymbol{b} := \Box(\boldsymbol{a} + \boldsymbol{b}), \qquad \bigwedge_{a \in T} \bigwedge_{\boldsymbol{a} \in S} a \boxdot \boldsymbol{a} := \Box(a \cdot \boldsymbol{a}).$$

(a) Then $\{S, T\}$ is a monotone screen vectoid of $\{V, R\}$.

(b) If the rounding $\Box : V \to S$ is downwardly (resp. upwardly) directed, then $\{S, T\}$ is a lower (resp. an upper) screen vectoid of $\{V, R\}$.

(c) If $\{V, R, \leq\}$ is a weakly ordered (resp. an ordered) vectoid, then $\{S, T, \leq\}$ is a weakly ordered (resp. an ordered) monotone screen vectoid.

*Proof*: The proof can be given in complete analogy to that of Lemma 3.4. Therefore, we simply list the properties which have to be shown.

(a): (V1), (V2), (VD1)–(VD4), (RG1), (RG2), (RG4).
(b): (RG3).
(c): (OV1), (OV2) (and (OV3) in the ordered case). ■

**Theorem 3.13:** Let $\{V, R\}$ be a multiplicative vectoid, $\{V, \leq\}$ a complete lattice, $\{S, \leq\}$ a symmetric screen (resp. a symmetric lower; resp. upper screen) of $\{V, \leq\}$, and $\{T, \oplus, \odot\}$ a screen ringoid of $\{R, +, \cdot\}$. Further, let $\square : V \to S$ be a semimorphism.

(a) Then $\{S, T\}$ is a multiplicative monotone screen vectoid of $\{V, R\}$.
(b) If the rounding $\square : V \to S$ is downwardly (resp. upwardly) directed, then $\{S, T\}$ is a lower (resp. an upper) screen vectoid of $\{V, R\}$.
(c) If $\{V, R, \leq\}$ is a weakly ordered (resp. an ordered) multiplicative vectoid, then $\{T, S, \leq\}$ is a weakly ordered (resp. an ordered) multiplicative screen vectoid.

*Proof*: The proof of this theorem can also be given analogously to the proofs of Lemmas 3.4 and 3.5. Therefore, we only list the properties that have to be shown.

(a): (V3)–(V5), (VD5), (RG1), and (RG2) for the multiplication.
(b): (RG3) for the multiplication.
(c): (OV4). ■

The last two theorems permit the definition of the arithmetic by means of semimorphisms in all rows in Fig. 22. As an example, one may again interpret $R$ as the linearly ordered ringoid of real numbers, $S$ as the subset of single-precision floating-point numbers, and $D$ as the set of double-precision floating-point numbers of a given computer. The arithmetic and structure in the product sets listed in Fig. 22 are defined by the theorems displayed on the connecting lines.

One may also extend the chain $R \supseteq D \supseteq S$ rightward by means of coarser symmetric screens. Theorems 3.12 and 3.13 then show that one again obtains vectoids in all rows of Fig. 22. In this sense the vectoid structures in all rows of Fig. 22 are invariant with respect to semimorphisms.

We mention finally that the arithmetic defined by semimorphisms in the sets listed in Fig. 22 is in principle different from that defined in the same sets by the vertical method. Within the structures in the rows beginning with $\{M_{mn}R, R, \leq\}$ and $\{V_nR, R, \leq\}$, however, both methods lead to the same operations.

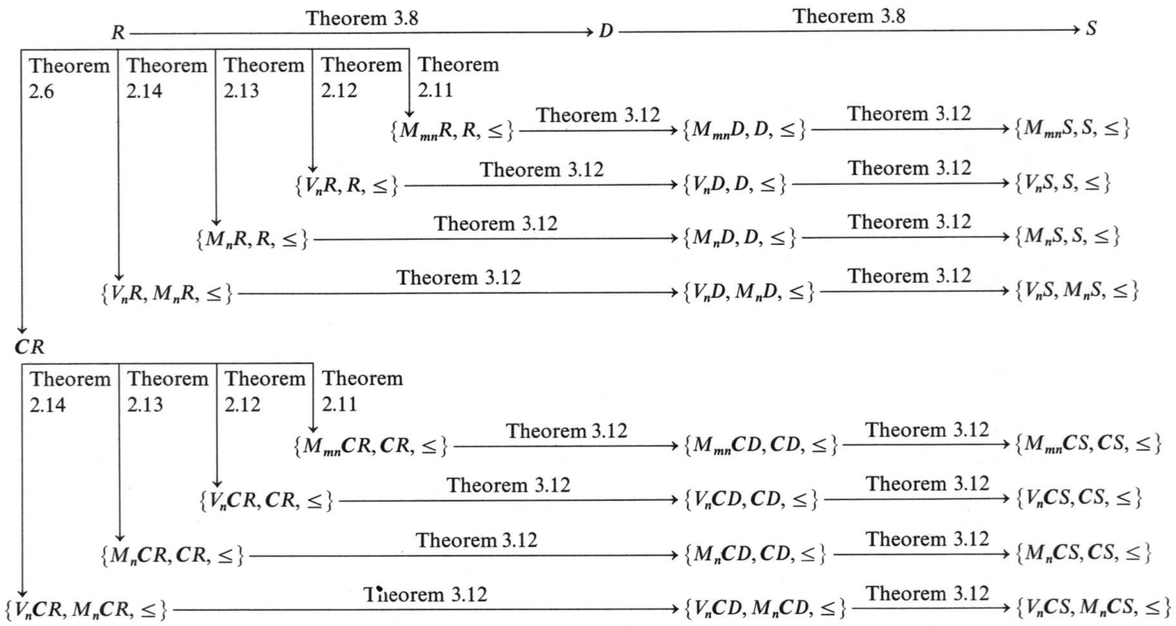

$$R \xrightarrow{\text{Theorem 3.8}} D \xrightarrow{\text{Theorem 3.8}} S$$

From $R$: Theorem 2.6 → $CR$; Theorem 2.14 → $\{V_nR, M_nR, \le\}$; Theorem 2.13 → $\{M_nR, R, \le\}$; Theorem 2.12 → $\{V_nR, R, \le\}$; Theorem 2.11 → $\{M_{mn}R, R, \le\}$

$$\{M_{mn}R, R, \le\} \xrightarrow{\text{Theorem 3.12}} \{M_{mn}D, D, \le\} \xrightarrow{\text{Theorem 3.12}} \{M_{mn}S, S, \le\}$$

$$\{V_nR, R, \le\} \xrightarrow{\text{Theorem 3.12}} \{V_nD, D, \le\} \xrightarrow{\text{Theorem 3.12}} \{V_nS, S, \le\}$$

$$\{M_nR, R, \le\} \xrightarrow{\text{Theorem 3.12}} \{M_nD, D, \le\} \xrightarrow{\text{Theorem 3.12}} \{M_nS, S, \le\}$$

$$\{V_nR, M_nR, \le\} \xrightarrow{\text{Theorem 3.12}} \{V_nD, M_nD, \le\} \xrightarrow{\text{Theorem 3.12}} \{V_nS, M_nS, \le\}$$

From $CR$: Theorem 2.14 → $\{V_nCR, M_nCR, \le\}$; Theorem 2.13 → $\{M_nCR, CR, \le\}$; Theorem 2.12 → $\{V_nCR, CR, \le\}$; Theorem 2.11 → $\{M_{mn}CR, CR, \le\}$

$$\{M_{mn}CR, CR, \le\} \xrightarrow{\text{Theorem 3.12}} \{M_{mn}CD, CD, \le\} \xrightarrow{\text{Theorem 3.12}} \{M_{mn}CS, CS, \le\}$$

$$\{V_nCR, CR, \le\} \xrightarrow{\text{Theorem 3.12}} \{V_nCD, CD, \le\} \xrightarrow{\text{Theorem 3.12}} \{V_nCS, CS, \le\}$$

$$\{M_nCR, CR, \le\} \xrightarrow{\text{Theorem 3.12}} \{M_nCD, CD, \le\} \xrightarrow{\text{Theorem 3.12}} \{M_nCS, CS, \le\}$$

$$\{V_nCR, M_nCR, \le\} \xrightarrow{\text{Theorem 3.12}} \{V_nCD, M_nCD, \le\} \xrightarrow{\text{Theorem 3.12}} \{V_nCS, M_nCS, \le\}$$

FIGURE 22. The definition of computer arithmetic by semimorphisms leading to vectoids.

## 5. A REMARK ABOUT ROUNDINGS

The results of this chapter show that the monotone and antisymmetric roundings are important roundings. When composed into a semimorphism, they leave invariant the ringoid and vectoid structure in all numerically interesting cases.

We recall, however, Theorem 1.28, which asserts that in a linearly ordered set every monotone rounding can be expressed by means of the monotone downwardly and upwardly directed roundings $\bigtriangledown$ and $\bigtriangleup$. Employing Theorem 3.3 in the case of a linearly ordered ringoid, we see that even the rounding $\bigtriangleup$ can be expressed by $\bigtriangledown$ and conversely. This leads to the result that in a linearly ordered ringoid (e.g., the real numbers) every monotone rounding can be expressed by the monotone downwardly directed rounding $\bigtriangledown$ alone. Apart from $\bigtriangleup$, no other monotone rounding has this property.

This implies that on a computer every monotone rounding can be performed if either the monotone downwardly or upwardly directed rounding $\bigtriangledown$ or $\bigtriangleup$ is available. Since nonmonotone roundings are of little interest, this points out the very central role of the monotone directed roundings $\bigtriangledown$ and $\bigtriangleup$ for all rounded computations.

Beyond this result, the monotone directed roundings $\bigtriangledown$ and $\bigtriangleup$ are needed for all interval computations as we shall see in Chapter 4. In spite of these facts, the monotone downwardly or upwardly directed rounding $\bigtriangledown$ or $\bigtriangleup$ is currently made available by hardware facilities on only very few computers.

We may interpolate the observation that the monotone downwardly directed rounding $\bigtriangledown$, for instance, is not at all difficult to perform. If negative numbers are represented by the $b$-complement within the rounding procedure, then for all $x \in \boldsymbol{R}$ the rounding $\bigtriangledown$ is identical with the truncation of the representation to finite length.

# Chapter 4 / INTERVAL ARITHMETIC

**Summary:** We start with the observation made in Chapter 2 that all the spaces listed in the leftmost element in every row in Fig. 1 are ringoids or vectoids. In particular, the spaces that are power sets are inclusion-isotonally ordered with respect to inclusion as an order relation. We shall define the operations in all interval spaces in Fig. 1 by means of semimorphisms. With these operations, each interval space becomes an inclusion-isotonally ordered monotone upper screen ringoid (resp. vectoid) of its left neighbor in Fig. 1. Furthermore, these ringoids and vectoids are ordered (resp. weakly ordered) with respect to the order relation $\leq$, which we shall define below.

For all operations to be defined in the interval sets listed in Fig. 1, we derive explicit formulas for the computation of the resulting interval in terms of the bounds of the interval operands.

To do this, we show that the structures $M_n\boldsymbol{I}R$ and $\boldsymbol{I}M_nR$ as well as $\{V_n\boldsymbol{I}R, \boldsymbol{I}R, \leq\}$ and $\{\boldsymbol{I}V_nR, \boldsymbol{I}R, \leq\}$—if the operations in $M_n\boldsymbol{I}R$ and $V_n\boldsymbol{I}R$ are defined by the vertical method—are isomorphic with respect to all algebraic operations and the order relation $\leq$. The same isomorphisms are shown to exist between the corresponding spaces, wherein $R$ is replaced by its complexification $\boldsymbol{C}R$.

The operations in all interval spaces in the columns listed under $D$ and $S$ in Fig. 1 are also defined via semimorphisms. The isomorphisms referred to here are the main tool for obtaining optimal (i.e., smallest appropriate interval) and computer executable formulas for the operations in these spaces.

## 1. INTERVAL SETS AND ARITHMETIC

Let $\{R, \leq\}$ be an ordered set and $\boldsymbol{P}R$ the power set of $R$. The intervals $[a_1, a_2] := \{x \in R \mid a_1, a_2 \in R, a_1 \leq x \leq a_2\}$ with $a_1 \leq a_2$ are special elements contained in $\boldsymbol{PR}$. Thus, denoting the set of all intervals over $R$ by $\boldsymbol{I}R$, we have $\boldsymbol{I}R \subset \boldsymbol{P}R$. In $\boldsymbol{I}R$ we define equality and the order relation $\leq$ by

$$[a_1, a_2] = [b_1, b_2] :\Leftrightarrow a_1 = b_1 \wedge a_2 = b_2,$$
$$[a_1, a_2] \leq [b_1, b_2] :\Leftrightarrow a_1 \leq b_1 \wedge a_2 \leq b_2.$$

With these definitions, we formulate the following theorem.

**Theorem 4.1:** If $\{R, \leq\}$ is a complete lattice, then $\{\boldsymbol{I}R, \leq\}$ is also a complete lattice. Moreover, with $A = [a_1, a_2]$

$$\bigwedge_{S \subseteq \boldsymbol{I}R} \inf S = \left[\inf_{A \in S} a_1, \inf_{A \in S} a_2\right] \wedge \sup S = \left[\sup_{A \in S} a_1, \sup_{A \in S} a_2\right]. \tag{1}$$

*Proof*: It is clear that $\{\boldsymbol{I}R, \leq\}$ is an ordered set. Since $\boldsymbol{I}R$ is a subset of the product set $R \times R$ and the order relation is defined componentwise, Theorem 1.10 implies that $\{\boldsymbol{I}R, \leq\}$ is a complete lattice and that (1) holds. ■

We remark that within $\boldsymbol{P}R$ the corresponding definition

$$A \leq B :\Leftrightarrow \inf_{a \in A} a \leq \inf_{b \in B} b \wedge \sup_{a \in A} a \leq \sup_{b \in B} b,$$

does not lead to an order relation because (O3) is obviously not valid.

If $\boldsymbol{R}^* := \boldsymbol{R} \cup \{-\infty\} \cup \{+\infty\}$, then the interval $[-\infty, -\infty]$ is the least element and the interval $[+\infty, +\infty]$ is the greatest element of $\{\boldsymbol{I}\boldsymbol{R}^*, \leq\}$.

If $\{R, \leq\}$ is a complete lattice, then $R$ is the greatest element of the power set $\{\boldsymbol{P}R, \subseteq\}$. Writing $R$ in the form $[o(R), i(R)]$, we see that it is also the greatest element of $\{\boldsymbol{I}R, \subseteq\}$. The empty set $\emptyset$ is the least element of $\{\boldsymbol{P}R, \subseteq\}$. According to the above definition, however, $\emptyset$ is not an element of $\boldsymbol{I}R$. In order to have a least element available whenever necessary, we adjoin the empty set to $\boldsymbol{I}R$ and define

$$\overline{\boldsymbol{I}R} := \boldsymbol{I}R \cup \{\emptyset\}.$$

In the following theorem we show that $\overline{\boldsymbol{I}R}$ serves as an upper screen for $\boldsymbol{P}R$ with respect to inclusion as an order relation.

**Theorem 4.2:** If $\{R, \leq\}$ is a complete lattice, then

(a) $\{\boldsymbol{I}R, \subseteq\}$ is a conditionally completely ordered set. For all $A \subseteq \boldsymbol{I}R$, which are bounded below, we have with $B := [b_1, b_2] \in \boldsymbol{I}R$

$$\inf_{\boldsymbol{I}R} A = \left[\sup_{B \in A} b_1, \inf_{B \in A} b_2\right] \wedge \sup_{\boldsymbol{I}R} A = \left[\inf_{B \in A} b_1, \sup_{B \in A} b_2\right],$$

i.e., the infimum is the intersection, and the supremum is the convex hull of the elements of $A$.

(b) $\{\overline{IR}, \subseteq\}$ is an upper screen of $\{PR, \subseteq\}$.

*Proof*: (a) Since $\{R, \leq\}$ is a complete lattice and $A$ is bounded below (i.e., $A$ contains an interval), the intervals

$$I := \left[\sup_{B \in A} b_1, \inf_{B \in A} b_2\right], \qquad S := \left[\inf_{B \in A} b_1, \sup_{B \in A} b_2\right]$$

exist, and for $B \in A$ we have $I \subseteq B \wedge B \subseteq S$, i.e., $I$ is a lower and $S$ an upper bound of $A$. If $I_1$ is any lower and $S_1$ any upper bound of $A$, then $I_1 \subseteq I$ and $S \subseteq S_1$, i.e., $I$ is the greatest lower and $S$ the least upper bound of $A$.

(b) Using (a), we see that $\{\overline{IR}, \subseteq\}$ is a complete lattice and therefore a complete subnet of $\{PR, \subseteq\}$. We still have to show that for every subset $A \subseteq \overline{IR}$, $\inf_{\overline{IR}} A = \inf_{PR} A$.

If $A$ is bounded below in $IR$, then by (a) the infimum in $IR$ is the intersection as it is in $PR$, and therefore $\inf_{IR} A = \inf_{\overline{IR}} A = \inf_{PR} A$.

If $A$ is not bounded below in $IR$, we consider three cases.

1. $A = \emptyset$. Then $\inf_{\overline{IR}} A = \inf_{PR} A = R = i(PR)$.
2. $A \neq \emptyset \wedge \emptyset \in A$. Then $\inf_{\overline{IR}} A = \emptyset = \inf_{PR} A$.
3. $\emptyset \notin A$. Since $A$ is not bounded below in $IR$, we obtain once again $\inf_{\overline{IR}} A = \emptyset = \inf_{PR} A$. ■

Now we consider the monotone upwardly directed rounding $\square : PR \Rightarrow \overline{IR}$, which is defined by the following properties:

(R1) $\bigwedge_{A \in \overline{IR}} \square A = A,$

(R2) $\bigwedge_{A,B \in PR} (A \subseteq B \Rightarrow \square A \subseteq \square B),$

(R3) $\bigwedge_{A \in PR} A \subseteq \square A.$

In the following theorem we develop properties of this rounding.

**Theorem 4.3:** Let $\{R, +, \cdot, \leq\}$ be a completely and weakly ordered ringoid with the neutral elements $o$ and $e$. Then

(a) $\{\overline{IR}, \subseteq\}$ is a symmetric upper screen of $\{PR, +, \cdot, \subseteq\}$, i.e., we have

(S3) $[o, o], [e, e] \in \overline{IR} \wedge \bigwedge_{A = [a_1, a_2] \in IR} - A = [-a_2, -a_1] \in IR.$

(b) The monotone upwardly directed rounding $\square : PR \to IR$ is antisymmetric, i.e.,

(R4) $\bigwedge_{\emptyset \neq A \in PR} \square(-A) = -\square A.$

(c) We have

$$\text{(R)} \quad \bigwedge_{\emptyset \neq A \in PR} \square A = \inf_{\overline{IR}}(U(A) \cap \overline{IR}) = \left[\inf_{a \in A} a, \sup_{a \in A} a\right]. \quad (3)$$

*Proof*: The assertions are shown in the order (a), (c), (b).

(a): With $A := [a_1, a_2] := \{x \in R \mid a_1 \leq x \leq a_2\}$, we get $-A := \{-e\} \cdot A := \{(-e) \cdot x \mid x \in A\} = \{-x \mid a_1 \leq x \leq a_2\} = \{x \mid a_1 \leq -x \leq a_2\} =_{(\text{OD2})_R} \{x \mid -a_2 \leq x \leq -a_1\} = [-a_2, -a_1] \in IR$.

(c): Theorem 1.24 implies $\square A = \inf(U(A) \cap \overline{IR})$. We show that $\square A = [\inf A, \sup A]$. For all $A \in PR$ such that $A \neq \emptyset$ we have $A \subseteq [\inf A, \sup A]$. For all $B = [b_1, b_2] \in IR$ with the property $A \subseteq B$ we obtain $\bigwedge_{a \in A}(b_1 \leq a \leq b_2) \Rightarrow b_1 \leq \inf A \wedge \sup A \leq b_2 \Rightarrow [\inf A, \sup A] \subseteq [b_1, b_2]$, i.e., $[\inf A, \sup A]$ is the least upper bound of $A$ in $\{IR, \subseteq\}$, which proves (3).

(b): With $\square A = [\inf A, \sup A]$, we get $\square(-A) = [\inf(-A), \sup(-A)]$, and with Theorem 2.2(y), $\square(-A) = [-\sup A, -\inf A]$. Now with (2) and (3) we obtain $\square(-A) = -\square A$. ■

Now we make use of the monotone upwardly directed rounding $\square : PR \to \overline{IR}$ in order to define operations $\boxed{*}$, $* \in \{+, \cdot, /\}$, in $IR$ by the corresponding semimorphism that has the following property:

$$\text{(RG)} \quad \bigwedge_{A,B \in IR} A \boxed{*} B := \square(A * B) = [\inf(a * b), \sup(a * b)]. \quad (4)$$

Here the infimum and supremum are taken over $a \in A$, $b \in B$, while the equality on the right-hand side is a simple consequence of (3). These operations have the following properties (to which by Theorem 1.30 (RG) is equivalent):

$$\text{(RG1)} \quad \bigwedge_{A,B \in IR} (A * B \in IR \Rightarrow A \boxed{*} B = A * B),$$

$$\text{(RG2)} \quad \bigwedge_{A,B,C,D \in IR} (A * B \subseteq C * D \Rightarrow A \boxed{*} B \subseteq C \boxed{*} D),$$

$$\text{(RG3)} \quad \bigwedge_{A,B \in IR} (A * B \subseteq A \boxed{*} B).$$

When in these formulas the operator $*$ represents division, we assume additionally that $B, D \notin \tilde{N}$. For the definition of $\tilde{N}$, see Theorem 4.4, which we now state.

**Theorem 4.4:** (a) Let $\{R, +, \cdot, \leq\}$ be a completely and weakly ordered ringoid (resp. $\{R, N, +, \cdot, /, \leq\}$ a completely and weakly ordered division ringoid) with the neutral elements $o$ and $e$. If operations are defined in $IR$ by the semimorphism $\square : PR \to \overline{IR}$, then $\{IR, \boxplus, \boxdot, \leq, \subseteq,\}$ becomes a completely and weakly ordered ringoid with respect to the order relation $\leq$.

The neutral elements are $[o, o]$ and $[e, e]$. (Likewise $\{\boldsymbol{IR}, \tilde{N}, \boxplus, \boxdot, \boxslash, \leq, \subseteq\}$ with $\tilde{N} := \{A \in \boldsymbol{IR} | A \cap N \neq \varnothing\}$ becomes a completely and weakly ordered division ringoid.) With respect to $\subseteq$, $\boldsymbol{IR}$ is an inclusion-isotonally ordered monotone upper screen (resp. screen division) ringoid of $\boldsymbol{PR}$.

(b) If additionally $\{R, +, \cdot, \leq\}$ is a completely ordered ringoid, then $\{\boldsymbol{IR}, \boxplus, \boxdot, \leq, \subseteq,\}$ is also a completely ordered ringoid (with respect to $\leq$). Moreover, for all $A = [a_1, a_2]$, $B = [b_1, b_2] \in \boldsymbol{IR}$, we have

(A) $A \boxplus B = [a_1 + b_1, a_2 + b_2]$, $A \boxminus B = A \boxplus (-B) = [a_1 - b_2, a_2 - b_1]$. These formulas (A) hold under the weaker assumption that $\{R, +, \cdot, \leq\}$ is only weakly ordered.

(B) $A \geq [o, o] \wedge B \geq [o, o] \Rightarrow A \boxdot B = [a_1 b_1, a_2 b_2]$.
(C) $A \leq [o, o] \wedge B \leq [o, o] \Rightarrow A \boxdot B = [a_2 b_2, a_1 b_1]$.
(D) $A \leq [o, o] \wedge B \geq [o, o] \Rightarrow A \boxdot B = [a_1 b_2, a_2 b_1]$.
$A \geq [o, o] \wedge B \leq [o, o] \Rightarrow A \boxdot B = [a_2 b_1, a_1 b_2]$.

(c) If $\{R, N, +, \cdot, /, \leq\}$ is a completely ordered division ringoid, then $\{\boldsymbol{IR}, \tilde{N}, \boxplus, \boxdot, \boxslash, \leq, \subseteq\}$ is also a completely ordered division ringoid in the sense that

(OD4) (a) $$\bigwedge_{A,B \in \boldsymbol{IR}} \bigwedge_{C \in \boldsymbol{IR} \setminus \tilde{N}} ([o, o] \leq A \leq B \wedge C > [o, o]$$
$$\Rightarrow [o, o] \leq A \boxslash C \leq B \boxslash C) \text{ and}$$

(b) $$\bigwedge_{A,B \in \boldsymbol{IR} \setminus \tilde{N}} \bigwedge_{C \in \boldsymbol{IR}} ([o, o] < A \leq B \wedge C \geq [o, o]$$
$$\Rightarrow C \boxslash A \geq C \boxslash B \geq [o, o]).$$

Moreover, for all $A = [a_1, a_2] \in \boldsymbol{IR}$ and $[b_1, b_2] \in \boldsymbol{IR} \setminus \tilde{N}$, we have

(E) $A \geq [o, o] \wedge B > [o, o] \Rightarrow A \boxslash B = [a_1/b_2, a_2/b_1]$.
(F) $A \geq [o, o] \wedge B < [o, o] \Rightarrow A \boxslash B = [a_2/b_2, a_1/b_1]$.
(G) $A \leq [o, o] \wedge B > [o, o] \Rightarrow A \boxslash B = [a_1/b_1, a_2/b_2]$.
(H) $A \leq [o, o] \wedge B < [o, o] \Rightarrow A \boxslash B = [a_2/b_1, a_1/b_2]$.

*Proof*: (a): Lemmas 3.4 and 3.5 directly imply the properties (D1)–(D4), (D7), and (D8), (D5), and (D9) for the interval $[-e, -e]$. The inclusion-isotony (OD5) is a simple consequence of (OD5) in $\boldsymbol{PR}$, of the monotonicity of the rounding $\square : \boldsymbol{PR} \to \overline{\boldsymbol{IR}}$, and of (RG). There remains to show (D6), (OD1) and (OD2).

(D6): We have to show that there exists only one element $X := [x, y]$ in $\boldsymbol{IR}$ that satisfies (D5). Every such $X$ must in particular fulfill (D5b, c):

(D5b) $X \boxdot X := \square(X \cdot X) = \square\{x\tilde{x} | x, \tilde{x} \in X\} = [e, e]$.

(D5c) $$\bigwedge_{A,B \in \boldsymbol{IR}} X \boxdot (A \boxdot B) = (X \boxdot A) \boxdot B = A \boxdot (X \boxdot B).$$

Since $\square$ is an upwardly directed rounding and the set $\{e\} = [e, e]$ consists only of one element, (D5b) implies

$$\bigwedge_{a,b \in X} a \cdot a = a \cdot b = b \cdot a = b \cdot b = e. \tag{5}$$

Substituting $A := [x, x]$ and $B := [y, y]$ in (D5c), we obtain

$$[x, y] \boxdot ([x, x] \boxdot [y, y]) = ([x, y] \boxdot [x, x]) \boxdot [y, y]$$
$$\Rightarrow [x, y] \boxdot \{\square([x, x] \cdot [y, y])\} = \{\square([x, y] \cdot [x, x])\} \boxdot [y, y].$$

Because of (5), both expressions contained within curly brackets here are equal to the interval $[e, e]$. Since $[e, e]$ is the neutral element of multiplication in $\boldsymbol{I}R$, we have $[x, y] = [y, y]$. Therefore $x = y$, i.e., every interval $X$ contains only one point $x \in R$, $X = [x, x]$.[†]

Let $\{\boldsymbol{I}R\}$ denote the subset of $\boldsymbol{I}R$ that consists of all point intervals $[a, a]$, $[b, b]$, .... Then the mapping $\psi : \{\boldsymbol{I}R\} \to R$ defined by $\bigwedge_{a \in R} [a, a] \to a$ is obviously an isomorphism. Because of this isomorphism, the properties (D5a)–(D5d) in $\{\boldsymbol{I}R\}$ are equivalent to (D5a)–(D5d) in $R$. Since $R$ is a ringoid, there exists only one such element, $x = -e$. Therefore $X = [-e, -e]$ is the only element in $\boldsymbol{I}R$ that fulfills (D5).

(OD1): We first demonstrate the relation

$$\bigwedge_{A,B \in \boldsymbol{I}R} (\inf A + \inf B = \inf(A + B) \wedge \sup A + \sup B = \sup(A + B)). \tag{6}$$

If $A = [a_1, a_2]$, $B = [b_1, b_2]$ and $C = [c_1, c_2]$, then

$$\bigwedge_{a \in A} \bigwedge_{b \in B} (\inf A \leq a \wedge \inf B \leq b$$
$$\underset{\text{Theorem 2.2(m)}}{\Rightarrow} \inf A + \inf B \leq a + b) \Rightarrow \inf A + \inf B \leq \inf(A + B).$$

On the other hand, $\inf(A + B) \leq a_1 + b_1 = \inf A + \inf B$ and therefore by (O3) $\inf A + \inf B = \inf(A + B)$. The proof of the second property in (6) is dual.

With (6) we now prove (OD1):

$$A \leq B :\Leftrightarrow a_1 \leq b_1 \wedge a_2 \leq b_2$$
$$\underset{(\text{OD1})_R}{\Rightarrow} a_1 + c_1 \leq b_1 + c_1 \wedge a_2 + c_2 \leq b_2 + c_2$$
$$\Rightarrow \inf A + \inf C \leq \inf B + \inf C \wedge \sup A + \sup C \leq \sup B + \sup C$$
$$\underset{(6)}{\Rightarrow} \inf(A + C) \leq \inf(B + C) \wedge \sup(A + C) \leq \sup(B + C)$$
$$\Rightarrow A \boxplus C \leq B \boxplus C.$$

[†] We call such an interval a point interval.

$$\text{(OD2)}: \quad A \le B :\Leftrightarrow a_1 \le b_1 \wedge a_2 \le b_2$$
$$\underset{(\mathrm{OD2})_R}{\Rightarrow} -b_1 \le -a_1 \wedge -b_2 \le -a_2$$
$$\underset{(2)}{\Rightarrow} -B \le -A \Rightarrow \boxminus B \le \boxminus A.$$

(b): Now we may assume that $\{R, +, \cdot, \le\}$ is an ordered ringoid. Instead of (B), (C), (D), we demonstrate the following properties for all $A, B \in IR$:

$$\text{(B')} \quad A \ge [o, o] \wedge B \ge [o, o] \Rightarrow \inf A \cdot \inf B = \inf(A \cdot B)$$
$$\wedge \sup A \cdot \sup B = \sup(A \cdot B).$$

$$\text{(C')} \quad A \le [o, o] \wedge B \le [o, o] \Rightarrow \inf A \cdot \inf B = \sup(A \cdot B)$$
$$\wedge \sup A \cdot \sup B = \inf(A \cdot B).$$

$$\text{(D')} \quad A \le [o, o] \wedge B \ge [o, o] \Rightarrow \sup A \cdot \inf B = \sup(A \cdot B)$$
$$\wedge \inf A \cdot \sup B = \inf(A \cdot B)$$
$$\wedge \inf B \cdot \sup A = \sup(B \cdot A)$$
$$\wedge \sup B \cdot \inf A = \inf(B \cdot A).$$

Applying (4), we get (A), (B), (C), (D) directly by using (6), (B′), (C′), (D′), respectively.

We only prove the first equality in each of (B′), (C′), and (D′), the proof of the second equality being always dual.

$$\text{(B')}: \quad A \ge [o, o] \wedge B \ge [o, o]$$
$$\Rightarrow \bigwedge_{a \in A} \bigwedge_{b \in B} o \le \inf A \le a \wedge o \le \inf B \le b$$
$$\underset{\text{Theorem 2.2(o)}}{\Rightarrow} \inf A \cdot \inf B \le ab \Rightarrow \inf A \cdot \inf B \le \inf(A \cdot B).$$

On the other hand,

$$\inf(A \cdot B) \le a_1 \cdot b_1 = \inf A \cdot \inf B \underset{(\mathrm{O3})}{\Rightarrow} \inf A \cdot \inf B = \inf(A \cdot B).$$

$$\text{(C')}: \quad A \le [o, o] \wedge B \le [o, o] \Rightarrow -A \ge [o, o] \wedge -B \ge [o, o]$$
$$\underset{(\mathrm{B'})}{\Rightarrow} \inf(-A) \cdot \inf(-B) \underset{\text{Theorem 2.2(y)}}{=} \sup A \cdot \sup B = \inf(A \cdot B).$$

$$\text{(D')}: \quad A \le [o, o] \wedge B \ge [o, o] \Rightarrow -A \ge [o, o] \wedge B \ge [o, o]$$
$$\underset{(\mathrm{B'})}{\Rightarrow} \inf(-A) \cdot \inf B = \inf((-A) \cdot B)$$
$$\underset{\text{Theorem 2.2(y)}}{\Rightarrow} \sup A \cdot \inf B = \sup(A \cdot B).$$

The proof of the remaining properties in (D′) is analogous.

Finally we prove the property (OD3):

$$[o,o] \le A \le B \wedge C \ge [o,o] \;\Rightarrow\; o \le a_1 \le b_1 \wedge o \le a_2 \le b_2 \wedge o \le c_1 \le c_2$$
$$\underset{(\mathrm{OD3})_R}{\Rightarrow} o \le a_1 c_1 \le b_1 c_1 \wedge o \le a_2 c_2 \le b_2 c_2$$
$$\underset{(B)}{\Rightarrow} [o,o] \le A \boxdot C \le B \boxdot C.$$

(c): Instead of (E), (F), (G), and (H), we demonstrate the following properties for all $A \in IR$ and all $B \in IR \backslash \tilde{N}$:

(E′) $A \ge [o,o] \wedge B > [o,o] \Rightarrow \inf A/\sup B = \inf(A/B) \wedge \sup A/\inf B = \sup(A/B)$.

(F′) $A \ge [o,o] \wedge B < [o,o] \Rightarrow \inf A/\inf B = \sup(A/B) \wedge \sup A/\sup B = \inf(A/B)$.

(G′) $A \le [o,o] \wedge B > [o,o] \Rightarrow \sup A/\sup B = \sup(A/B) \wedge \inf A/\inf B = \inf(A/B)$.

(H′) $A \le [o,o] \wedge B < [o,o] \Rightarrow \sup A/\inf B = \inf(A/B) \wedge \inf A/\sup B = \sup(A/B)$.

As before, we then apply (4) to obtain (E), (F), (G), and (H) directly by using (E′), (F′), (G′), and (H′), respectively.

(E′): First we remark that the property (OD4) in $R$ can obviously be extended to the form

$$\bigwedge_{a,b,c \in R} (o \le a \le b \wedge c > o \Rightarrow o \le a/c \le b/c) \wedge \tag{7}$$

$$\bigwedge_{a,b,c \in R} (o < a \le b \wedge c \ge o \Rightarrow c/a \ge c/b \ge o). \tag{8}$$

Since $o \in N$ and $B \in IR \backslash \tilde{N}$, we have $o \notin B$. Therefore,

$$A \ge [o,o] \wedge B > [o,o] \Rightarrow \bigwedge_{a \in A} \bigwedge_{b \in B} o \le \inf A \le a \wedge o < b \le \sup B$$

$$\underset{(7),(8)}{\Rightarrow} \inf A/\sup B \le a/\sup B \le a/b \Rightarrow \inf A/\sup B \le \inf(A/B).$$

On the other hand, $\inf(A/B) \le a_1/b_2 = \inf A/\sup B \Rightarrow \inf A/\sup B = \inf(A/B)$. The second property is dual.

(F′): $A \ge [o,o] \wedge B < [o,o] \Rightarrow A \ge [o,o] \wedge -B > [o,o]$

$$\underset{(\mathrm{E}')}{\Rightarrow} \inf A/\inf B = \sup(A/B) \wedge \sup A/\sup B = \inf(A/B).$$

(G′): $A \leq [o, o) \wedge B > [o, o] \Rightarrow -A \geq [o, o] \wedge B > [o, o]$

$$\underset{(E')}{\Rightarrow} \sup A/\sup B = \sup(A/B) \wedge \inf A/\inf B = \inf(A/B).$$

(H′): $A \leq [o, o] \wedge B < [o, o] \Rightarrow -A \geq [o, o] \wedge -B > [o, o]$

$$\underset{(E')}{\Rightarrow} \sup A/\inf B = \inf(A/B) \wedge \inf A/\sup B = \sup(A/B).$$

There remains only to prove (OD4).

(OD4) (a): $C \in IR \backslash \tilde{N} \wedge C > [o, o]$

$$\Rightarrow o \notin C$$
$$\Rightarrow o \leq a_1 \leq b_1 \wedge o \leq a_2 \leq b_2 \wedge o < c_1 < c_2$$
$$\Rightarrow_{(7)} o \leq a_1/c_2 \leq b_1/c_2 \wedge o \leq a_2/c_1 \leq b_2/c_1$$
$$\Rightarrow_{(E)} [o, o] \leq A \boxed{/} C \leq B \boxed{/} C.$$

(b): $A, B \in IR \backslash \tilde{N} \wedge [o, o] < A \leq B$

$$\Rightarrow o < a_1 \leq b_1 \wedge o < a_2 \leq b_2 \wedge o \leq c_1 \leq c_2$$
$$\Rightarrow_{(8)} c_1/a_2 \geq c_1/b_2 \geq o \wedge c_2/a_1 \geq c_2/b_1 \geq o$$
$$\Rightarrow_{(E)} C \boxed{/} A \geq C \boxed{/} B \geq [o, o]. \blacksquare$$

The rules (A)–(H) in Theorem 4.4 cover all cases where both of the operand intervals are comparable with zero with respect to the order relation $\leq$. In all of these cases the result of an operation $A \circledast B$ can be expressed in terms of the bounds of the operand intervals. The remaining cases are those in which one or both operands are not comparable with $[o, o]$. We shall deal with these cases in the next section under the additional hypothesis that the set $\{R, \leq\}$ is linearly ordered.

We already know that the real numbers $\{\mathbf{R}, \{0\}, +, \cdot, /, \leq\}$ are a completely ordered division ringoid if we assume that the least and greatest elements $-\infty$ and $+\infty$ are included in $\mathbf{R}$. By Theorem 4.4, therefore, $\{\mathbf{IR}, N, \boxplus, \boxdot, \boxed{/}, \leq, \subseteq\}$, $N := \{A \in \mathbf{IR} \mid 0 \in A\}$ is a completely ordered division ringoid with respect to $\leq$, while with respect to inclusion $\subseteq$ it is, moreover, an inclusion-isotonally ordered monotone upper screen ringoid of the power set $\mathbf{PR}$.

Since $\{\mathbf{C}, \{0\}, +, \cdot, /, \leq\}$ is a completely and weakly ordered division ringoid, we obtain the same properties for the interval set $\{\mathbf{IC}, N, \boxplus, \boxdot, \boxed{/}, \leq, \subseteq\}$, $N := \{A \in \mathbf{IC} \mid 0 \in A\}$ with the exception that the latter is only weakly ordered with respect to $\leq$.

We also know that $\{M_n\mathbf{R}, +, \cdot, \leq\}$ is a completely ordered ringoid and that $\{M_n\mathbf{C}, +, \cdot, \leq\}$ is a completely and weakly ordered ringoid. By Theorem 4.4 therefore $\{\mathbf{I}M_n\mathbf{R}, \boxplus, \boxdot, \leq, \subseteq\}$ is a completely ordered ringoid and $\{\mathbf{I}M_n\mathbf{C}, \boxplus, \boxdot, \leq, \subseteq\}$ is a completely and weakly ordered ringoid with respect to $\leq$. With respect to inclusion $\subseteq$, both ringoids are inclusion-isotonally ordered and monotone upper screen ringoids of $\boldsymbol{PM_nR}$ resp. $\boldsymbol{PM_nC}$.

We now consider the case of vectoids. The corresponding concepts can be developed quite similarly. Let $\{V, \leq\}$ be a complete lattice, $\boldsymbol{I}V$ the set of intervals over $V$, and $\overline{\boldsymbol{I}V} := \boldsymbol{I}V \cup \{\varnothing\}$. Then $\{\overline{\boldsymbol{I}V}, \subseteq\}$ is an upper screen of $\{\boldsymbol{P}V, \subseteq\}$. The monotone upwardly directed rounding is defined by the following properties:

(R1) $\bigwedge_{A \in \boldsymbol{I}V} \square A = A.$

(R2) $\bigwedge_{A,B \in \boldsymbol{P}V} (A \subseteq B \Rightarrow \square A \subseteq \square B).$

(R3) $\bigwedge_{A \in \boldsymbol{P}V} A \subseteq \square A.$

The following theorem characterizes the set $\overline{\boldsymbol{I}V}$ and rounding $\square$ just introduced.

**Theorem 4.5:** Let $\{V, R, \leq\}$ be a completely and weakly ordered vectoid with the neutral element $\boldsymbol{o}$. Then

(a) $\{\overline{\boldsymbol{I}V}, \subseteq\}$ is symmetric upper screen of $\{\boldsymbol{P}V, \boldsymbol{P}R, \subseteq\}$, i.e., we have

(S3) $[\boldsymbol{o}, \boldsymbol{o}]$ (and $[\boldsymbol{e}, \boldsymbol{e}])^{\dagger} \in \overline{\boldsymbol{I}V} \wedge \bigwedge_{A = [\boldsymbol{a}_1, \boldsymbol{a}_2] \in \boldsymbol{I}V} -A = [-\boldsymbol{a}_2, -\boldsymbol{a}_1] \in \boldsymbol{I}V;$

(b) the monotone upwardly directed rounding $\square : \boldsymbol{P}V \to \overline{\boldsymbol{I}V}$ is antisymmetric, i.e.,

(R4) $\bigwedge_{\varnothing \neq A \in \boldsymbol{P}V} \square(-A) = -\square A;$

(c) we have

(R) $\bigwedge_{\varnothing \neq A \in \boldsymbol{P}V} \square A = \inf_{\overline{\boldsymbol{I}V}}(U(A) \cap \boldsymbol{I}V) = \left[\inf_{\boldsymbol{a} \in A} \boldsymbol{a}, \sup_{\boldsymbol{a} \in A} \boldsymbol{a}\right].$

*Proof*: By employing corresponding properties of the vectoid, the proof can be given in a manner similar to the proof of Theorem 4.3. ■

Now we employ the monotone upwardly directed rounding $\square : \boldsymbol{P}V \to \overline{\boldsymbol{I}V}$ to define inner and outer operations in $\boldsymbol{I}V$ by the corresponding semimorphism. This semimorphism has the following properties:

$$
\begin{aligned}
&\text{(RG)} \quad \bigwedge_{A,B \in \boldsymbol{I}V} A \boxast B := \square(A * B) = [\inf(A * B), \sup(A * B)], * \in \{+, \cdot\}, \\
&\qquad \bigwedge_{A \in \boldsymbol{I}R} A \boxdot \boldsymbol{A} := \square(A \cdot \boldsymbol{A}) = [\inf(A * \boldsymbol{A}), \sup(A * \boldsymbol{A})].
\end{aligned} \tag{9}
$$

† If $V$ is multiplicative.

The operations defined by (RG) have the following properties (to which by Theorems 1.30 and 1.34, (RG) is equivalent):

(RG1) $\bigwedge_{\mathbf{A},\mathbf{B} \in \boldsymbol{IV}} (\mathbf{A} * \mathbf{B} \in \boldsymbol{IV} \Rightarrow \mathbf{A} \boxed{*} \mathbf{B} = \mathbf{A} * \mathbf{B}),$

$\bigwedge_{A \in \boldsymbol{IR}} \bigwedge_{\mathbf{A} \in \boldsymbol{IV}} (A \cdot \mathbf{A} \in \boldsymbol{IV} \Rightarrow A \boxdot \mathbf{A} = A \cdot \mathbf{A}).$

(RG2) $\bigwedge_{\mathbf{A},\mathbf{B},\mathbf{C},\mathbf{D} \in \boldsymbol{IV}} (\mathbf{A} * \mathbf{B} \subseteq \mathbf{C} * \mathbf{D} \Rightarrow \mathbf{A} \boxed{*} \mathbf{B} \subseteq \mathbf{C} \boxed{*} \mathbf{D}),$

$\bigwedge_{A,B \in \boldsymbol{IR}} \bigwedge_{\mathbf{A},\mathbf{B} \in \boldsymbol{IV}} (A \cdot \mathbf{A} \subseteq B \cdot \mathbf{B} \Rightarrow A \boxdot \mathbf{A} \subseteq B \boxdot \mathbf{B}).$

(RG3) $\bigwedge_{\mathbf{A},\mathbf{B} \in \boldsymbol{IV}} (\mathbf{A} * \mathbf{B} \subseteq \mathbf{A} \boxed{*} \mathbf{B}),$ $\bigwedge_{A \in \boldsymbol{IR}} \bigwedge_{\mathbf{A} \in \boldsymbol{IV}} (A \cdot \mathbf{A} \subseteq A \boxdot \mathbf{A}).$

In the following theorem we characterize interval vectoids and derive explicit representations for certain interval operations.

**Theorem 4.6:** (a) Let $\{V, R, \leq\}$ be a completely and weakly ordered vectoid with the neutral elements $\boldsymbol{o}$ and $\boldsymbol{e}$ (the latter if a multiplication exists), and let $\{\boldsymbol{IR}, \boxplus\ \boxdot, \leq, \subseteq\}$ be the monotone upper screen ringoid of $\boldsymbol{PR}$. If operations in $\boldsymbol{I}V$ are defined by the semimorphism $\Box : \boldsymbol{P}V \to \overline{\boldsymbol{I}V}$, then $\{\boldsymbol{I}V, \boldsymbol{IR}, \leq, \subseteq\}$ is a completely and weakly ordered vectoid with respect to the order relation $\leq$. The neutral elements are $[\boldsymbol{o}, \boldsymbol{o}]$ and $[\boldsymbol{e}, \boldsymbol{e}]$. With respect to inclusion $\subseteq$, $\boldsymbol{I}V$ is an inclusion-isotonally ordered monotone upper screen vectoid of $\{\boldsymbol{P}V, \boldsymbol{P}R\}$. It is multiplicative if $\{V, R, \leq\}$ is.

(b) If in addition $\{V, R, \leq\}$ is completely ordered, then $\{\boldsymbol{I}V, \boldsymbol{I}R, \leq, \subseteq\}$ is also completely ordered with respect to $\leq$. Moreover, for all $A = [a_1, a_2] \in \boldsymbol{I}R$ and all $\mathbf{A} = [\mathbf{a}_1, \mathbf{a}_2]$, $\mathbf{B} = [\mathbf{b}_1, \mathbf{b}_2] \in \boldsymbol{I}V$, we have

(A) $\mathbf{A} \boxplus \mathbf{B} = [\mathbf{a}_1 + \mathbf{b}_1, \mathbf{a}_2 + \mathbf{b}_2]$, $\mathbf{A} \boxminus \mathbf{B} = \mathbf{A} \boxplus (-\mathbf{B}) = [\mathbf{a}_1 - \mathbf{b}_2, \mathbf{a}_2 - \mathbf{b}_1]$.
These formulas (A) hold under the weaker assumption that $\{V, R, \leq\}$ is only weakly ordered.

(B) $A \geq [o, o] \wedge \mathbf{A} \geq [\boldsymbol{o}, \boldsymbol{o}] \Rightarrow A \boxdot \mathbf{A} = [a_1 \mathbf{a}_1, a_2 \mathbf{a}_2]$.
(C) $A \leq [o, o] \wedge \mathbf{A} \leq [\boldsymbol{o}, \boldsymbol{o}] \Rightarrow A \boxdot \mathbf{A} = [a_2 \mathbf{a}_2, a_1 \mathbf{a}_1]$.
(D) $A \geq [o, o] \wedge \mathbf{A} \leq [\boldsymbol{o}, \boldsymbol{o}] \Rightarrow A \boxdot \mathbf{A} = [a_2 \mathbf{a}_1, a_1 \mathbf{a}_2]$.
(E) $A \leq [o, o] \wedge \mathbf{A} \geq [\boldsymbol{o}, \boldsymbol{o}] \Rightarrow A \boxdot \mathbf{A} = [a_1 \mathbf{a}_2, a_2 \mathbf{a}_1]$.

If a multiplication exists in $V$, we also obtain:

(F) $\mathbf{A} \geq [\boldsymbol{o}, \boldsymbol{o}] \wedge \mathbf{B} \geq [\boldsymbol{o}, \boldsymbol{o}] \Rightarrow \mathbf{A} \boxdot \mathbf{B} = [\mathbf{a}_1 \mathbf{b}_1, \mathbf{a}_2 \mathbf{b}_2]$.
(G) $\mathbf{A} \leq [\boldsymbol{o}, \boldsymbol{o}] \wedge \mathbf{B} \leq [\boldsymbol{o}, \boldsymbol{o}] \Rightarrow \mathbf{A} \boxdot \mathbf{B} = [\mathbf{a}_2 \mathbf{b}_2, \mathbf{a}_1 \mathbf{b}_1]$.
(H) $\mathbf{A} \leq [\boldsymbol{o}, \boldsymbol{o}] \wedge \mathbf{B} \geq [\boldsymbol{o}, \boldsymbol{o}] \Rightarrow \mathbf{A} \boxdot \mathbf{B} = [\mathbf{a}_1 \mathbf{b}_2, \mathbf{a}_2 \mathbf{b}_1]$
$\wedge\, \mathbf{B} \boxdot \mathbf{A} = [\mathbf{b}_2 \mathbf{a}_1, \mathbf{b}_1 \mathbf{a}_2]$.

*Proof*: (a): Theorems 3.12 and 3.13 imply that $\{\boldsymbol{IV}, \boldsymbol{IR}, \leq, \subseteq\}$ is a monotone upper screen vectoid of $\{\boldsymbol{PV}, \boldsymbol{PR}\}$, which is multiplicative if $\{V, R, \leq\}$ is. The inclusion-isotony is a simple consequence of (OV5) in $\boldsymbol{PV}$, of the monotonicity of the rounding $\square : \boldsymbol{PV} \to \overline{\boldsymbol{IV}}$, and of (RG). There remains to prove (OV1) and (OV2) (analogous to (6) in the proof of Theorem 4.4). To do this, we use the following property:

$$(\mathrm{A}') \quad \bigwedge_{\boldsymbol{A},\boldsymbol{B} \in \boldsymbol{IV}} (\inf \boldsymbol{A} + \inf \boldsymbol{B} = \inf(\boldsymbol{A} + \boldsymbol{B}) \wedge \sup \boldsymbol{A} + \sup \boldsymbol{B} = \sup(\boldsymbol{A} + \boldsymbol{B})),$$

the proof of which is straightforward and omitted. With (A′), (OV1) and (OV2) can be proved in a manner analogous to the proof of (OD1, 2) in Theorem 4.4.

(b): If $\{V, R, \leq\}$ is a completely ordered vectoid, we proceed by demonstrating the following properties for all $A \in \boldsymbol{IR}$ and all $\boldsymbol{A}, \boldsymbol{B} \in \boldsymbol{IV}$:

$$(\mathrm{B}') \quad A \geq [o, o] \wedge \boldsymbol{A} \geq [\boldsymbol{o}, \boldsymbol{o}] \Rightarrow \inf A \cdot \inf \boldsymbol{A} = \inf(A \cdot \boldsymbol{A}) \wedge \sup A \cdot \sup \boldsymbol{A} = \sup(A \cdot \boldsymbol{A}).$$

$$(\mathrm{C}') \quad A \leq [o, o] \wedge \boldsymbol{A} \leq [\boldsymbol{o}, \boldsymbol{o}] \Rightarrow \inf A \cdot \inf \boldsymbol{A} = \sup(A \cdot \boldsymbol{A}) \wedge \sup A \cdot \sup \boldsymbol{A} = \inf(A \cdot \boldsymbol{A}).$$

$$(\mathrm{D}') \quad A \geq [o, o] \wedge \boldsymbol{A} \leq [\boldsymbol{o}, \boldsymbol{o}] \Rightarrow \inf A \cdot \sup \boldsymbol{A} = \sup(A \cdot \boldsymbol{A}) \wedge \sup A \cdot \inf \boldsymbol{A} = \inf(A \cdot \boldsymbol{A}).$$

$$(\mathrm{E}') \quad A \leq [o, o] \wedge \boldsymbol{A} \geq [\boldsymbol{o}, \boldsymbol{o}] \Rightarrow \sup A \cdot \inf \boldsymbol{A} = \sup(A \cdot \boldsymbol{A}) \wedge \inf A \cdot \sup \boldsymbol{A} = \inf(A \cdot \boldsymbol{A}).$$

$$(\mathrm{F}') \quad \boldsymbol{A} \geq [\boldsymbol{o}, \boldsymbol{o}] \wedge \boldsymbol{B} \geq [\boldsymbol{o}, \boldsymbol{o}] \Rightarrow \inf \boldsymbol{A} \cdot \inf \boldsymbol{B} = \inf(\boldsymbol{A} \cdot \boldsymbol{B}) \wedge \sup \boldsymbol{A} \cdot \sup \boldsymbol{B} = \sup(\boldsymbol{A} \cdot \boldsymbol{B}).$$

$$(\mathrm{G}') \quad \boldsymbol{A} \leq [\boldsymbol{o}, \boldsymbol{o}] \wedge \boldsymbol{B} \leq [\boldsymbol{o}, \boldsymbol{o}] \Rightarrow \inf \boldsymbol{A} \cdot \inf \boldsymbol{B} = \sup(\boldsymbol{A} \cdot \boldsymbol{B}) \wedge \sup \boldsymbol{A} \cdot \sup \boldsymbol{B} = \inf(\boldsymbol{A} \cdot \boldsymbol{B}).$$

$$(\mathrm{H}') \quad \boldsymbol{A} \leq [\boldsymbol{o}, \boldsymbol{o}] \wedge \boldsymbol{B} \geq [\boldsymbol{o}, \boldsymbol{o}] \Rightarrow \sup \boldsymbol{A} \cdot \inf \boldsymbol{B} = \sup(\boldsymbol{A} \cdot \boldsymbol{B}) \wedge \inf \boldsymbol{A} \cdot \sup \boldsymbol{B} = \inf(\boldsymbol{A} \cdot \boldsymbol{B}) \wedge \inf \boldsymbol{B} \cdot \sup \boldsymbol{A} = \sup(\boldsymbol{B} \cdot \boldsymbol{A}) \wedge \sup \boldsymbol{B} \cdot \inf \boldsymbol{A} = \inf(\boldsymbol{B} \cdot \boldsymbol{A}).$$

The proofs of these properties, which we omit, are analogous to the proof of the corresponding properties of Theorem 4.4.

Employing (9), we obtain (A)–(H) immediately from (A′)–(H′), respectively.

The proof of (OV3) is analogous to that of (OD3) in Theorem 4.4. ■

As before, the rules (A)–(H) in Theorem 4.5 cover all cases where both operand intervals are comparable with zero with respect to $\leq$. In all these cases the results of the operation $\boldsymbol{A} \boxdot \boldsymbol{A}$ or $\boldsymbol{A} \boxdot \boldsymbol{B}$ can be expressed in terms of the bounds of the operand intervals. The remaining cases are those in which one or both operands are not comparable with the zero interval. In Section 3 of this chapter we shall describe a method that furnishes explicit formulas for the resulting interval in these cases.

We know that $\{\boldsymbol{R}, +, \cdot, \leq\}$ is a completely ordered ringoid and that $\{\boldsymbol{C}, +, \cdot, \leq\}$ is a completely and weakly ordered ringoid. By Theorem 4.6, therefore, $\{\boldsymbol{IV_nR}, \boldsymbol{IR}, \leq, \subseteq\}$ and $\{\boldsymbol{IV_nR}, \boldsymbol{IM_nR}, \leq, \subseteq\}$ are completely ordered vectoids with respect to $\leq$. $\{\boldsymbol{IM_nR}, \boldsymbol{IR}, \leq, \subseteq\}$ is a completely ordered multiplicative vectoid. Further, $\{\boldsymbol{IV_nC}, \boldsymbol{IC}, \leq, \subseteq\}$ and $\{\boldsymbol{IV_nC}, \boldsymbol{IM_nC}, \leq, \subseteq\}$ are completely and weakly ordered vectoids with respect to $\leq$, while $\{\boldsymbol{IM_nC}, \boldsymbol{IC}, \leq, \subseteq\}$ is a completely and weakly ordered multiplicative vectoid. With respect to the inclusion $\subseteq$, all these vectoids are inclusion-isotonally ordered monotone upper screen vectoids of the vectoids in the corresponding power sets.

## 2. INTERVAL ARITHMETIC OVER A LINEARLY ORDERED SET

We begin with a characterization of incomparable intervals.

**Lemma 4.7:** In a linearly ordered set $\{R, \leq\}$, two intervals $A = [a_1, a_2]$, $B = [b_1, b_2] \in \boldsymbol{IR}$ are incomparable with respect to $\leq$ if and only if

$$a_1 < b_1 \leq b_2 < a_2 \vee b_1 < a_1 \leq a_2 < b_2.$$

*Proof*: It is clear that two intervals with this property are incomparable. If $a_1 = b_1$, then $A$ and $B$ are comparable. Therefore $A \| B \Rightarrow a_1 < b_1 \vee b_1 < a_1$. If in the first case $a_2 \leq b_2$, then $A$ and $B$ are comparable. Therefore $b_2 < a_2$. The second case is dual. ■

Thus far we have defined all interval operations by means of the formula (RG). This definition, however, is not directly executable on computers. In certain cases in Theorem 4.4, we have expressed the result of an operation $A \circledast B$ in terms of the bounds of the interval operands. In the cases of multiplication and division, this was only possible whenever both of the operands were comparable with $[o, o]$ with respect to $\leq$. The remaining cases are those in which one or both operands are incomparable with $[o, o]$.

By Lemma 4.7 we see that in a linearly ordered set an interval $A$ is incomparable with $[o, o]$ with respect to $\leq$ if and only if $o$ is an interior point of $A$, i.e., $o \in \mathring{A} := \{x \in R \mid a_1 < x < a_2\}$. Therefore, the remaining cases are those in which one or both interval operands have the element $o \in R$ as an interior point. Therefore, in the case of multiplication $A \boxdot B$, we still have to consider the cases

(a) $A \geq [o, o] \wedge o \in \mathring{B}$,
(b) $A \leq [o, o] \wedge o \in \mathring{B}$,
(c) $o \in \mathring{A} \wedge B \geq [o, o]$,
(d) $o \in \mathring{A} \wedge B \leq [o, o]$,
(e) $o \in \mathring{A} \wedge o \in \mathring{B}$,

and in the case of division $A \boxed{/} B$, with $A \in IR$ and $B \in IR \backslash \tilde{N}$, the cases

(f) $o \in \mathring{B}$,
(g) $o \in \mathring{A} \wedge B > [o, o]$,
(h) $o \in \mathring{A} \wedge B < [o, o]$

need to be considered.

All these cases will be covered by Theorem 4.9 to follow. In order to prove Theorem 4.9, we require the following lemma:

**Lemma 4.8:** Let $\{R, \leq\}$ be a linearly ordered complete lattice, and let $A_i \subseteq R$, $i = 1(1)n$. If $A := \bigcup_{i=1}^{n} A_i$, then

$$\inf A = \min_{i=1(1)n} (\inf A_i) \wedge \sup A = \max_{i=1(1)n} (\sup A_i).$$

*Proof*: Suppose $A$ consists of two sets $A = A_1 \cup A_2$. Then either $\inf A_1 = \inf A_2$ or by relabeling if necessary $\inf A_1 < \inf A_2$. In the first case $\inf A = \inf A_1 = \inf A_2$, and in the second case $\inf A = \inf A_1 < \inf A_2$. Thus the assertion of the lemma is correct in either case. The balance of the proof is a consequence of the fact that the union of sets is associative: $A = (\cdots (A_1 \cup A_2) \cup \cdots \cup A_{n-1}) \cup A_n$. ■

**Theorem 4.9:** Let $\{R, N, +, \cdot, /, \leq\}$ be a completely and linearly ordered division ringoid with the neutral elements $o$ and $e$, and let $\{IR, \tilde{N}, \boxplus, \boxdot, \boxed{/}, \leq\}$, $\tilde{N} := \{A \in IR \mid A \cap N \neq \varnothing\}$ be the completely ordered division ringoid of intervals over $R$. Then for all $A = [a_1, a_2]$, $B = [b_1, b_2] \in IR$ the following properties hold:

(a) $A \geq [o, o] \wedge o \in B \Rightarrow A \boxdot B = [a_2 b_1, a_2 b_2]$,
(b) $A \leq [o, o] \wedge o \in B \Rightarrow A \boxdot B = [a_1 b_2, a_1 b_1]$,
(c) $o \in A \wedge B \geq [o, o] \Rightarrow A \boxdot B = [a_1 b_2, a_2 b_2]$,
(d) $o \in A \wedge B \leq [o, o] \Rightarrow A \boxdot B = [a_2 b_1, a_1 b_1]$,
(e) $o \in A \wedge o \in B \Rightarrow A \boxdot B = [\min(a_1 b_2, a_2 b_1), \max(a_1 b_1, a_2 b_2)]$.

Moreover, for all $A = [a_1, a_2] \in IR$ and $B = [b_1, b_2] \in IR \backslash \tilde{N}$:

(f) $o \in B \Rightarrow A \boxslash B$ undefined,
(g) $o \in A \wedge B > [o, o] \Rightarrow A \boxslash B = [a_1/b_1, a_2/b_1]$,
(h) $o \in A \wedge B < [o, o] \Rightarrow A \boxslash B = [a_2/b_2, a_1/b_2]$.

*Proof*: The operations are defined by (4):

$$A \boxast B := \square(A * B) = [\inf(A * B), \sup(A * B)].$$

If $o \in B = [b_1, b_2]$ so that $b_1 \leq o \leq b_2$ and if $B_1 := [b_1, o]$ and $B_2 := [o, b_2]$ so that $B = B_1 \cup B_2$, then $A \cdot B = \{a \cdot b | a \in A \wedge b \in B\} = \{a \cdot b | a \in A \wedge b \in B_1\} \cup \{a \cdot b | a \in A \wedge b \in B_2\}$.

By Lemma 4.8 we get

$$\inf(A \cdot B) = \min(\inf(A \cdot B_1), \inf(A \cdot B_2)).$$

With this relation we use Theorem 4.4 to obtain

(a) $A \geq [o, o] \wedge o \in B$
$$\Rightarrow A \boxdot B = [\min(a_2 b_1, o), \max(o, a_2 b_2)] = [a_2 b_1, a_2 b_2],$$

(b) $A \leq [o, o] \wedge o \in B$
$$\Rightarrow A \boxdot B = [\min(o, a_1 b_2), \max(a_1 b_1, o)] = [a_1 b_2, a_1 b_1],$$

(c) $o \in A \wedge B \geq [o, o]$
$$\Rightarrow A \boxdot B = [\min(a_1 b_2, o), \max(o, a_2 b_2)] = [a_1 b_2, a_2 b_2],$$

(d) $o \in A \wedge B \leq [o, o]$
$$\Rightarrow A \boxdot B = [\min(o, a_2 b_1), \max(a_1 b_1, o)] = [a_2 b_1, a_1 b_1],$$

(e) $o \in A \wedge o \in B$
$$\Rightarrow A \boxdot B = [\min(o, o, a_1 b_2, a_2 b_1), \max(a_1 b_1, a_2 b_2, o, o)]$$
$$= [\min(a_1 b_2, a_2 b_1), \max(a_1 b_1, a_2 b_2)],$$

(f) $o \in B \Rightarrow A \boxbslash B$ undefined,

(g) $o \in A \wedge B > [o, o]$
$$\Rightarrow A \boxslash B = [\min(a_1/b_1, o), \max(o, a_2/b_1)] = [a_1/b_1, a_2/b_1],$$

(h) $o \in A \wedge B < [o, o] \Rightarrow A \boxslash B = [\min(o, a_2/b_2), \max(a_1/b_2, o)]$
$$= [a_2/b_2, a_1/b_2]. \quad \blacksquare$$

As the result of Theorems 4.4 and 4.9, we can state that in a linearly ordered set even in the cases of multiplication and division, the result of an interval operation $A \boxast B$ can be expressed in terms of the bounds of the interval operands. In order to get each of these bounds, typically only one multipli-

TABLE 1

EXECUTION OF MULTIPLICATION

| | $A = [a_1, a_2]$ | $B = [b_1, b_2]$ | $A \boxdot B = [\inf(AB), \sup(AB)]$ |
|---|---|---|---|
| 1 | $A \geq [o, o]$ | $B \geq [o, o]$ | $[a_1b_1, a_2b_2]$ |
| 2 | $A \geq [o, o]$ | $B \leq [o, o]$ | $[a_2b_1, a_1b_2]$ |
| 3 | $A \geq [o, o]$ | $o \in B$ | $[a_2b_1, a_2b_2]$ |
| 4 | $A \leq [o, o]$ | $B \geq [o, o]$ | $[a_1b_2, a_2b_1]$ |
| 5 | $A \leq [o, o]$ | $B \leq [o, o]$ | $[a_2b_2, a_1b_1]$ |
| 6 | $A \leq [o, o]$ | $o \in B$ | $[a_1b_2, a_1b_1]$ |
| 7 | $o \in A$ | $B \geq [o, o]$ | $[a_1b_2, a_2b_2]$ |
| 8 | $o \in A$ | $B \leq [o, o]$ | $[a_2b_1, a_1b_1]$ |
| 9 | $o \in A$ | $o \in B$ | $[\min(a_1b_2, a_2b_1), \max(a_1b_1, a_2b_2)]$ |

TABLE 2

EXECUTION OF DIVISION

| | $A = [a_1, a_2] \in IR$ | $B = [b_1, b_2] \in IR \backslash \tilde{N}$ | $A \boxslash B =$ $[\inf(A/B), \sup(A/B)]$ |
|---|---|---|---|
| 1 | $A \geq [o, o]$ | $B > [o, o]$ | $[a_1/b_2, a_2/b_1]$ |
| 2 | $A \geq [o, o]$ | $B < [o, o]$ | $[a_2/b_2, a_1/b_1]$ |
| 3 | $A \geq [o, o]$ | $o \in B$ | undefined |
| 4 | $A \leq [o, o]$ | $B > [o, o]$ | $[a_1/b_1, a_2/b_2]$ |
| 5 | $A \leq [o, o]$ | $B < [o, o]$ | $[a_2/b_1, a_1/b_2]$ |
| 6 | $A \leq [o, o]$ | $o \in B$ | undefined |
| 7 | $o \in A$ | $B > [o, o]$ | $[a_1/b_1, a_2/b_1]$ |
| 8 | $o \in A$ | $B < [o, o]$ | $[a_2/b_2, a_1/b_2]$ |
| 9 | $o \in A$ | $o \in B$ | undefined |

cation or division is necessary. Only in the case of Theorem 4.9(e), $o \in A$ and $o \in B$, do two products have to be calculated and compared. All these formulas are easily implemented on a computer.

In the cases of multiplication and division we summarize these results in Tables 1 and 2. The formulas therein hold, in particular, for computations with real intervals.

The results of Theorems 4.4 and 4.9 are summarized by the following corollary.

**Corollary 4.10:** Let $\{R, \{o\}, +, \cdot, /, \leq\}$ be a completely and linearly ordered division ringoid. For intervals $A = [a_1, a_2]$ and $B = [b_1, b_2]$ of $IR$, the following formula holds:

$$A \circledast B := \square(A * B) = \left[\min_{i,j=1,2} (a_i * b_j), \max_{i,j=1,2} (a_i * b_j)\right]$$

for operations $* \in \{+, -, \cdot, /\}$. ∎

Whenever in Tables 1 and 2 both operands are comparable with the interval $[o, o]$, the result of the interval operation $A \boxast B$ contains both bounds of $A$ as well as those of $B$. If one or both of the operands $A$ or $B$, however, contains zero, then the result $A \boxast B$ is always expressed by only three of the four bounds of $A$ and $B$. In all these cases in Table 1, the bound which is missing in the expression for the result can be shifted toward zero and in general even beyond zero without changing the result of the operation $A \boxast B$. Similarly, in cases 7 and 8 in Table 2, the bound of $B$, which is missing in the expression for the resulting interval, can be shifted toward $\infty$ (resp. $-\infty$) without changing the result of the operation.

This shows a certain lack of sensitivity of interval arithmetic whenever in the cases of multiplication and division one of the operands contains zero.

In all such cases—cases 3, 6, 7, 8, 9 of Table 1 and 7, 8 of Table 2—the result $A \boxast B$ also contains zero, and the formulas show that the result tends toward the zero interval if the operands that contain zero do likewise. In the limit case when the operand that contains zero has become the zero interval, no such imprecision is left. This suggests that within arithmetic expressions interval operands that contain zero should be made as small in diameter as possible.

We conclude this section with the following observation concerning $\boldsymbol{IR}$.

**Remark 4.11:** It is an interesting fact, but not essential for the development of this theory, that the intervals $\boldsymbol{IR}$ over the real numbers $\boldsymbol{R}$ form a closed subdivision ringoid of the power set $\boldsymbol{PR}$, i.e.,

$$\bigwedge_{A,B \in \boldsymbol{IR}} A \boxast B := \square(A * B) = A * B, \qquad \text{for all} \quad * \in \{+, -, \cdot, /\}.$$

Here in the case of division we assume that $0 \notin B$.

In order to see this, we recall that $A \boxast B := \square(A * B)$ is the least interval that contains $A * B := \{a * b \mid a \in A \wedge b \in B\}$. It is well known from real analysis that for all $* \in \{+, -, \cdot, /\}$, $a * b$ is a continuous function of both variables. $A * B$ is the range of this function over the product set $A \times B$. Since $A$ and $B$ are closed intervals, $A \times B$ is a simply connected, bounded, and closed subset of $\boldsymbol{R}^2$. In such a region the continuous function $a * b$ takes a maximum and a minimum as well as all values in between. Therefore,

$$\begin{aligned} A * B &= \left[ \min_{a \in A, b \in B} (a * b), \max_{a \in A, b \in B} (a * b) \right] \\ &= \left[ \inf_{a \in A, b \in B} (a * b), \sup_{a \in A, b \in B} (a * b) \right] \\ &= A \boxast B. \quad \blacksquare \end{aligned}$$

## 3. INTERVAL MATRICES

Let $\{R, +, \cdot, \leq\}$ be a completely (resp. weakly) ordered ringoid with the neutral elements $o$ and $e$, and $\{M_nR, +, \cdot, \leq\}$ the ordered (resp. weakly ordered) ringoid of matrices over $R$ with the neutral elements

$$\boldsymbol{O} = \begin{pmatrix} o & o & \cdots & o \\ o & o & \cdots & o \\ \vdots & \vdots & & \vdots \\ o & o & \cdots & o \end{pmatrix}, \qquad \boldsymbol{E} = \begin{pmatrix} e & o & \cdots & o \\ o & e & \cdots & o \\ \vdots & \vdots & & \vdots \\ o & o & \cdots & e \end{pmatrix}.$$

The power set $\{\boldsymbol{P}M_nR, +, \cdot, \subseteq\}$ is also a ringoid. If $\boldsymbol{I}M_nR$ denotes the set of intervals over $\{M_nR, \leq\}$, then according to Theorem 4.3, $\overline{\boldsymbol{I}M_nR} := \boldsymbol{I}M_nR \cup \{\varnothing\}$ is a symmetric upper screen of $\{\boldsymbol{P}M_nR, \subseteq\}$, and the monotone upwardly directed rounding $\square : \boldsymbol{P}M_nR \to \overline{\boldsymbol{I}M_nR}$ is antisymmetric. We consider its semimorphism, which in $\boldsymbol{I}M_nR$ defines operations $\boxast$, $* \in \{+, \cdot\}$, by

$$\text{(RG)} \quad \bigwedge_{\boldsymbol{A},\boldsymbol{B} \in \boldsymbol{I}M_nR} \boldsymbol{A} \boxast \boldsymbol{B} := \square(\boldsymbol{A} * \boldsymbol{B}) = [\inf(\boldsymbol{A} * \boldsymbol{B}), \sup(\boldsymbol{A} * \boldsymbol{B})].$$

Then according to Theorem 4.4, $\{\boldsymbol{I}M_nR, \boxplus, \boxdot, \leq, \subseteq\}$ is also a completely ordered (resp. weakly ordered) ringoid with respect to $\leq$ and an inclusion-isotonally ordered monotone upper screen ringoid of $\boldsymbol{P}M_nR$ with respect to $\subseteq$.

By Theorem 4.4(A), the result of an addition or subtraction in $\boldsymbol{I}M_nR$ can always be expressed in terms of the bounds of the operands. In case of a multiplication this is only possible if $\{R, +, \cdot, \leq\}$ is an ordered ringoid and the operands are comparable with the interval $[\boldsymbol{O}, \boldsymbol{O}] \in \boldsymbol{I}M_nR$ with respect to $\leq$. We are now going to derive explicit formulas that are simple to implement for all multiplications in $\boldsymbol{I}M_nR$.

In order to do this, we consider the set of $n \times n$ matrices $M_n\boldsymbol{I}R$. The elements of this set have components that are intervals over $R$. If $\{R, +, \cdot, \leq\}$ is a completely ordered (resp. weakly ordered) ringoid, then by Theorem 4.4, $\{\boldsymbol{I}R, \boxplus, \boxdot, \leq\}$ is such a structure also. With the operations and order relation of the latter, we define operations $\circledast$, $* \in \{+, \cdot\}$, and an order relation $\leq$ in $M_n\boldsymbol{I}R$ by employing the vertical method:

$$\bigwedge_{\boldsymbol{A}=(A_{ij}), \boldsymbol{B}=(B_{ij}) \in M_n\boldsymbol{I}R} \left( \boldsymbol{A} \oplus \boldsymbol{B} := (A_{ij} \boxplus B_{ij}) \wedge \boldsymbol{A} \odot \boldsymbol{B} := \left( \boxed{\sum_{\nu=1}^{n}} A_{i\nu} \boxdot B_{\nu j} \right) \right) \tag{1}$$

$$(A_{ij}) \leq (B_{ij}) :\Leftrightarrow \bigwedge_{i=1(1)n} \bigwedge_{j=1(1)n} A_{ij} \leq B_{ij}.$$

Here $\boxed{\Sigma}$ denotes the repeated summation in $\boldsymbol{I}R$.

By Theorem 2.5, we deduce directly that $\{M_n\boldsymbol{I}R, \oplus, \odot, \leq\}$ is a completely ordered (resp. weakly ordered) ringoid. We shall see that under certain further assumptions the ringoids $\{M_n IR, \oplus, \odot, \leq\}$ and $\{IM_n R, \boxplus, \boxdot, \leq\}$ are isomorphic.

To this end, we define a mapping

$$\chi: M_n IR \to IM_n R,$$

which for matrices $A = (A_{ij}) \in M_n IR$ with $A_{ij} = [a_{ij}^{(1)}, a_{ij}^{(2)}] \in IR$, $i, j = 1(1)n$, has the property

$$\chi A = \chi(A_{ij}) = \chi([a_{ij}^{(1)}, a_{ij}^{(2)}]) := [(a_{ij}^{(1)}), (a_{ij}^{(2)})]. \tag{2}$$

Obviously $\chi$ is a one-to-one mapping of $M_n IR$ onto $IM_n R$ and an order isomorphism with respect to $\leq$. To characterize the algebraic operations, we prove Lemmas 4.12 and 4.13. In certain cases we find that $\chi$ is also an algebraic isomorphism.

**Lemma 4.12:** Let $\{R, +, \cdot, \leq\}$ be a completely and weakly ordered ringoid, and consider the semimorphism defined by the monotone upwardly directly rounding $\square: PR \to \overline{IR}$. If the formula

$$\bigwedge_{A_\nu, B_\nu \in IR} \left( \sum_{\nu=1}^{n} A_\nu \boxdot B_\nu \subseteq \square \left( \sum_{\nu=1}^{n} A_\nu \cdot B_\nu \right) \right) \tag{3}$$

between the operations in $\{PR, +, \cdot\}$ and $\{IR, \boxplus, \boxdot, \leq\}$ holds, then the mapping $\chi$ establishes an isomorphism between the completely and weakly ordered ringoids $\{M_n IR, \oplus, \odot, \leq\}$ and $\{IM_n R, \boxplus, \boxdot, \leq\}$ with respect to the algebraic operations and the order relation.

*Proof*: The isomorphism of the order structures was already noted. Let $A, B \in M_n IR$, where

$$A := (A_{ij}) := ([a_{ij}^{(1)}, a_{ij}^{(2)}]), \qquad B := (B_{ij}) := ([b_{ij}^{(1)}, b_{ij}^{(2)}]).$$

Then for the addition of the images,

$$\chi A = [(a_{ij}^{(1)}), (a_{ij}^{(2)})], \qquad \chi B = [(b_{ij}^{(1)}), (b_{ij}^{(2)})] \in IM_n R,$$

we obtain by Theorem 4.4(A) that

$$\begin{aligned} \chi A \boxplus \chi B := \square(\chi A + \chi B) &= [(a_{ij}^{(1)}) + (b_{ij}^{(1)}), (a_{ij}^{(2)}) + (b_{ij}^{(2)})] \\ &= [(a_{ij}^{(1)} + b_{ij}^{(1)}), (a_{ij}^{(2)} + b_{ij}^{(2)})]. \end{aligned} \tag{4}$$

In $M_n IR$ we obtain also by Theorem 4.4(A) that

$$A \oplus B := (A_{ij} \boxplus B_{ij}) = ([a_{ij}^{(1)} + b_{ij}^{(1)}, a_{ij}^{(2)} + b_{ij}^{(2)}]). \tag{5}$$

Equations (4) and (5) imply

$$\chi A \boxplus \chi B = \chi(A \oplus B),$$

which proves the isomorphism of addition. For the case of subtraction the proof can be given analogously.

As a next step, we prove that for multiplication in $\boldsymbol{IM}_nR \subset \boldsymbol{PM}_nR$, we have

$$\chi\boldsymbol{A} \cdot \chi\boldsymbol{B} \subseteq \chi(\boldsymbol{A} \odot \boldsymbol{B}). \tag{6}$$

Here inclusion is in general proper. We consider the case $n = 2$. Let

$$\boldsymbol{A} = (A_{ij}), \qquad \boldsymbol{B} = (B_{ij}) \in M_2\boldsymbol{IR} \qquad \text{and} \qquad \underset{\cdot}{\boldsymbol{A}} = (a_{ij}), \qquad \underset{\cdot}{\boldsymbol{B}} = (b_{ij}) \in M_2R.$$

Then

$$\begin{aligned}
\chi\boldsymbol{B} \cdot \chi\boldsymbol{B} &:= \left\{ \begin{pmatrix} a_{11}b_{11} + a_{12}b_{21} & a_{11}b_{12} + a_{12}b_{22} \\ a_{21}b_{11} + a_{22}b_{21} & a_{21}b_{12} + a_{22}b_{22} \end{pmatrix} \middle| \, \underset{\cdot}{\boldsymbol{A}} \in \chi\boldsymbol{A} \wedge \underset{\cdot}{\boldsymbol{B}} \in \chi\boldsymbol{B} \right\} \\
&\subseteq \chi \begin{pmatrix} A_{11} \boxdot B_{11} \boxplus A_{12} \boxdot B_{21} & A_{11} \boxdot B_{12} \boxplus A_{12} \boxdot B_{22} \\ A_{21} \boxdot B_{11} \boxplus A_{22} \boxdot B_{21} & A_{21} \boxdot B_{12} \boxplus A_{22} \boxdot B_{22} \end{pmatrix} \\
&= \chi(\boldsymbol{A} \odot \boldsymbol{B}).
\end{aligned}$$

In general the strict inclusion sign holds in this inequality since if $\underset{\cdot}{\boldsymbol{A}} := (a_{ij}) \in \chi\boldsymbol{A}$, $\underset{\cdot}{\boldsymbol{B}} := (b_{ij}) \in \chi\boldsymbol{B}$, and $\underset{\cdot}{\boldsymbol{C}} := (c_{ij}) \in \chi\boldsymbol{B}$, then $\underset{\cdot}{\boldsymbol{A}} \cdot \underset{\cdot}{\boldsymbol{B}} \in \chi\boldsymbol{A} \cdot \chi\boldsymbol{B}$ and $\underset{\cdot}{\boldsymbol{A}} \cdot \underset{\cdot}{\boldsymbol{C}} \in \chi\boldsymbol{A} \cdot \chi\boldsymbol{B}$. However, in general

$$\underset{\cdot}{\boldsymbol{D}} := \begin{pmatrix} a_{11}b_{11} + a_{12}b_{21} & a_{11}b_{12} + a_{12}b_{22} \\ a_{21}c_{11} + a_{22}c_{21} & a_{21}c_{12} + a_{22}c_{22} \end{pmatrix} \notin \chi\boldsymbol{A} \cdot \chi\boldsymbol{B},$$

while $\underset{\cdot}{\boldsymbol{D}} \in \chi(\boldsymbol{A} \odot \boldsymbol{D})$. In (6) the equality sign may occur. It holds, for instance, whenever both of the matrices $\chi\boldsymbol{A}$ and $\chi\boldsymbol{B}$ contain only one element of $M_nR$.

Since $\square : \boldsymbol{PM}_nR \to \overline{\boldsymbol{IM}_nR}$ is a monotone upwardly directed rounding, then because of (6) we obtain

$$\chi\boldsymbol{A} \cdot \chi\boldsymbol{B} \underset{(\mathrm{R3})}{\subseteq} \square(\chi\boldsymbol{A} \cdot \chi\boldsymbol{B}) \underset{(\mathrm{RG})}{=} \chi\boldsymbol{A} \boxdot \chi\boldsymbol{B} \underset{(\mathrm{R1})}{\subseteq} \chi(\boldsymbol{A} \odot \boldsymbol{B}). \tag{7}$$

Further,

$$(3) \underset{(\mathrm{R1,2})}{\Rightarrow} \bigwedge_{\substack{A_{iv}, B_{vj} \in \boldsymbol{IR} \\ i,j,v = 1(1)n}} \square\left( \sum_{v=1}^{n} A_{iv} \boxdot B_{vj} \right) \subseteq \square\left( \sum_{v=1}^{n} A_{iv} \cdot B_{vj} \right). \tag{8}$$

Now let $C_{ijv} := [C_{ijv}^{(1)}, C_{ijv}^{(2)}] := A_{iv} \boxdot B_{vj}$, $i, j, v = 1(1)n$. Then

$$\begin{aligned}
\boldsymbol{A} \odot \boldsymbol{B} := \left( \boxed{\sum_{v=1}^{n}} A_{iv} \boxdot B_{vj} \right) &= \left( \boxed{\sum_{v=1}^{n}} C_{ijv} \right) \\
&\underset{\text{Theorem 4.4(A)}}{=} \left( \left[ \sum_{v=1}^{n} c_{ijv}^{(1)}, \sum_{v=1}^{n} c_{ijv}^{(2)} \right] \right) \\
&= \left( \square \sum_{v=1}^{n} C_{ijv} \right) = \left( \square \sum_{v=1}^{n} A_{iv} \boxdot B_{vj} \right).
\end{aligned}$$

If we now apply the mapping $\chi$, we get with $\underset{\bullet}{A} = (a_{ij}) \in \chi A$ and $\underset{\bullet}{B} = (b_{ij}) \in \chi B$

$$\chi(A \odot B) = \chi\left(\square \sum_{\nu=1}^{n} A_{i\nu} \boxdot B_{\nu j}\right) \underset{(8)}{\subseteq} \chi\left(\square \sum_{\nu=1}^{n} A_{i\nu} B_{\nu j}\right)$$

$$= \chi\left(\left[\inf \sum_{\nu=1}^{n} A_{i\nu} B_{\nu j}, \sup \sum_{\nu=1}^{n} A_{i\nu} B_{\nu j}\right]\right)$$

$$= \chi\left(\left[\inf_{\underset{\bullet}{A} \in \chi A, \underset{\bullet}{B} \in \chi B} \sum_{\nu=1}^{n} a_{i\nu} b_{\nu j}, \sup_{\underset{\bullet}{A} \in \chi A, \underset{\bullet}{B} \in \chi B} \sum_{\nu=1}^{n} a_{i\nu} b_{\nu j}\right]\right)$$

$$= \left[\inf_{\underset{\bullet}{A} \in \chi A, \underset{\bullet}{B} \in \chi B} \left(\sum_{\nu=1}^{n} a_{i\nu} b_{\nu j}\right), \sup_{\underset{\bullet}{A} \in \chi A, \underset{\bullet}{B} \in \chi B} \left(\sum_{\nu=1}^{n} a_{i\nu} b_{\nu j}\right)\right]$$

$$\Rightarrow \chi(A \odot B) \subseteq \square(\chi A \cdot \chi B) = \chi A \boxdot \chi B. \tag{9}$$

By (7) and (9) we obtain finally that $\chi A \boxdot \chi B = \chi(A \odot B)$, which proves the isomorphism in the case of multiplication. ■

**Lemma 4.13:** Let $\{R, +, \cdot, \leq\}$ be a completely and linearly ordered ringoid and $\square : PR \to \overline{IR}$ the monotone upwardly directed rounding. Then

$$\bigwedge_{A_\nu, B_\nu \in IR} \left(\sum_{\nu=1}^{n} A_\nu \boxdot B_\nu \subseteq \square\left(\sum_{\nu=1}^{n} A_\nu \cdot B_\nu\right)\right).$$

*Proof*: By Corollary 4.10, we have for all $A = [a_1, a_2]$ and $B = [b_1, b_2] \in IR$

$$A \boxast B = \left[\min_{i,j=1,2} (a_i * b_j), \max_{i,j=1,2} (a_i * b_j)\right]. \tag{10}$$

Let $A_i = [a_{i1}, a_{i2}]$ and $B_i = [b_{i1}, b_{i2}]$, $i = 1(1)n$. Then

$$x \in \sum_{i=1}^{n} A_i \boxdot B_i \;\Rightarrow\; \bigwedge_{i=1(1)n} \bigvee_{x_i \in A_i \boxdot B_i} x = \sum_{i=1}^{n} x_i$$

$$\Rightarrow \bigwedge_{i=1(1)n} \bigvee_{x_i} \min_{\mu,\nu=1,2} a_{i\mu} b_{i\nu} \leq x_i \leq \max_{\mu,\nu=1,2} a_{i\mu} b_{i\nu} \wedge x = \sum_{i=1}^{n} x_i$$

$$\underset{(\mathrm{OD1})_R}{\Rightarrow} \sum_{i=1}^{n} \min_{\mu,\nu=1,2} a_{i\mu} \cdot b_{i\nu} \leq x = \sum_{i=1}^{n} x_i \leq \sum_{i=1}^{n} \max_{\mu,\nu=1,2} a_{i\mu} \cdot b_{i\nu}$$

$$\Rightarrow x \in \square\left(\sum_{i=1}^{n} A_i \cdot B_i\right). \quad ■$$

Lemmas 4.12 and 4.13 imply the following theorem, which establishes the isomorphic character of $\chi$ in case of a linearly ordered ringoid.

**Theorem 4.14:** Let the structure $\{R, +, \cdot, \leq\}$ be a completely and linearly ordered ringoid. Then the mapping $\chi$ establishes an isomorphism between the completely ordered ringoids $\{M_n IR, \oplus, \odot, \leq\}$ and $\{IM_n R, \boxplus, \boxdot, \leq\}$ with respect to the algebraic and the order structure. ■

Whenever the two structures $\{M_n IR, \oplus, \odot, \leq\}$ and $\{IM_n R, \boxplus, \boxdot, \leq\}$ are isomorphic corresponding elements can be identified with each other. This allows us to define an inclusion relation even for elements $\boldsymbol{A} = (A_{ij})$, $\boldsymbol{B} = (B_{ij}) \in M_n IR$ by

$$\boldsymbol{A} \subseteq \boldsymbol{B} :\Leftrightarrow \bigwedge_{i,j=1(1)n} A_{ij} \subseteq B_{ij}$$

$$\text{and} \qquad A = (a_{ij}) \in \boldsymbol{A} = (A_{ij}) :\Leftrightarrow \bigwedge_{i,j=1(1)n} a_{ij} \in A_{ij}.$$

This convenient definition allows for the interpretation that a matrix $\boldsymbol{A} = (A_{ij}) \in M_n IR$ also represents a set of matrices as demonstrated by the following identity:

$$\boldsymbol{A} = (A_{ij}) \equiv \{(a_{ij}) \mid a_{ij} \in A_{ij},\ i,j = 1(1)n\}.$$

Both matrices contain the same elements.

In the case of a linearly ordered ringoid $\{R, +, \cdot, \leq\}$ and for all operations in the ringoid $\{IR, \boxplus, \boxdot, \leq\}$, we have derived explicit and easily implementable formulas in Section 2. The formulas (1) for the operation $\circledast$, $* \in \{+, \cdot\}$, in $\{M_n IR, \oplus, \odot, \leq\}$, therefore, are also easily implementable. This is not the case for the operations $\boxdot$ originally defined in $IM_n R$. By means of the isomorphism, the operation $\odot$ can be used in order to perform the multiplication $\boxdot$.

The isomorphism shows that both the vertical definition of the operations in $M_n IR$ in terms of those in $IR$ and the horizontal definition of the operations in $IM_n R$ by means of the semimorphism $\square : PM_n R \to \overline{IM_n R}$ lead to the same result. Figure 23 illustrates the connection.

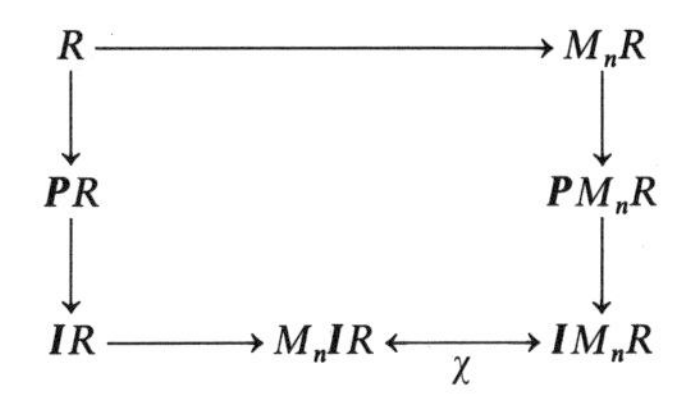


FIGURE 23. Illustration of Theorem 4.14.

## 4. INTERVAL VECTORS

Similarly to the considerations of Section 3, we are now going to derive easily implementable formulas for the interval operations occurring in vectoids.

Again let $\{R, +, \cdot, \leq\}$ be a completely ordered or weakly ordered ringoid. Then $\{V_nR, R, \leq\}$, $\{M_nR, R, \leq\}$, and $\{V_nR, M_nR, \leq\}$, where the operations and the order relation are defined by the usual formulas, are completely ordered (resp. weakly ordered) vectoids. $\{M_nR, R, \leq\}$ is in particular multiplicative.

Moreover, by Theorem 2.10 the three power sets $\{\boldsymbol{P}V_nR, \boldsymbol{P}R, \subseteq\}$, $\{\boldsymbol{P}M_nR, \boldsymbol{P}R, \subseteq\}$, and $\{\boldsymbol{P}V_nR, \boldsymbol{P}M_nR, \subseteq\}$ are also inclusion-isotonally ordered vectoids.

Let us now consider the semimorphism $\square: \boldsymbol{P}V_nR \to \overline{\boldsymbol{I}V_nR} := \boldsymbol{I}V_nR \cup \varnothing$. With this mapping and by Theorem 4.6, $\{\boldsymbol{I}V_nR, \boldsymbol{I}R, \leq, \subseteq\}$, $\{\boldsymbol{I}M_nR, \boldsymbol{I}R, \leq, \subseteq\}$, and $\{\boldsymbol{I}V_nR, \boldsymbol{I}M_nR, \leq, \subseteq\}$ become completely ordered (resp. weakly ordered) vectoids with respect to $\leq$ and inclusion-isotonally ordered monotone upper screen vectoids of the corresponding power sets with respect to $\subseteq$. Moreover, $\boldsymbol{I}M_nR$ is multiplicative.

By Theorem 4.6, the result of an addition or subtraction in $\boldsymbol{I}V_nR$ and $\boldsymbol{I}M_nR$ can always be expressed in terms of the bounds of the interval operands. In the case of an outer multiplication, this is possible only if the vectoid is ordered and if the operands are comparable with the corresponding zero element with respect to $\leq$. (The case of the inner multiplication in $\boldsymbol{I}M_nR$ was already considered in Section 3.) We are now going to derive explicit and easily implementable formulas for all outer multiplications in the vectoids $\{\boldsymbol{I}V_nR, \boldsymbol{I}R\}$, $\{\boldsymbol{I}M_nR, \boldsymbol{I}R\}$, and $\{\boldsymbol{I}V_nR, \boldsymbol{I}M_nR\}$.

In order to do this, we also consider the sets $V_n\boldsymbol{I}R$ and $M_n\boldsymbol{I}R$. The elements of these sets have components that are intervals over $R$. If $\{R, +, \cdot, \leq\}$ is a completely ordered (resp. weakly ordered) ringoid, then by Theorem 4.4, $\{\boldsymbol{I}R, \boxplus, \boxdot, \leq\}$ also has these properties. Employing the operations and order relation of the latter, we define operations and an order relation $\leq$ in $V_n\boldsymbol{I}R$ and $M_n\boldsymbol{I}R$ by means of the vertical method:

$$\bigwedge_{\boldsymbol{a}=(A_i),\boldsymbol{b}=(B_i)\in V_n\boldsymbol{I}R} \boldsymbol{a} \oplus \boldsymbol{b} := (A_i \boxplus B_i),$$

$$\bigwedge_{A\in \boldsymbol{I}R}\ \bigwedge_{\boldsymbol{a}=(A_i)\in V_n\boldsymbol{I}R} A \odot \boldsymbol{a} := (A \boxdot A_i),$$

$$\bigwedge_{A\in \boldsymbol{I}R}\ \bigwedge_{\boldsymbol{A}=(A_{ij})\in M_n\boldsymbol{I}R} A \odot \boldsymbol{A} := (A \boxdot A_{ij}), \tag{11}$$

$$\bigwedge_{\boldsymbol{A}=(A_{ij})\in M_n\boldsymbol{I}R}\ \bigwedge_{\boldsymbol{a}=(A_i)\in V_n\boldsymbol{I}R} \boldsymbol{A} \odot \boldsymbol{a} := \left(\boxed{\sum_{\nu=1}^{n}} A_{i\nu} \boxdot A_\nu\right),$$

$$(A_i) \leq (B_i) := \bigwedge_{i=1(1)n} A_i \leq B_i.$$

Here as before $\boxed{\Sigma}$ denotes the repeated summation in $\boldsymbol{I}R$.

For $M_n\boldsymbol{I}R$, we take the definition of the inner operations $\oplus$, $\odot$, and the order relation $\leq$ as given in (1) in Section 3.

Using Theorems 2.12, 2.13, and 2.14, we obtain directly that $\{V_n IR, IR, \leq\}$, $\{M_n IR, IR, \leq\}$, and $\{V_n IR, M_n IR, \leq\}$ are completely ordered (resp. weakly ordered) vectoids. Moreover, $\{M_n IR, IR, \leq\}$ is multiplicative.

Under certain further assumptions, we shall see that isomorphisms exist between the following pairs of vectoids:

$$\{V_n IR, IR, \leq\} \leftrightarrow \{IV_n R, IR, \leq\},$$
$$\{M_n IR, IR, \leq\} \leftrightarrow \{IM_n R, IR, \leq\},$$
$$\{V_n IR, M_n IR, \leq\} \leftrightarrow \{IV_n R, IM_n, R, \leq\}.$$

In order to see this for vectors and matrices

$$\boldsymbol{a} = (A_i) \in V_n IR \quad \text{with} \quad A_i = [a_i^{(1)}, a_i^{(2)}] \in IR,$$
$$\boldsymbol{A} = (A_{ij}) \in M_n IR \quad \text{with} \quad A_{ij} = [a_{ij}^{(1)}, a_{ij}^{(2)}] \in IR,$$

we define the mappings $\psi$ and $\chi$:

$$\psi : V_n IR \to IV_n R, \tag{12}$$

where

$$\psi \boldsymbol{a} = \psi(A_i) = \psi([a_i^{(1)}, a_i^{(2)}]) := [(a_i^{(1)}), (a_i^{(2)})],$$

and

$$\chi : M_n IR \to IM_n R, \tag{13}$$

where

$$\chi \boldsymbol{a} = \chi(A_{ij}) = \chi([a_{ij}^{(1)}, a_{ij}^{(2)}]) := [(a_{ij}^{(1)}), (a_{ij}^{(2)})].$$

Obviously, the mappings $\psi$ and $\chi$ are one-to-one and order isomorphisms with respect to $\leq$. The following theorem characterizes properties of $\psi$ and $\chi$ with respect to the algebraic operations.

**Theorem 4.15:** Let $\{R, +, \cdot, \leq\}$ be a completely and weakly ordered ringoid and $\{IR, \boxplus, \boxdot, \leq\}$ be the completely and weakly ordered ringoid of intervals over $R$. Then we have

(a) The mapping $\psi$ establishes an isomorphism between $\{V_n IR, IR, \leq\}$ and $\{IV_n R, IR, \leq\}$ with respect to the algebraic and the order structure.

(b) If $\square : PR \to \overline{IR}$ denotes the monotone upwardly directed rounding and if for the operations in $PR$ and $IR$ the formula

$$\bigwedge_{A_\nu, B_\nu \in IR} \sum_{\nu=1}^{n} A_\nu \boxdot B_\nu \subseteq \square\left(\sum_{\nu=1}^{n} A_\nu \cdot B_\nu\right) \tag{3}$$

holds, then the mapping $\chi$ establishes an isomorphism between the completely and weakly ordered multiplicative vectoids $\{M_n IR, IR, \leq\}$ and $\{IM_n R, IR, \leq\}$ with respect to the algebraic and the order structure.

(c) Under the assumption (3), the mappings $\psi$ and $\chi$ establish an isomorphism between the completely and weakly ordered vectoids $\{V_n IR, M_n IR, \leq\}$ and $\{IV_n R, IM_n R, \leq\}$ with respect to the algebraic and the order structure.

*Proof*: The isomorphism of the order structures has already been noted immediately preceding the statement of this theorem.

(a): The isomorphism of the addition can be proved as in Lemma 4.12. For the outer multiplication, we get with $A \in IR$ and $\boldsymbol{a} = (A_i) \in V_n IR$, $A_i \in IR$, $i = 1(1)n$:

$$\begin{aligned}\psi(A \odot \boldsymbol{a}) &:= \psi(A \boxdot A_i) = \psi(\square(A \cdot A_i)) \\ &= \psi([\inf(A \cdot A_i), \sup(A \cdot A_i)]) = [(\inf(A \cdot A_i)), (\sup(A \cdot A_i))] \\ &= [\inf(A \cdot A_i), \sup(A \cdot A_i)] = \square(A \cdot \psi\boldsymbol{a}) = A \boxdot \psi\boldsymbol{a}.\end{aligned}$$

(b): The isomorphism of the inner operations is proved in Lemma 4.12 above. The proof for the outer multiplication is analogous to (a).

(c): Lemma 4.12 demonstrated the isomorphism of the ringoids $\{M_n IR, \oplus, \odot, \leq\}$ and $\{IM_n R, \boxplus, \boxdot, \leq\}$. The isomorphism of addition in $V_n IR$ and $IV_n R$ was proved under (a). Therefore only the isomorphism of the outer multiplication remains to be shown. With $A = (A_{ij}) \in M_n IR$ and $\boldsymbol{a} = (A_i) \in V_n IR$, we get

$$\chi A \cdot \psi\boldsymbol{a} := \{\underset{\cdot}{A} \cdot \underset{\cdot}{\boldsymbol{a}} \mid \underset{\cdot}{A} \in \chi A \wedge \underset{\cdot}{\boldsymbol{a}} \in \psi\boldsymbol{a}\} \subseteq \psi\left(\boxed{\Sigma}_{\nu=1}^{n} A_{i\nu} \boxdot A_\nu\right) = \psi(A \odot \boldsymbol{a}).$$

Applying the monotone upwardly directed rounding $\square : PV_n R \to \overline{IV_n R}$ to this inequality, we get

$$\chi A \cdot \psi\boldsymbol{a} \subseteq \square(\chi A \cdot \psi\boldsymbol{a}) = \chi A \boxdot \psi\boldsymbol{a} \subseteq \psi(A \odot \boldsymbol{a}). \tag{14}$$

Using (3), we may show as in the proof of Lemma 4.2 that

$$\psi(A \odot \boldsymbol{a}) \subseteq \square(\chi A \cdot \psi\boldsymbol{a}) = \chi A \boxdot \psi\boldsymbol{a}. \tag{15}$$

With (14) and (15) we finally obtain $\chi A \boxdot \psi\boldsymbol{a} = \psi(A \odot \boldsymbol{a})$. ■

Combining this theorem with Lemma 4.13, we obtain the following theorem which further characterizes the isomorphisms induced by the mappings $\psi$ and $\chi$.

**Theorem 4.16:** Let $\{R, +, \cdot, \leq\}$ be a completely and linearly ordered ringoid. Then

(a) The mapping $\chi$ establishes an isomorphism between the completely ordered vectoids $\{M_n IR, IR, \leq\}$ and $\{IM_n R, IR, \leq\}$ with respect to the algebraic and the order structure.

(b) The mapping $\psi$ and $\chi$ establish an isomorphism between the two completely ordered vectoids $\{V_n IR, M_n IR, \leq\}$ and $\{IV_n R, IM_n R, \leq\}$ with respect to the algebraic and the order structure. ■

Whenever the structures $\{V_n IR, IR, \leq\}$ and $\{IV_n R, IR, \leq\}$ (resp. $\{V_n IR, M_n IR, \leq\}$ and $\{IV_n R, IM_n R, \leq\}$) are isomorphic, corresponding elements can be identified with one another. This allows us to define an inclusion relation even for elements $\boldsymbol{a} = (A_i)$, $\boldsymbol{b} = (B_i) \in V_n IR$ by

$$\boldsymbol{a} \subseteq \boldsymbol{b} :\Leftrightarrow \bigwedge_{i=1(1)n} A_i \subseteq B_i,$$

$$\underset{\cdot}{a} = (a_i) \in \boldsymbol{a} = (A_i) :\Leftrightarrow \bigwedge_{i=1(1)n} a_i \in A_i.$$

By means of the following identity, this definition allows for the interpretation that a vector $\boldsymbol{a} = (A_i) \in V_n IR$ also represents a set of vectors

$$\boldsymbol{a} = (A_i) \equiv \{(a_i) \mid a_i \in A_i,\ i = 1(1)n\}.$$

Both vectors contain the same elements.

In the case of a linearly ordered ringoid $\{R, +, \cdot, \leq\}$ and for all operations in the ringoid $\{IR, \boxplus, \boxdot, \leq\}$, we have derived explicit and easily implementable formulas in Section 2 of this chapter. Formulas (11) for the operations $\circledast$, therefore, are also easily implementable. This is not the case for the operations $\boxast$ originally defined, for instance, in $\{IV_n R, IM_n R, \leq\}$. Appealing to the isomorphism, we see that the operations $\circledast$ can be used in order to execute the operations $\boxast$.

## 5. INTERVAL ARITHMETIC ON A SCREEN

Let $\{R, \leq\}$ be a complete lattice and $\{S, \leq\}$ a screen of $\{R, \leq\}$. Now we consider intervals over $R$ with endpoints in $S$:

$$[a_1, a_2] = \{x \in R \mid a_1, a_2 \in S, a_1 \leq x \leq a_2\} \qquad \text{with} \quad a_1 \leq a_2.$$

We denote the set of all such intervals by $IS$. Then $IS \subseteq IR$ and $\overline{IS} := IS \cup \{\varnothing\} \subseteq \overline{IR} := IR \cup \{\varnothing\}$.

Several properties of $IS$ are described in the following theorem.

**Theorem 4.17:** Let $\{R, \leq\}$ be a complete lattice and $\{S, \leq\}$ a screen of $\{R, \leq\}$. Then

(a) $\{IS, \subseteq\}$ is a conditionally completely ordered set. For all nonempty subsets $A$ of $IS$, which are bounded below, we have with $B = [b_1, b_2] \in IS$:

$$\inf_{IS} A = \left[\sup_{B \in A} b_1, \inf_{B \in A} b_2\right] \wedge \sup_{IS} A = \left[\inf_{B \in A} b_1, \sup_{B \in A} b_2\right],$$

i.e., the infimum is the intersection, and the supremum is the convex hull of the elements of $A$.

(b) $\{\overline{IS}, \subseteq\}$ is a screen of $\{\overline{IR}, \subseteq\}$.

*Proof*: (a) Since $\{S, \leq\}$ is a complete lattice and $A$ is bounded below, the intervals

$$I := \left[\sup_{B \in A} b_1, \inf_{B \in A} b_2\right] \wedge S := \left[\inf_{B \in A} b_1, \sup_{B \in A} b_2\right]$$

exist, and for all $B \in A$, we have $I \subseteq B \wedge B \subseteq S$, i.e., $I$ is a lower and $S$ an upper bound of $A$. If $I_1$ is any lower and $S_1$ any upper bound of $A$, then $I_1 \subseteq I$ and $S \subseteq S_1$, i.e., $I$ is the greatest lower and $S$ the least upper bound of $A$.

(b) $\{\overline{IS}, \subseteq\}$ is a complete lattice and therefore a complete subnet of $\{\overline{IR}, \subseteq\}$. We still have to show that for every subset $A \subseteq \overline{IS}$, $\inf_{\overline{IS}} A = \inf_{\overline{IR}} A$ and $\sup_{\overline{IS}} A = \sup_{\overline{IR}} A$.

If $A$ is bounded below in $\overline{IS}$, then by (a), the infimum is the intersection and the supremum is the convex hull as in $\overline{IR}$. If $A$ is not bounded below in $\overline{IS}$, we consider the cases

1. $A = \varnothing$. Then $\inf_{\overline{IS}} A = \inf_{\overline{IR}} A = i(\overline{IS}) = i(\overline{IR}) = R$ and $\sup_{\overline{IS}} A = \sup_{\overline{IR}} A = o(\overline{IS}) = o(\overline{IR}) = \varnothing$.

2. $A \neq \varnothing$. Then $\sup_{\overline{IS}} A = \sup_{\overline{IR}} A = [\inf_{B \in A} b_1, \sup_{B \in A} b_2]$.

2.1 $\varnothing \in A$. Then $\inf_{\overline{IS}} A = \varnothing = \inf_{\overline{IR}} A$.

2.2 $\varnothing \notin A$. Since $A$ is not bounded below in $IS$, we get again $\inf_{IS} A = \varnothing = \inf_{\overline{IR}} A$. ■

In general, interval calculations are employed to determine sets that include the solution of a given problem. If $\{R, \leq\}$ is an ordered set and the operations for the intervals of $IR$ cannot be performed precisely on a computer, one has to approximate them on a screen $\{S, \leq\}$ of $\{R, \leq\}$. This approximation will be required to have the following qualitative properties, which we shall make precise later:

(a) The result of any computation in the subset $IS$ always has to include the result of the corresponding calculation in $IR$.

(b) The result of the computation in $IS$ has to be as close as possible to the result of the corresponding calculation in $IR$.

We may seek to achieve these qualitative requirements by defining the operations in $IS$ by means of the semimorphism associated with the monotone upwardly directed rounding $\Diamond: IR \to \overline{IS}$. We are now going to describe this process.

The monotone upwardly directed rounding $\lozenge: \overline{IR} \to \overline{IS}$ is defined by the properties

(R1) $\bigwedge_{A \in \overline{IS}} \lozenge A = A.$

(R2) $\bigwedge_{A,B \in \overline{IR}} (A \subseteq B \Rightarrow \lozenge A \subseteq \lozenge B).$

(R3) $\bigwedge_{A \in \overline{IR}} A \subseteq \lozenge A.$

In the following theorem we develop some properties of this rounding.

**Theorem 4.18:** Let $\{R, +, \cdot, \leq\}$ be a completely and weakly ordered ringoid with the neutral elements $o$ and $e$, and $\{S, \leq\}$ be a symmetric screen of $\{R, +, \cdot, \leq\}$. Further, let $\{IR, \boxplus, \boxdot\}$ be the ringoid defined by the semimorphism $\square: PR \to \overline{IR}$. Then

(a) $\{\overline{IS}, \subseteq\}$ is a symmetric screen of $\{IR, \boxplus, \boxdot\}$, i.e., we have

$$\text{(S3)} \quad [o, o], [e, e] \in \overline{IS} \wedge \bigwedge_{A = [a_1, a_2] \in IS} \boxminus A = [-a_2, -a_1] \in IS. \tag{1}$$

(b) The monotone upwardly directed rounding $\lozenge: \overline{IR} \to \overline{IS}$ is antisymmetric, i.e.,

(R4) $\bigwedge_{A \in IR} \lozenge(\boxminus A) = \boxminus(\lozenge A).$

(c) We have

$$\text{(R)} \quad \bigwedge_{A = [a_1, a_2] \in IR} \lozenge A = \inf(U(A) \cap IS) = [\triangledown a_1, \triangle a_2]. \tag{2}$$

*Proof*: (a) With $A = [a_1, a_2] \in IS$ we obtain

$$\begin{aligned} \boxminus A \quad &:= \quad [-e, -e] \boxdot A := \square(\{-e\} \cdot A) = \square(-A) \\ &\underset{(S3)_{IR}, (R1)}{=} -A \underset{\text{Theorem 4.3(a)}}{=} [-a_2, -a_1] \in IS. \end{aligned}$$

As the next step, we prove (c).

(c) With $B = [b_1, b_2] \in IS$, we get for all $A = [a_1, a_2] \in IR$:

$$\begin{aligned} \inf_{IR}(U(A) \cap IS) &\underset{\text{Theorem 4.17(b)}}{=} \inf_{IS}(U(A) \cap IS) \\ &\underset{\text{Theorem 4.17(a)}}{=} \left[ \sup_{B \in U(A) \cap IS} b_1, \inf_{B \in U(A) \cap IS} b_2 \right]. \end{aligned}$$

In general we have

$$[a_1, a_2] \subseteq [b_1, b_2] \Leftrightarrow b_1 \leq a_1 \wedge a_2 \leq b_2.$$

With this and Theorem 1.24, we get

$$\Diamond A = \inf_{IS}(U(A) \cap IS)$$
$$= \left[ \sup_{b_1 \in S \wedge b_1 \le a_1} b_1, \inf_{b_2 \in S \wedge a_2 \le b_2} b_2 \right]$$
$$= [\sup(L(a_1) \cap S), \inf(U(a_2) \cap S)] = [\nabla a_1, \Delta a_2].$$

(b) For all $A = [a_1, a_1] \in IR$ we have

$$\Diamond(\boxminus A) \underset{(1)}{=} \Diamond[-a_2, -a_1] \underset{(2)}{=} [\nabla(-a_2), \Delta(-a_1)]$$
$$\underset{\text{Theorem 3.3}}{=} [-\Delta a_2, -\nabla a_1] \underset{(1)}{=} \boxminus[\nabla a_1, \Delta a_2] \underset{(2)}{=} \boxminus(\Diamond A). \quad \blacksquare$$

The monotone upwardly directed rounding $\Diamond: \overline{IR} \to \overline{IS}$ can be employed as a semimorphism in order to define operations $\diamond\!\!\!\!\ast$, $* \in \{+, \cdot, /\}$, in $IS$ as

$$\text{(RG)} \quad \bigwedge_{A,B \in IS} A \,\diamond\!\!\!\!\ast\, B := \Diamond(A \boxast B) := \Diamond(\Box(A * B))$$
$$\underset{1,(4)}{=} \Diamond[\inf(A * B), \sup(A * B)]$$
$$\underset{(2)}{=} [\nabla \inf(A * B), \Delta \sup(A * B)]. \quad (3)$$

These operations have the following properties which by Theorem 1.30 also define them:

$$\text{(RG1)} \quad \bigwedge_{A,B \in IS} (A \boxast B \in IS \Rightarrow A \,\diamond\!\!\!\!\ast\, B = A \boxast B),$$

$$\text{(RG2)} \quad \bigwedge_{A,B,C,D \in IS} (A \boxast B \subseteq C \boxast D \Rightarrow A \,\diamond\!\!\!\!\ast\, B \subseteq C \,\diamond\!\!\!\!\ast\, D),$$

$$\text{(RG3)} \quad \bigwedge_{A,B \in IS} (A \boxast B \subseteq A \,\diamond\!\!\!\!\ast\, B).$$

In these formulas we assume additionally, in the case of division, that $B$, $D \notin \overline{N}$. For the definition of $\overline{N}$, see Lemma 4.19(d), to which we now turn.

**Lemma 4.19:** Let $\{R, +, \cdot, \le\}$ be a completely and weakly ordered ringoid with the neutral elements $o$ and $e$, and let $\{S, \le\}$ be a symmetric screen of $\{R, +, \cdot, \le\}$. Consider the semimorphisms $\Box: PR \to \overline{IR}$ and $\Diamond: \overline{IR} \to \overline{IS}$. Then

(a) Every element $X \in IS$ that fulfills (D5) in $\{IS, \diamond\!\!\!\!+, \diamond\!\!\!\!\cdot\}$ is a point interval, i.e., $X = [x, x]$.

(b) If every element of $IS$ that fulfills both (D5) and (OD2) belongs to a linearly ordered subset $\{L, \le\}$ of $\{IS, \le\}$, then $\{IS, \diamond\!\!\!\!+, \diamond\!\!\!\!\cdot, \le\}$ is a weakly ordered ringoid.

(c) If $\{R, +, \cdot, \leq\}$ is an ordered ringoid, then under the hypothesis in (b), $\{\boldsymbol{IS}, \diamondplus, \diamonddot, \leq\}$ also is an ordered ringoid.

(d) If $\{R, N, +, \cdot, /, \leq\}$ is a completely ordered (resp. weakly ordered) division ringoid, then under the hypothesis in (b), $\{\boldsymbol{IS}, \bar{N}, \diamondplus, \diamonddot, \diamonddiv, \leq\}$ with $\bar{N} := \{A \in \boldsymbol{IS} \mid A \cap N \neq \emptyset\}$ is also a completely ordered (resp. weakly ordered) division ringoid.

*Proof*: (a): By Theorem 4.4 $\{\boldsymbol{IR}, \boxplus, \boxdot\}$ is a ringoid, and by Lemma 3.4, $\{\boldsymbol{IS}, \diamondplus, \diamonddot\}$ has the properties (D1), (D2) for $[o, o]$, (D3) for $[e, e]$, (D4) and (D5) for $[-e, -e]$. Assume that $X = [x, y]$ is any element of $\boldsymbol{IS}$ that fulfills (D5). Then, in particular, (D5b, c) hold:

$$\text{(D5b)} \quad X \diamonddot X := \Diamond(X \boxdot X) := \Diamond(\square(X \cdot X)) = \Diamond(\square\{x \cdot \tilde{x} \mid x, \tilde{x} \in X\}) = [e, e].$$

$$\text{(D5c)} \quad X \diamonddot (A \diamonddot B) = (X \diamonddot A) \diamonddot B = A \diamonddot (X \diamonddot B) \qquad \text{for all} \quad A, B \in \boldsymbol{IS}.$$

Since $\Diamond$ and $\square$ are upwardly directed roundings with respect to $\subseteq$ and the set $[e, e]$ consists only of one element, (D5b) implies

$$\bigwedge_{a, b \in X} a \cdot a = a \cdot b = b \cdot a = b \cdot b = e. \tag{4}$$

Substituting $A = [x, x]$ and $B = [y, y]$ in (D5c), we get in particular

$$\begin{aligned}[x, y] \diamonddot ([x, x] \diamonddot [y, y]) &= ([x, y] \diamonddot [x, x]) \diamonddot [y, y] \\ &= [x, y] \diamonddot (\Diamond(\square([x, x] \cdot [y, y]))) \\ &= (\Diamond(\square([x, y] \cdot [x, x]))) \diamonddot [y, y].\end{aligned}$$

Because of (4) and (R1) for the roundings $\Diamond$ and $\square$, all parenthetic expressions here equal the interval $[e, e]$. Since $[e, e]$ is the neutral element of multiplication in $\boldsymbol{IS}$, we have $[x, y] = [y, y]$ and therefore $x = y$, i.e., $X = [x, x]$.

(b): By (a) every interval that fulfills (D5) is of the form $X = [x, x]$. Assume that there are two intervals $X = [x, x]$ and $Y = [y, y]$ that fulfill (D5) and (OD2). Then by hypothesis, we may take $X \leq Y$ without loss of generality. If we multiply this inequality by $X$ (resp. $Y$), we get because of $X \diamonddot X = Y \diamonddot Y = [e, e]$ and (OD2) that

$$X \diamonddot Y \leq [e, e] \wedge [e, e] \leq Y \diamonddot X. \tag{5}$$

From (D5c) we get with $B = [e, e]$ directly that $X \diamonddot A = A \diamonddot X$ for all $A \in \boldsymbol{IS}$. In (5), therefore, $X \diamonddot Y = Y \diamonddot X$, and by (O3) we obtain $X \diamonddot Y = [e, e]$. If we now multiply this equality by $X$, we get (because of (D5c) and

(D5b)) that $Y = X$. Since $[-e, -e]$ satisfies (D5) and (OD2) in $IS$, we have $X = Y = [-e, -e]$.

(c) and (d): To show (OD1)–(OD4) by Lemmas 3.4(c) and 3.5(c), it suffices to show that the rounding $\Diamond : \overline{IR} \to \overline{IS}$ is also monotone with respect to $\leq$:

(R2): With $A = [a_1, a_2]$ and $B = [b_1, b_2] \in IR$, we have

$$A \leq B :\Leftrightarrow a_1 \leq b_1 \wedge a_2 \leq b_2 \underset{(R2)_{\nabla,\Delta}}{\Rightarrow} \nabla a_1 \leq \nabla b_1 \wedge \Delta a_2 \leq \Delta b_2$$

$$\Rightarrow [\nabla a_1, \Delta a_2] \leq [\nabla b_1, \Delta b_2] \underset{(2)}{\Rightarrow} \Diamond A \leq \Diamond B.$$

Lemma 3.5(a) implies (D7)–(D9). ■

We shall apply this lemma to various interval sets. The first application, given in the following theorem, deals with the case of intervals on a screen of a linearly ordered division ringoid. We recall that the real numbers $\boldsymbol{R}$ furnish an example of such a structure.

**Theorem 4.20:** Let $\{R, +, \cdot, \leq\}$ be a completely and linearly ordered ringoid with the neutral elements $o$ and $e$, and let $\{S, \leq\}$ be a symmetric screen of $\{R, +, \cdot, \leq\}$. Consider the semimorphisms $\square : PR \to \overline{IR}$ and $\Diamond : \overline{IR} \to \overline{IS}$. Then $\{IS, \diamondplus, \diamonddot, \leq, \subseteq\}$ is a completely ordered ringoid with respect to $\leq$. The neutral elements are $[o, o]$ and $[e, e]$. With respect to $\subseteq$, $IS$ is an inclusion-isotonally ordered monotone upper screen ringoid of $IR$.

If in addition, $\{R, \{o\}, +, \cdot, /, \leq\}$ is a division ringoid, then $\{IS, \bar{N}, \diamondplus, \diamonddot, \diamondslash, \leq, \subseteq\}$ with $\bar{N} := \{A \in IS \mid o \in A\}$ is an ordered division ringoid with respect to $\leq$ and an inclusion-isotonally ordered monotone upper screen division ringoid of $IR$ with respect to $\subseteq$. ■

*Proof*: By Lemma 4.19(a) all intervals that fulfill (D5) in $IS$ are point intervals. The subset of point intervals of $IS$ is order-isomorphic to the set $\{S, \leq\}$ and therefore is linearly ordered. By Lemma 4.19(b)–(d), we obtain that $\{IS, \bar{N}, \diamondplus, \diamonddot, \diamondslash, \leq\}$ is a completely ordered division ringoid. By Lemmas 3.4 and 3.5, it is an inclusion-isotonally ordered monotone upper screen division ringoid of $IR$ with respect to $\subseteq$.

The central application of Theorem 4.20 is, of course, the set of intervals on a screen of the real numbers $\boldsymbol{R}$. Because of the great importance of this model, we derive explicit formulas for the operations in $IS$.

Let $\{R, \{o\}, +, \cdot, /, \leq\}$ be a completely and linearly ordered division ringoid and $A = [a_1, a_2], B = [b_1, b_2] \in IR$. Corollary 4.10 summarizes the result of all operations $\boxast, * \in \{+, -, \cdot, /\}$ in $IR$ by means of the formula

$$A \boxast B := \square(A * B) = \left[ \min_{i,j=1,2} (a_i * b_j), \max_{i,j=1,2} (a_i * b_j) \right].$$

By the definition of the operations in ***IS*** and by (2), we obtain directly for all $A = [a_1, a_2]$, $B = [b_1, b_2] \in \boldsymbol{IS}$ that

$$A \diamond_{*} B := \diamond (A \boxed{*} B) = \left[ \nabla \min_{i,j=1,2} (a_i * b_j), \triangle \max_{i,j=1,2} (a_i * b_j) \right].$$

Since $\nabla : R \to S$ and $\triangle : R \to S$ are monotone mappings, we obtain

$$A \diamond_{*} B := \diamond (A \boxed{*} B) = \left[ \min_{i,j=1,2} (a_i \nabla_{*} b_j), \max_{i,j=1,2} (a_i \triangle_{*} b_j) \right].$$

For execution of the operations $\diamond_{*}$, $* \in \{+, -, \cdot, /\}$, in ***IS*** on a computer, we display the following formulas and Tables 3 and 4. These are obtained by employing the preceding equation, Theorems 4.4 and 4.9, and especially Tables 1 and 2.

TABLE 3

EXECUTION OF THE MULTIPLICATION IN *IS*

| | $A = [a_1, a_2]$ | $B = [b_1, b_2]$ | $A \diamond_{\cdot} B = [\min_{i,j=1,2} (a_i \nabla_{\cdot} b_j), \max_{i,j=1.2} (a_i \triangle_{\cdot} b_j)]$ |
|---|---|---|---|
| 1 | $A \geq [o, o]$ | $B \geq [o, o]$ | $[a_1 \nabla_{\cdot} b_1, a_2 \triangle_{\cdot} b_2]$ |
| 2 | $A \geq [o, o]$ | $B \leq [o, o]$ | $[a_2 \nabla_{\cdot} b_1, a_1 \triangle_{\cdot} b_2]$ |
| 3 | $A \geq [o, o]$ | $o \in B$ | $[a_2 \nabla_{\cdot} b_1, a_2 \triangle_{\cdot} b_2]$ |
| 4 | $A \leq [o, o]$ | $B \geq [o, o]$ | $[a_1 \nabla_{\cdot} b_2, a_2 \triangle_{\cdot} b_1]$ |
| 5 | $A \leq [o, o]$ | $B \leq [o, o]$ | $[a_2 \nabla_{\cdot} b_2, a_1 \triangle_{\cdot} b_1]$ |
| 6 | $A \leq [o, o]$ | $o \in B$ | $[a_1 \nabla_{\cdot} b_2, a_1 \triangle_{\cdot} b_1]$ |
| 7 | $o \in A$ | $B \geq [o, o]$ | $[a_1 \nabla_{\cdot} b_2, a_2 \triangle_{\cdot} b_2]$ |
| 8 | $o \in A$ | $B \leq [o, o]$ | $[a_2 \nabla_{\cdot} b_1, a_1 \quad b_1]$ |
| 9 | $o \in A$ | $o \in B$ | $[\min(a_1 \nabla_{\cdot} b_2, a_2 \nabla_{\cdot} b_1), \max(a_1 \triangle_{\cdot} b_1, a_2 \triangle_{\cdot} b_2)]$ |

TABLE 4

EXECUTION OF THE DIVISION IN *IS*

| | $A = [a_1, a_2] \in \boldsymbol{IS}$ | $B = [b_1, b_2] \in IS \backslash \bar{N}$ | $A \diamond_{/} B = [\min_{i,j=1,2} (a_i \nabla_{/} b_j), \max_{i,j=1,2} (a_i \triangle_{/} b_j)]$ |
|---|---|---|---|
| 1 | $A \geq [o, o]$ | $B > [o, o]$ | $[a_1 \nabla_{/} b_2, a_2 \triangle_{/} b_1]$ |
| 2 | $A \geq [o, o]$ | $B < [o, o]$ | $[a_2 \nabla_{/} b_2, a_1 \triangle_{/} b_1]$ |
| 3 | $A \geq [o, o]$ | $o \in B$ | undefined |
| 4 | $A \leq [o, o]$ | $B > [o, o]$ | $[a_1 \nabla_{/} b_1, a_2 \triangle_{/} b_2]$ |
| 5 | $A \leq [o, o]$ | $B < [o, o]$ | $[a_2 \nabla_{/} b_1, a_1 \triangle_{/} b_2]$ |
| 6 | $A \leq [o, o]$ | $o \in B$ | undefined |
| 7 | $o \in A$ | $B > [o, o]$ | $[a_1 \nabla_{/} b_1, a_2 \triangle_{/} b_1]$ |
| 8 | $o \in A$ | $B < [o, o]$ | $[a_2 \nabla_{/} b_2, a_1 \triangle_{/} b_2]$ |
| 9 | $o \in A$ | $o \in B$ | undefined |

These formulas show, in particular, that the operations $\mathbin{\Diamond\!\!\!\!\!\ast}$ , $* \in \{+, -, \cdot, /\}$, in $\boldsymbol{IS}$ are executable on a computer if the operations $\mathbin{\nabla\!\!\!\!\!\ast}$ and $\mathbin{\Delta\!\!\!\!\!\ast}$, $* \in \{+, -, \cdot, /\}$, for elements of $S$ are available. These latter operations are defined by the following formulas:

$$\bigwedge_{a,b \in S} a \mathbin{\nabla\!\!\!\!\!\ast} b := \nabla(a * b) \wedge \bigwedge_{a,b \in S} a \mathbin{\Delta\!\!\!\!\!\ast} b := \Delta(a * b).$$

Here in the case of a division we additionally assume that $b \neq o$.

In order to perform the eight operations $\mathbin{\nabla\!\!\!\!\!\ast}$ and $\mathbin{\Delta\!\!\!\!\!\ast}$, $* \in \{+, -, \cdot, /\}$, on a computer, it is sufficient if three of them, for instance, $\mathbin{\nabla\!\!\!\!\!+}$, $\mathbin{\nabla\!\!\!\!\!\cdot}$, $\mathbin{\nabla\!\!\!\!\!/}$, or an equivalent triple, are available. This is a consequence of the fact that subtraction can be expressed by addition and the result of Theorem 3.3, which asserts that

$$\bigwedge_{a \in R} \nabla a = -\Delta(-a) \wedge \bigwedge_{a \in R} \Delta a = -\nabla(-a).$$

For instance, we obtain the following list of formulas:

$$\begin{aligned}
a \mathbin{\nabla\!\!\!\!\!+} b &= \nabla(a + b), \\
a \mathbin{\nabla\!\!\!\!\!-} b &= \nabla(a + (-b)) = a \mathbin{\nabla\!\!\!\!\!+} (-b), \\
a \mathbin{\nabla\!\!\!\!\!\cdot} b &= \nabla(a \cdot b), \\
a \mathbin{\nabla\!\!\!\!\!/} b &= \nabla(a/b), \\
a \mathbin{\Delta\!\!\!\!\!+} b &= -\nabla(-(a + b)) = -((-a) \mathbin{\nabla\!\!\!\!\!+} (-b)), \\
a \mathbin{\Delta\!\!\!\!\!-} b &= -\nabla(-(a + (-b))) = -((-a) \mathbin{\nabla\!\!\!\!\!+} b), \\
a \mathbin{\Delta\!\!\!\!\!\cdot} b &= -\nabla(-(a \cdot b)) = -((-a) \mathbin{\nabla\!\!\!\!\!\cdot} b), \\
a \mathbin{\Delta\!\!\!\!\!/} b &= -\nabla(-(a/b)) = -((-a) \mathbin{\nabla\!\!\!\!\!/} b).
\end{aligned}$$

We have already discussed the meaning of the monotone directed roundings $\nabla$ and $\Delta$ in Section 3.5. Theorems 1.28 and 3.3 show, in particular, that in a linearly ordered ringoid every monotone rounding can be expressed in terms of either $\nabla$ or $\Delta$. The formula system that we have derived above shows additionally that in order to perform all interval operations in $\boldsymbol{IS}$ either $\nabla$ or $\Delta$ suffices.

Addition and multiplication in $\boldsymbol{R}$ are associative. Consequently, addition and multiplication in $\boldsymbol{PR}$ and, referring to Remark 4.11, those in $\boldsymbol{IR}$ are also associative. These associativity properties, however, are no longer valid in $\boldsymbol{IS}$. We show this by means of a simple example in case of addition.

Without loss of generality we may assume that $S$ is a floating-point system with the number representation $m \cdot b^e$ with $m = -0.9(0.1)0.9$ and $e \in \{-1, 0, 1\}$. Throughout the example, $A, B \in \boldsymbol{IS}$.

Let $A := [0.3, 0.6]$, $B := [0.3, 0.5]$, $C := [0.3, 0.4]$. Then $A \diamondplus (B \diamondplus C) = [0.3, 0,6] \diamondplus [0.6, 0.9] = [0.9, 0.2 \cdot 10], (A \diamondplus B) \diamondplus C = (\Diamond[0.6, 1.1]) \diamondplus [0.3, 0.4] = [0.6, 0.2 \cdot 10] \diamondplus [0.3, 0.4] = \Diamond[0.9, 2.4] = [0.9, 0.3 \cdot 10]$, i.e., $(A \diamondplus B) \diamondplus C \neq A \diamondplus (B \diamondplus C)$.

These characteristics of the operations in $IS$ are simple consequences of the fact that the associative laws fail already in $\{S, \overset{\ast}{\triangledown}\}$ and $\{S, \overset{\ast}{\triangle}\}$, $\ast \in \{+ . \cdot\}$.

All the properties that we demonstrated in Theorem 1.31 hold for the rounding $\Diamond: \overline{IR} \to IS$ and the operations in $\{IS, \bar{N}, \diamondplus, \diamonddot, \diamondslash, \leq, \subseteq,\}$. In particular, the inequality

$$\Diamond(A \boxast B) \subseteq (\Diamond A) \diamondast (\Diamond B)$$

is valid for all $A, B \in IR$. We show by means of a simple example that the strict inclusion sign can occur. Thus there exists no homomorphism between the operations $\boxast$ in $IR$ and $\diamondast$ in $IS$ for all $\ast \in \{+, -, \cdot, /\}$. For the example we use the floating-point system defined above.

Let $A := [0.36, 0.54]$ and $B := [0.35, 0.45] \in IR$. Then $A \boxplus B = [0.71, 0.99]$. By (2) we get $\Diamond A = [0.3, 0.6], \Diamond B = [0.3, 0.5]$, and $\Diamond(A \boxplus B) = [0.7, 0.1 \cdot 10]$. Therefore, $(\Diamond A) \diamondplus (\Diamond B) = \Diamond((\Diamond A) \boxplus (\Diamond B)) = \Diamond[0.6, 1.1] = [0.6, 0.2 \cdot 10]$ and $\Diamond(A \boxplus B) = [0.7, 0.1 \cdot 10] \subset [0.6, 0.2 \cdot 10] = (\Diamond A) \diamondplus (\Diamond B)$.

## 6. INTERVAL MATRICES AND INTERVAL VECTORS ON A SCREEN

We begin with the following characterization of $IM_nS$.

**Theorem 4.21:** Let $\{R, +, \cdot, \leq\}$ be a completely and linearly ordered ringoid with the neutral elements $o$ and $e$, and let $\{S, \leq\}$ be a symmetric screen of $\{R, +, \cdot, \leq\}$. Consider the completely ordered ringoid of matrices $\{M_nR, +, \cdot, \leq\}$ with the neutral elements $O$ and $E$ and the semimorphisms $\square: PM_nR \to \overline{IM_nR}$ and $\Diamond : \overline{IM_nR} \to \overline{IM_nS}$. Then $\{IM_nS, \diamondplus, \diamonddot, \leq, \subseteq\}$ is a completely ordered ringoid with respect to $\leq$. The neutral elements are $[O, O]$ and $[E, E]$. With respect to $\subseteq$, $IM_nS$ is an inclusion-isotonally ordered monotone upper screen ringoid of $IM_nR$.

*Proof*: Theorem 3.9 implies that $M_nS$ is a symmetric screen of $M_nR$. By Theorem 4.4, $\{IM_nR, \diamondplus, \diamonddot\}$ is a ringoid, and by Lemma 4.19(a), all elements in $IM_nS$ that fulfill (D5) are point intervals $X = [X, X]$ with $X = (x_{ij})$. Each such interval, in particular, fulfills (D5a) in $IM_nS$:

(D5a) $[X, X] \diamondplus [E, E] = [O, O]$,

which implies $X + E = O$, i.e., $X$ is a diagonal matrix, and $x_{ij} = 0$ for all $i \neq j$.

By (D5c) we get with $\boldsymbol{B} = [E, E]$ and for all $\boldsymbol{A} \in \boldsymbol{I}M_nS$ that

$$\boldsymbol{X} \diamondsuit\!\!\!\!\cdot\;\, \boldsymbol{A} = \boldsymbol{A} \diamondsuit\!\!\!\!\cdot\;\, \boldsymbol{X}$$

Now we choose $\boldsymbol{A} := [(a_{ij}), (a_{ij})]$ with $a_{ij} = e$ for all $i, j = 1(1)n$. Since $\boldsymbol{A}$ and $\boldsymbol{X}$ are point matrices, their product which is also a point matrix satisfies the equality $(x_{ij}) \cdot (a_{ij}) = (a_{ij}) \cdot (x_{ij})$. This implies that $x_{11} = x_{22} = \ldots = x_{nn}$, i.e., $X$ is a diagonal matrix with a constant diagonal element.

The set of all diagonal point matrices of $\boldsymbol{I}M_nS$ with a constant diagonal element forms a linearly ordered subset of $\{\boldsymbol{I}M_nS, \leq\}$, which is order isomorphic to the set $\{S, \leq\}$. By Lemma 4.19(b)–(d), therefore, we obtain that $\{\boldsymbol{I}M_nS, \diamondsuit\!\!\!\!+\;, \diamondsuit\!\!\!\!\cdot\;, \leq, \subseteq\}$ is a completely ordered ringoid. By Lemma 3.5 it is an inclusion-isotonally ordered monotone upper screen ringoid of $\boldsymbol{I}M_nR$. ■

The operations $\diamondsuit\!\!\!\!*\;$ in $\boldsymbol{I}M_nS$ are not executable on a computer. Therefore, we are now going to express these operations in terms of computer executable formulas. In order to do this, let us once more consider the ringoid $\{M_n\boldsymbol{I}R, \oplus, \odot, \leq, \subseteq\}$. According to Theorem 4.14, this ringoid is isomorphic to $\{\boldsymbol{I}M_nR, \boxplus, \boxdot, \leq, \subseteq\}$. The isomorphism is expressed by the equality

$$\bigwedge_{\boldsymbol{A},\boldsymbol{B} \in M_n\boldsymbol{I}R} \chi\boldsymbol{A} \boxast \chi\boldsymbol{B} = \chi(\boldsymbol{A} \circledast \boldsymbol{B}), \qquad * \in \{+, \cdot\}. \tag{1}$$

Here the mapping $\chi: M_n\boldsymbol{I}R \to \boldsymbol{I}M_nS$ is defined by (2) in Section 3.

Using the rounding $\diamondsuit : \overline{\boldsymbol{I}R} \to \overline{\boldsymbol{I}S}$ and appealing to Theorem 3.10, a rounding $\diamondsuit: M_n\boldsymbol{I}R \to M_n\boldsymbol{I}S$ and operations in $M_n\boldsymbol{I}S$ are defined by

$$\bigwedge_{\boldsymbol{A}=(\boldsymbol{A}_{ij}) \in M_n\boldsymbol{I}R} \diamondsuit\boldsymbol{A} := (\diamondsuit\boldsymbol{A}_{ij}),$$

$$\bigwedge_{\boldsymbol{A},\boldsymbol{B} \in M_n\boldsymbol{I}S} \boldsymbol{A} \diamondsuit\!\!\!\!*\;\, \boldsymbol{B} := \diamondsuit\,(\boldsymbol{A} \circledast \boldsymbol{B}), \qquad * \in \{+, \cdot\}.$$

For further purposes, let us denote the matrix $\boldsymbol{A} \circledast \boldsymbol{B} \in M_n\boldsymbol{I}R$ by

$$(C_{ij}) := \boldsymbol{A} \circledast \boldsymbol{B} \qquad \text{with} \quad C_{ij} = [c_{ij}^1, c_{ij}^2] \qquad \text{and} \qquad c_{ij}^1, c_{ij}^2 \in R. \tag{2}$$

We are interested in the expression $\chi(\boldsymbol{A} \diamondsuit\!\!\!\!*\;\, \boldsymbol{B})$ and obtain for it

$$\chi(\boldsymbol{A} \diamondsuit\!\!\!\!*\;\, \boldsymbol{B}) = \chi(\diamondsuit C_{ij}) \underset{\text{Theorem 4.18(c)}}{=} \chi(\inf(U(C_{ij}) \cap \boldsymbol{I}S)).$$

If we exchange the componentwise infima with the infimum of a matrix, we obtain

$$\chi(\boldsymbol{A} \diamondsuit\!\!\!\!*\;\, \boldsymbol{B}) = \chi\{\inf(UC_{ij}) \cap \boldsymbol{I}S)\}.$$

Since the inclusion and intersection in $M_nIR$ are defined componentwise, we may exchange the matrix parentheses with the upper bounds and obtain

$$\chi(A \circledast B) = \chi\{\inf U\{([c^1_{ij}, c^2_{ij}]) \cap M_nIS\}\}.$$

If we now apply the mapping $\chi$, we obtain stepwise

$$\begin{aligned}\chi(A \circledast B) &= \inf U\{[(c^1_{ij}), (c^2_{ij})] \cap IM_nS\} \\ &\underset{(2)}{=} \inf U\{\chi(A \circledast B) \cap IM_nS\} \\ &\underset{(1)}{=} \inf U\{(\chi A \boxast \chi B) \cap IM_nS\}.\end{aligned}$$

Since the rounding $\lozenge : \overline{IM_nR} \to \overline{IM_nS}$ has the property expressed by Theorem 4.18(c), we obtain

$$\chi(A \circledast B) = \lozenge(\chi A \boxast \chi B).$$

By the definition of the operation $\circledast$ in $IM_nS$, this finally leads to

$$\chi(A \circledast B) = \chi A \circledast \chi B. \tag{3}$$

This states the fact that the mapping $\chi$ also represents an isomorphism between the ringoid $\{M_nIS, \oplus, \odot, \leq, \subseteq\}$ defined by Theorem 3.10 and the ringoid $\{IM_nS, \oplus, \odot, \leq, \subseteq\}$ studied in Theorem 4.21.

The isomorphism (3) reduces the operations in $IM_nS$, which are not computer executable, to the operations in $M_nIS$, which have the properties

$$A \oplus B = (\lozenge(A_{ij} \boxplus B_{ij})) = (A_{ij} \oplus B_{ij}), \tag{4}$$

$$A \odot B = \left(\lozenge \boxed{\sum_{\nu=1}^{n}} A_{i\nu} \boxdot B_{\nu j}\right). \tag{5}$$

These operations are executable on a computer. The componentwise addition can be performed by means of addition in $IS$. The multiplication can be executed using Table 1 and one of the algorithms for scalar products with maximal accuracy. See Chapter 6.

Since a semimorphism of a semimorphism is a semimorphism, this result can be used not only to describe the structure on a screen but also on one or more coarser screens. Accordingly, the passage that we made from $IM_nR$ to $IM_nS$ can also be performed from $IM_nR$ to $IM_nD$ or from $IM_nD$ to $IM_nS$. This results in isomorphisms between the sets $IM_nR$, $IM_nD$, $IM_nS$ and $M_nIR$, $M_nID$, $M_nIS$, respectively.

We are now going to illustrate these results. As an example, let $R$ denote the linearly ordered set of real numbers and $S$ the subset of single precision floating-point numbers of a given computer. Although these sets comprise the most important applications, we stress that our considerations are not

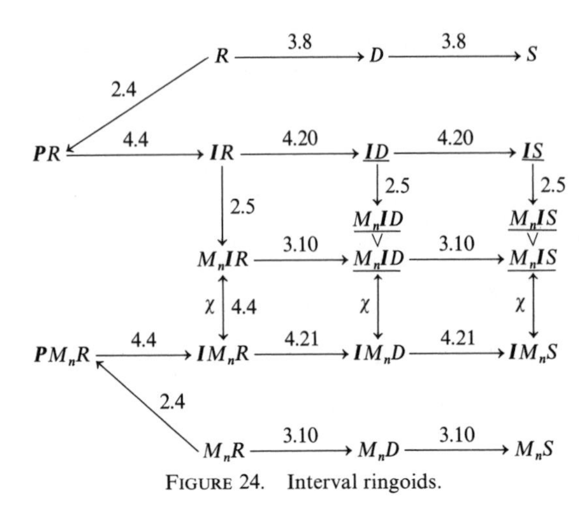


FIGURE 24. Interval ringoids.

restricted to this case. Figure 24 shows the spaces that were shown to be ringoids and lists the theorems that provide the required properties. In particular, we display the isomorphism between the ringoids $\boldsymbol{IM_nR}$, $\boldsymbol{IM_nD}$, $\boldsymbol{IM_nS}$ and $M_n\boldsymbol{IR}$, $M_n\boldsymbol{ID}$, $M_n\boldsymbol{IS}$, respectively. We have already noted that the operations in the latter ringoids defined by Theorem 3.10 are executable on a computer.

Within the sets $M_nID$ and $M_nIS$, operations and a ringoid can also be defined by the vertical method using the operations in $ID$ (resp. $IS$). We show that addition in these ringoids is identical to the addition defined by Theorem 3.10. The multiplication defined by Theorem 2.5 delivers upper bounds of the result of the multiplication defined by Theorem 3.10.

Using (4) and (5), these observations can be directly verified by means of the following formulas, wherein $\boldsymbol{A} = (A_{ij})$, $\boldsymbol{B} = (B_{ij}) \in M_n\boldsymbol{IS}$:

$$\boldsymbol{A} \mathbin{\diamond\!\!\!+} \boldsymbol{B} = (\Diamond(A_{ij} \boxplus B_{ij})) = (A_{ij} \mathbin{\diamond\!\!\!+} B_{ij}),$$

$$\boldsymbol{A} \mathbin{\diamond\!\!\!\cdot} \boldsymbol{B} = \left( \Diamond \boxed{\sum_{\nu=1}^{n}} A_{i\nu} \boxdot B_{\nu j} \right) \subseteq \left( \mathop{\Diamond\!\!\!\!\Sigma}_{\nu=1}^{n} A_{i\nu} \mathbin{\diamond\!\!\!\cdot} B_{\nu j} \right). \tag{6}$$

The last inequality is a consequence of (R3) applied to the rounding $\Diamond : \overline{\boldsymbol{IR}} \to \overline{\boldsymbol{IS}}$ and the inclusion-isotony of all operations in $IS$. The expressions on the right-hand side of these formulas represent the sum (resp. the product) defined in $M_n\boldsymbol{IS}$ by the vertical method (Theorem 2.5). On computers practical calculations in $M_n\boldsymbol{ID}$ and $M_n\boldsymbol{IS}$ are, for simplicity, often done by use of the operations defined by the vertical method. The operations defined by semimorphisms, on the other hand, provide optimal accuracy for each execution.

In Fig. 24 double-headed arrows indicate isomorphisms, horizontal arrows denote semimorphisms, and vertical arrows indicate a vertical definition of the arithmetic operations. A slanting arrow shows where the powerset operations come from. The sign $\vee$ used in Fig. 24 is intended to illustrate the inequality (6). Interval sets, the operations of which are executable on a computer, are undefined in Fig. 24. The numbers on the arrows drawn in the figure indicate the theorems that provide the corresponding properties.

We are now going to consider vectoids. Let $\{V, \leq\}$ be a complete lattice, $\{S, \leq\}$ a screen of $\{V, \leq\}$, and $\{IS, \subseteq\}$ the set of intervals over $V$ with bounds in $S$ of the form

$$A = [a_1, a_2] = \{\chi \in V \mid a_1, a_2 \in S, a_1 \leq \chi \leq a_2\} \qquad \text{with} \quad a_1 \leq a_2.$$

Then $IS \subseteq IV$ and $\overline{IS} := IS \cup \{\varnothing\} \subseteq \overline{IV} := IV \cup \{\varnothing\}$. By Theorem 4.17, $\{\overline{IS}, \subseteq\}$ is a screen of $\{\overline{IV}, \subseteq\}$. We consider the monotone upwardly directed rounding $\Diamond: \overline{IV} \to \overline{IS}$, which is defined by the properties

(R1) $\bigwedge_{A \in \overline{IS}} \Diamond A = A,$

(R2) $\bigwedge_{A,B \in \overline{IV}} (A \subseteq B \Rightarrow \Diamond A \subseteq \Diamond B),$

(R3) $\bigwedge_{A \in \overline{IV}} A \subseteq \Diamond A.$

For this rounding, we prove the following theorem.

**Theorem 4.22:** Let $\{V, R, \leq\}$ be a completely and weakly ordered ringoid, and let $\{S, \leq\}$ be a symmetric screen of $\{V, R, \leq\}$. Further, let $\{IV, IR, \leq, \subseteq\}$ be the vectoid defined by the semimorphism $\square: PV \to \overline{IV}$. Then setting $A = [a_1, a_2]$, we have

(a) $\{\overline{IS}, \subseteq\}$ is a symmetric screen of $\{\overline{IV}, \subseteq\}$, i.e., we have

(S3) $[o, o]([e, e]) \in \overline{IS} \wedge \bigwedge_{A=[a_1,a_2] \in IS} \boxminus A = [-a_2, -a_1] \in IS.$

(b) The monotone upwardly directed rounding $\Diamond: \overline{IV} \to \overline{IS}$ is antisymmetric, i.e.,

(R4) $\bigwedge_{A \in IV} \Diamond(\boxminus A) = \boxminus(\Diamond A).$

(c)

(R) $\bigwedge_{A \in IV} \Diamond A = \inf(U(A) \cap IS) = [\nabla a_1, \Delta a_2].$ (7)

*Proof:* The proof of this theorem is completely analogous to that of Theorem 4.18. ∎

Now we use the monotone upwardly directed rounding $\diamondsuit : \overline{IV} \to \overline{IS}$ in order to define inner and outer operations $\diamondsuit\!\!\!\!\ast$ in $IS$ in terms of its associated semimorphism which, by Theorem 4.5 and (1), has the following properties:

$$\text{(RG)} \quad \bigwedge_{A,B \in IS} A \mathbin{\diamondsuit\!\!\!\!\ast} B := \diamondsuit(A \mathbin{\boxed{\ast}} B) = \diamondsuit(\square(A * B))$$

$$\underset{(7)}{=} [\nabla \inf(A * B), \triangle \sup(A * B)] \qquad \text{for} \quad * \in \{+, \cdot\},$$

and with $IT \subseteq IR$,

$$\bigwedge_{A \in IT} \bigwedge_{A \in IS} A \mathbin{\diamondsuit\!\!\!\!\cdot} A := \diamondsuit(A \mathbin{\boxdot} A) = \diamondsuit(\square(A \cdot A))$$

$$\underset{(7)}{=} [\nabla \inf(A \cdot A), \triangle \sup(A \cdot A)].$$

The following theorem gives a characterization of $\{IS, IT, \leq, \subseteq\}$.

**Theorem 4.23:** Let $\{V, R, \leq\}$ be a completely and weakly ordered vectoid with the neutral elements $o$ and $e$ if a multiplication exists, and let $\{S, \leq\}$ be a symmetric screen of $\{V, R, \leq\}$. Further, let $\{IT, \mathbin{\diamondsuit\!\!\!\!+}, \mathbin{\diamondsuit\!\!\!\!\cdot}\}$ be a monotone upper screen ringoid of $\{IR, \boxplus, \boxdot\}$ with respect to $\subseteq$. Consider the two semimorphisms $\square : PV \to \overline{IV}$ and $\diamondsuit : \overline{IV} \to \overline{IS}$. Then $\{IS, IT, \leq, \subseteq\}$ is a weakly ordered vectoid with respect to $\leq$. It is multiplicative if $\{V, R, \leq\}$ is. The neutral elements are $[o, o]$ and $[e, e]$. With respect to $\subseteq$, $\{IS, IT, \leq, \subseteq\}$ is an inclusion-isotonally ordered monotone upper screen vectoid of $\{IV, IR, \leq, \subseteq\}$.

If $\{V, R, \leq\}$ is completely ordered, then $\{IS, IT, \leq, \subseteq\}$ is also completely ordered with respect to $\leq$.

*Proof*: The theorem is an immediate consequence of Theorems 3.12 and 3.13. ■

A special application of this theorem is the set of interval vectors on a screen of a linearly ordered ringoid with outer multiplication by elements of the ringoids $\{IS, \mathbin{\diamondsuit\!\!\!\!+}, \mathbin{\diamondsuit\!\!\!\!\cdot}, \leq, \subseteq\}$ or $\{IM_nS, \mathbin{\diamondsuit\!\!\!\!+}, \mathbin{\diamondsuit\!\!\!\!\cdot}, \leq, \subseteq\}$. The following corollary describes this application.

**Corollary 4.24:** Let $\{R, +, \cdot, \leq\}$ be a completely and linearly ordered ringoid, and $\{S, \leq\}$ a symmetric screen of $\{R, +, \cdot, \leq\}$. Consider the completely ordered vectoids $\{V_nR, R, \leq\}$, $\{M_nR, R, \leq\}$, and $\{V_nR, M_nR, \leq\}$ as well as the semimorphisms $\square : PV_nR \to \overline{IV_nR}$, $\diamondsuit : \overline{IV_nR} \to \overline{IV_nS}$ and the semimorphisms $\square : PM_nR \to \overline{IM_nR}$, $\diamondsuit : \overline{IM_nR} \to \overline{IM_nS}$. Then $\{IV_nS, IS, \leq, \subseteq\}$, $\{IM_nS, IS, \leq, \subseteq\}$, and $\{IV_nS, IM_nS, \leq, \subseteq\}$ are completely ordered vectoids with respect to $\leq$. $IM_nS$ is multiplicative. With respect to $\subseteq$, all of these vectoids are inclusion-isotonally ordered and monotone upper screen vectoids of $\boldsymbol{I}V_nR$ resp. $\boldsymbol{I}M_nR$. ■

In practical applications of this corollary, of course, $R$ is the set of real numbers $\boldsymbol{R}$.

The principle situation is similar to that which arose in the case of matrix structures. The elements of the vectoids $\{\boldsymbol{I}V_nR, \boldsymbol{I}R\}$, $\{\boldsymbol{I}M_nR, \boldsymbol{I}R\}$, $\{\boldsymbol{I}V_nR, \boldsymbol{I}M_nR\}$ cannot be represented on a computer, while the operations within the vectoids $\{\boldsymbol{I}V_nS, \boldsymbol{I}S\}$, $\{\boldsymbol{I}M_nS, \boldsymbol{I}S\}$, $\{\boldsymbol{I}V_nS, \boldsymbol{I}M_nS\}$ cannot in general even be executed on a computer since they are based on the corresponding powerset operations.

Therefore, we are now going to express these operations in terms of computer executable formulas. By Theorem 4.16 we saw that under the assumption that $R$ is a completely and linearly ordered ringoid, there exist isomorphisms between the following pairs of vectoids:

$$\{V_n\boldsymbol{I}R, \boldsymbol{I}R, \leq\} \leftrightarrow \{\boldsymbol{I}V_nR, \boldsymbol{I}R, \leq\},$$
$$\{M_n\boldsymbol{I}R, \boldsymbol{I}R, \leq\} \leftrightarrow \{\boldsymbol{I}M_nR, \boldsymbol{I}R, \leq\},$$
$$\{V_n\boldsymbol{I}R, M_n\boldsymbol{I}R, \leq\} \leftrightarrow \{\boldsymbol{I}V_nR, \boldsymbol{I}M_nR, \leq\}.$$

The algebraic isomorphisms are expressed by the relations

$$\begin{aligned} \psi(A \odot \boldsymbol{a}) &= A \boxdot \psi\boldsymbol{a}, & \psi(\boldsymbol{a} \oplus \boldsymbol{b}) &= \psi\boldsymbol{a} \boxplus \psi\boldsymbol{b}, & & \\ \chi(A \odot \boldsymbol{A}) &= A \boxdot \chi\boldsymbol{A}, & \chi(\boldsymbol{A} \circledast \boldsymbol{B}) &= \chi\boldsymbol{A} \boxast \chi\boldsymbol{B}, & * &\in \{+, \cdot\}, \\ \psi(\boldsymbol{A} \odot \boldsymbol{a}) &= \chi\boldsymbol{A} \boxdot \psi\boldsymbol{a}, & \psi(\boldsymbol{a} \oplus \boldsymbol{b}) &= \psi\boldsymbol{a} \boxplus \psi\boldsymbol{b}. & & \end{aligned}$$

Here the mappings $\psi: V_n\boldsymbol{I}R \to \boldsymbol{I}V_nR$ and $\chi: M_n\boldsymbol{I}R \to \boldsymbol{I}M_nR$ are defined by (12) and (13) in Section 4.4.

The operations $\boxast$ in $\boldsymbol{I}V_nR$ and $\boldsymbol{I}M_nR$ are not performable because of their definition in terms of powerset operations. The isomorphisms $\psi$ and $\chi$ reduce these operations to the explicit operations $\circledast$ in $V_n\boldsymbol{I}R$ resp. $M_n\boldsymbol{I}R$.

Because of the isomorphism $\psi$, corresponding elements in $V_n\boldsymbol{I}R$ and $\boldsymbol{I}V_nR$ can be identified with each other. This shows that an inclusion relation for elements $\boldsymbol{a} = (A_i)$, $\boldsymbol{b} = (B_i) \in V_n\boldsymbol{I}R$ can be defined as in Section 4.4 by

$$\boldsymbol{a} \subseteq \boldsymbol{b} :\Leftrightarrow \bigwedge_{i=1(1)n} A_i \subseteq B_i.$$

We observed above that the (inner and outer) operations $\diamond\!\!\!*$ in $\boldsymbol{I}V_nS$, which are defined by the semimorphism $\Diamond: \overline{\boldsymbol{I}V_nR} \to \overline{\boldsymbol{I}V_nS}$, are not executable on the computer. We now express these operations in terms of executable formulas. Using the monotone upwardly directed rounding $\Diamond: \overline{\boldsymbol{I}R} \to \overline{\boldsymbol{I}S}$, a rounding $\Diamond: V_n\boldsymbol{I}R \to V_n\boldsymbol{I}S$ resp. $\Diamond: M_n\boldsymbol{I}R \to M_n\boldsymbol{I}S$ and operations in $V_n\boldsymbol{I}S$

(resp. $M_n IS$) can be defined by the formulas

$$\bigwedge_{a=(A_i)\in V_n IS} \Diamond a := (\Diamond A_i) \quad \text{resp.} \quad \bigwedge_{A=(A_{ij})\in M_n IS} \Diamond A := (\Diamond A_{ij}),$$

$$\bigwedge_{a,b\in V_n IS} a \mathbin{\hat{+}} b := \Diamond(a \oplus b),$$

$$\bigwedge_{A\in IS} \bigwedge_{a\in V_n IS} A \mathbin{\hat{\cdot}} a := \Diamond(A \odot a),$$

$$\bigwedge_{A\in IS} \bigwedge_{A\in M_n IS} A \mathbin{\hat{\cdot}} A := \Diamond(A \odot A),$$

$$\bigwedge_{A\in M_n IS} \bigwedge_{a\in V_n IS} A \mathbin{\hat{\cdot}} a := \Diamond(A \odot a).$$

Similar to the matrix case treated above, it can be shown that the mappings $\psi$ and $\chi$ establish isomorphisms between the ordered algebraic structures

$$\{V_n IS, IS, \leq\} \leftrightarrow \{IV_n S, IS, \leq\},$$
$$\{M_n IS, IS, \leq\} \leftrightarrow \{IM_n S, IS, \leq\},$$
$$\{V_n IS, M_n IS, \leq\} \leftrightarrow \{IV_n S, IM_n S, \leq\}.$$

With $a = (A_i)$ and $A = (A_{ij})$, we obtain the following relations for the operations defined above:

$$a \mathbin{\hat{+}} b = \Diamond(a \oplus b) = \Diamond(A_i \boxplus B_i) = (A_i \mathbin{\hat{+}} B_i),$$
$$A \mathbin{\hat{\cdot}} a = \Diamond(A \odot a) = \Diamond(A \boxdot A_i) = (A \mathbin{\hat{\cdot}} A_i),$$
$$A \mathbin{\hat{\cdot}} A = \Diamond(A \odot A) = \Diamond(A \boxdot A_{ij}) = (A \mathbin{\hat{\cdot}} A_{ij}),$$
$$A \mathbin{\hat{\cdot}} a = \Diamond(A \odot a) = \Diamond\left(\boxed{\sum_{\nu=1}^{n}} A_{i\nu} \boxdot A_\nu\right) = \left(\Diamond \boxed{\sum_{\nu=1}^{n}} A_{i\nu} \boxdot A_\nu\right).$$

All these operations furnish the smallest appropriate interval (i.e., are of optimal accuracy). Moreover, they all can be executed on a computer. The formulas also show that the operations $a \mathbin{\hat{+}} b$, $A \mathbin{\hat{\cdot}} a$, and $A \mathbin{\hat{\cdot}} A$ are identical, whether defined by the vertical or the horizontal method. The horizontal and the vertical definitions of the outer multiplication in $\{V_n IS, M_n IS, \leq\}$, however, lead to the inequality

$$A \mathbin{\hat{\cdot}} a = \left(\Diamond \boxed{\sum_{\nu=1}^{n}} A_{i\nu} \boxdot A_\nu\right) \subseteq \left(\hat{\sum_{\nu=1}^{n}} A_{i\nu} \mathbin{\hat{\cdot}} A_\nu\right). \tag{8}$$

Both sides of the last inequality are computer-executable formulas. The expression on the left-hand side is of optimal accuracy. In practice calculations on computers, however, are for simplicity often performed using the operations defined by the vertical method, which occurs on the right-hand side of (8).

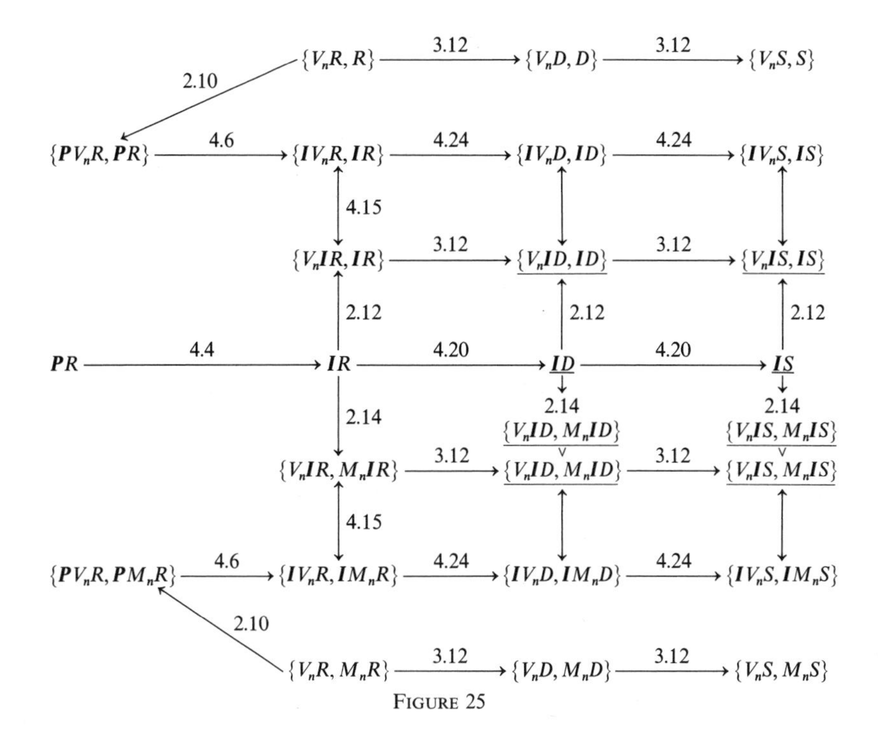

FIGURE 25

Since a semimorphism of a semimorphism is a semimorphism, these results can be used not only to describe the structure on a screen but also on one or more coarser screens as well.

Figure 25 shows a diagram displaying the interval vectoids discussed in this chapter. In Fig. 25 arrows are used in correspondence to those in the diagram of Fig. 24. Double arrows indicate isomorphisms. Horizontal arrows denote semimorphisms. Vertical arrows indicate a vertical definition of the arithmetic operations. The numbers on the arrows drawn in the figure indicate the theorems that provide the corresponding properties. The sign $\vee$ used in Fig. 25 indicates the inequality (8). All structures, the operations of which are executable on a computer, are underlined in Fig. 25.

## 7. COMPLEX INTERVAL ARITHMETIC

To complete our discussion of all spaces displayed in Fig. 1, there remains to consider the complex interval spaces, which we now proceed to do.

Let $\{R, N, +, \cdot, /, \leq\}$ be a completely and linearly ordered division ringoid with the neutral elements $o$ and $e$, and let $\boldsymbol{C}R := \{(x, y) \mid x, y \in R\}$ be the set of pairs over $R$. If in $\boldsymbol{C}R$ we define equality, the order relation $\leq$

and the operations $* \in \{+, \cdot, /\}$ by means of the usual formulas, we know by Theorem 2.6 that

$$\{CR, \bar{N}, +, \cdot, /, \leq\} \text{ with } \bar{N} := \{\gamma \in CR \mid \gamma = (c_1, c_2) \wedge c_1 c_1 + c_2 c_2 \in N\}$$

is a completely and weakly ordered division ringoid with the neutral elements $(o, o)$ and $(e, o)$. Then by Theorem 2.4, the power set $\{PCR, \tilde{N}, +, \cdot, /\}$ with $\tilde{N} := \{\Phi \in PCR \mid \{\Phi \cap \bar{N} \neq \varnothing\}$ is also a division ringoid.

Now let $ICR$ denote the set of intervals over $CR$ of the form

$$\Phi = [\varphi_1, \varphi_2] = \{\varphi \in CR \mid \varphi_1, \varphi_2 \in CR, \varphi_1 \leq \varphi \leq \varphi_2\} \qquad \text{with} \quad \varphi_1 \leq \varphi_2.$$

Then by Theorems 4.2 and 4.3, $\overline{ICR} := ICR \cup \{\varnothing\}$ is a symmetric upper screen of $PCR$, and the monotone upwardly directed rounding $\square$: $PCR \to \overline{ICR}$ is antisymmetric. In $ICR$ we define operations by employing the upwardly directed rounding $\square$ as a semimorphism:

$$\text{(RG)} \quad \bigwedge_{\Phi, \Psi \in ICR} \Phi \boxast \Psi := \square(\Phi * \Psi)$$
$$= \square\{\varphi * \psi \mid \varphi \in \Phi \wedge \psi \in \Psi\}, \qquad * \in \{+, \cdot, /\},$$

where for division, we assume that $\Psi \cap \tilde{N} = \varnothing$. For the intervals $\Phi = [\varphi_1, \varphi_2]$ and $\Psi = [\psi_1, \psi_2] \in ICR$, we further define an order relation

$$\Phi \leq \Psi :\Leftrightarrow (\varphi_1 \leq \psi_1 \wedge \varphi_2 \leq \psi_2).$$

Then by Theorem 4.4, $\{ICR, N^*, \boxplus, \boxdot, \boxslash, \leq, \subseteq\}$ with $N^* := \{\Phi \in ICR \mid \Phi \cap \tilde{N} \neq \varnothing\}$ is a completely and weakly ordered division ringoid with respect to $\leq$. With respect to $\subseteq$ it is an inclusion-isotonally ordered monotone upper screen division ringoid of $PCR$.

The operations in $ICR$ are defined by those in $PCR$, and are therefore not executable in practice. In order to obtain executable formulas, we are going to study the connection between these operations and those in the set $CIR$.

We start once more with the completely and linearly ordered division ringoid $\{R, N, +, \cdot, /, \leq\}$. By Theorem 4.4, $\{IR, N', \boxplus, \boxdot, \boxslash, \leq, \subseteq\}$ is a completely ordered division ringoid. Theorems 4.4 and 4.9 prescribe explicit and simply implementable formulas for all operations in $IR$. If $A = [a_1, a_2]$ and $B = [b_1, b_2]$, these formulas are summarized in Corollary 4.10 by the relation

$$\bigwedge_{A, B \in IR} A \boxast B := \square(A * B) = \left[\min_{i,j=1,2} (a_i * b_j), \max_{i,j=1,2} (a_i * b_j)\right], * \in \{+, -, \cdot, /\}.$$

Now consider the set $\boldsymbol{CIR} := \{(X, Y) | X, Y \in \boldsymbol{IR}\}$. For elements $\Phi = (X_1, X_2)$, $\Psi = (Y_1, Y_2)$ we define equality $=$, the order relation $\leq$, and the operations $\circledast$, $* \in \{+, \cdot, /\}$, as in Theorem 2.6 by

$$\Phi = \Psi :\Leftrightarrow X_1 = Y_1 \wedge X_2 = Y_2,$$
$$\Phi \leq \Psi :\Leftrightarrow X_1 \leq Y_1 \wedge X_2 \leq Y_2,$$
$$\bigwedge_{\Phi,\Psi \in \boldsymbol{CIR}} (\Phi \oplus \Psi := (X_1 \boxplus Y_1, X_2 \boxplus Y_2),$$
$$\Phi \odot \Psi := (X_1 \boxdot Y_1 \boxminus X_2 \boxdot Y_2, X_1 \boxdot Y_2 \boxplus X_2 \boxdot Y_1)),$$
$$\bigwedge_{\Phi \in \boldsymbol{CIR}} \bigwedge_{\Phi \in \boldsymbol{CIR} \setminus N''} \Phi \oslash \Psi := ((X_1 \boxdot Y_1 \boxplus X_2 \boxdot Y_2) \boxslash (Y_1 \boxdot Y_1 \boxplus Y_2 \boxdot Y_2),$$
$$(X_2 \boxdot Y_1 \boxminus X_1 \boxdot Y_2) \boxslash (Y_1 \boxdot Y_1 \boxplus Y_2 \boxdot Y_2)),$$

with $N'' := \{\Phi \in \boldsymbol{CIR} | \Phi = (X_1, X_2) \wedge X_1 \boxdot X_1 \boxplus X_2 \boxdot X_2 \in N')$.

Thus all of these operations $\circledast$ are easily implementable in terms of the operations $\boxast$ in $\boldsymbol{IR}$. By Theorem 2.6, $\{\boldsymbol{CIR}, N'', \oplus, \odot, \oslash, \leq\}$ is a completely and weakly ordered division ringoid.

We are interested in relating the operations in $\boldsymbol{ICR}$ and $\boldsymbol{CIR}$. Thus we consider the mapping

$$\tau: \boldsymbol{CIR} \to \boldsymbol{ICR} \quad \text{with} \quad \tau([x_1, x_2], [y_1, y_2]) := [(x_1, y_1), (x_2, y_2)].$$

Here $\tau$ obviously is a one-to-one mapping of $\boldsymbol{CIR}$ onto $\boldsymbol{ICR}$ and an order isomorphism.

We show that $\tau$ also is a ringoid isomorphism. In order to do this, we prove that addition and multiplication in $\boldsymbol{ICR}$ and in $\boldsymbol{CIR}$ are isomorphic. Let $\Phi = (X_1, X_2)$, $\Psi = (Y_1, Y_2) \in \boldsymbol{CIR}$. Then

$$\tau\Phi \boxplus \tau\Psi := \square(\tau\Phi + \tau\Psi)$$
$$= \left[\inf_{\varphi \in \Phi, \psi \in \Psi} (\varphi + \psi), \sup_{\varphi \in \Phi, \psi \in \Psi} (\varphi + \psi)\right]$$
$$= [\inf(x_1 + y_1, x_2 + y_2), \sup(x_1 + y_1, x_2 + y_2)]$$
$$= [(\inf(x_1 + y_1), \inf(x_2 + y_2)), (\sup(x_1 + y_1), \sup(x_2 + y_2))],$$

where in the last two lines displayed here, the infimima and the suprema are taken over $x_i \in X_i$, $y_i \in Y_i$, $i = 1, 2$. On the other hand, we have

$$\Phi \oplus \Psi := (X_1 \boxplus Y_1, X_2 \boxplus Y_2)$$
$$= (\square(X_1 + Y_1), \square(X_2 + Y_2))$$
$$= ([\inf(x_1 + y_1), \sup(x_1 + y_1)], [\inf(x_2 + y_2), \sup(x_2 + y_2)]),$$

where in the last line displayed here, the infima and the suprema are taken over $x_1 \in X_1$, $y_1 \in Y_1$ or over $x_2 \in X_2$, $y_2 \in Y_2$, as the case may be. These formulas show that

$$\tau\Phi \boxplus \tau\Psi = \tau(\Phi \oplus \Psi),$$

which demonstrates the isomorphism in the case of the addition.

In the case of multiplication we have

$$\bigwedge_{\varphi=(x_1,x_2)\in\tau\Phi} \bigwedge_{\psi=(y_1,y_2)\in\tau\Psi} \varphi\cdot\psi = (x_1y_1 - x_2y_2, x_1y_2 + x_2y_1)$$

$$\in \tau(X_1 \boxdot Y_1 \boxminus X_2 \boxdot Y_2, X_1 \boxdot X_2 \boxplus X_2 \boxdot Y_1)$$
$$= \tau(\Phi \odot \Psi),$$

and therefore $\tau\Phi \cdot \tau\Psi \subseteq \tau(\Phi \odot \Psi)$. If we apply the monotone upwardly directed rounding $\square : \boldsymbol{PCR} \to \overline{\boldsymbol{ICR}}$ to this inequality, we obtain

$$\tau\Phi \cdot \tau\Psi \underset{(\mathrm{R3})}{\subseteq} \square(\tau\Phi \cdot \tau\Psi) \underset{(\mathrm{RG})}{=} \tau\Phi \boxdot \tau\Psi \underset{(\mathrm{R1,2})}{\subseteq} \tau(\Phi \odot \Psi).$$

On the other hand, by employing the explicit formulas for the operations in the ringoid $\boldsymbol{IR}$, we obtain

$$X_1 \boxdot Y_1 \boxminus X_2 \boxdot Y_2 = \left[\min_{i,j=1,2} (x_1{}^i y_1{}^j), \max_{i,j=1,2} (x_1{}^i y_1{}^j)\right]$$
$$\boxminus \left[\min_{i,j=1,2} (x_2{}^i y_2{}^j), \max_{i,j=1,2} (x_2{}^i y_2{}^j)\right]$$
$$= \left[\min_{i,j=1,2} (x_1{}^i y_1{}^j) - \max_{i,j=1,2} (x_2{}^i y_2{}^j),\right.$$
$$\left. \max_{i,j=1,2} (x_1{}^i y_1{}^j) - \min_{i,j=1,2} (x_2{}^i y_2{}^j)\right],$$

$$X_1 \boxdot Y_2 \boxplus X_2 \boxdot Y_1 = \left[\min_{i,j=1,2} (x_1{}^i y_2{}^j) + \min_{i,j=1,2} (x_2{}^i y_1{}^j),\right.$$
$$\left. \max_{i,j=1,2} (x_1{}^i y_2{}^j) + \max_{i,j=1,2} (x_2{}^i y_1{}^j)\right],$$

i.e., the bounds of $\tau(\Phi \odot \Psi)$ consist of inner points of

$$\square(\tau\Phi \cdot \tau\Psi) = \tau\Phi \boxdot \tau\Psi$$

or

$$\tau(\Phi \odot \Psi) \subseteq \square(\tau\Phi \cdot \tau\Psi) = \tau\Phi \boxdot \tau\Psi.$$

If we take both inequalities together, we obtain $\tau(\Phi \odot \Psi) = \tau\Phi \boxdot \tau\Psi$, which demonstrates the ringoid isomorphism.

In case of the division, we similarly obtain

$$\tau\Phi/\tau\Psi \subseteq \square(\tau\Phi/\tau\Psi) = \tau\Phi \boxed{/} \tau\Psi \subseteq \tau(\Phi \oslash \Psi).$$

Equality, however, is not provable. We show this by means of a simple example. Let $\{\boldsymbol{R}, \{0\}, +, \cdot, /, \leq\}$ be the completely and linearly ordered division ringoid of the real numbers. Then with $\Phi = (X_1, X_2) = ([0,0], [2,4])$ and $\Psi = (Y_1, Y_2) = ([1,2]), [0,0])$ we obtain

$$\begin{aligned}\tau\Phi \boxed{/} \tau\Psi &= \square(\tau\Phi/\tau\Psi)\\ &= [(\inf((x_1y_1 + x_2y_2)/(y_1{}^2 + y_2{}^2)), \inf((x_2y_1 - x_1y_2)/(y_1{}^2 + y_2{}^2))),\\ &\qquad (\sup(x_1y_1 + x_2y_2)/(y_1{}^2 + y_2{}^2)), \sup((x_2y_1 - x_1y_2)/(y_1{}^2 + y_2{}^2)))]\\ &= [(0, \tfrac{1}{2}), (0,4)],\end{aligned}$$

$$\begin{aligned}\Phi \oslash \Psi &= ((X_1 \boxdot Y_1 \boxplus X_2 \boxdot Y_2 \boxed{/} (Y_1 \boxdot Y_1 \boxplus Y_2 \boxdot Y_2),\\ &\qquad (X_2 \boxdot Y_1 \boxminus X_1 \boxdot Y_2) \boxed{/} (Y_1 \boxdot Y_1 \boxplus Y_2 \boxdot Y_2))\\ &= ([0,0] \boxed{/} [1,4], [2,8] \boxed{/} [1,4] = ([0,0], [\tfrac{1}{2}, 8]).\end{aligned}$$

This effect is not at all surprising. It is well known from interval analysis. In the complex quotient, the quantities $y_1$ and $y_2$ occur more than once. The expression $\tau(\Phi \oslash \Psi)$, therefore, only delivers an overestimate of the range of the complex function $\varphi/\psi$. In interval analysis, therefore, other formulas are occasionally used in order to compute the complex quotient.

We illustrate the ringoid isomorphism derived above in a diagram. See Fig. 26.

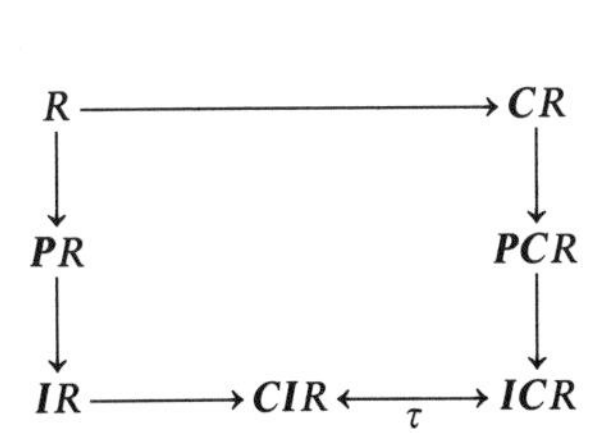


FIGURE 26. Illustration of the ringoid isomorphism $\tau : \boldsymbol{CIR} \to \boldsymbol{ICR}$.

Since the ringoids $\boldsymbol{CIR}$ and $\boldsymbol{ICR}$ are isomorphic, we may identify corresponding elements with each other. This allows us to define an inclusion for elements $\Phi = (X_1, X_2)$, $\Psi = (Y_1, Y_2) \in \boldsymbol{CIR}$ as

$$\Phi \subseteq \Psi :\Leftrightarrow X_1 \subseteq Y_1 \wedge X_2 \subseteq Y_2$$

and

$$\varphi = (x_1, x_2) \in \Phi = (X_1, X_2) :\Leftrightarrow x_1 \in X_1 \wedge x_2 \in X_2.$$

These definitions permit the interpretation that an element $\Phi = (X_1, X_2) \in \boldsymbol{CIR}$ represents a set of pairs of elements of $\boldsymbol{CR}$ by means of the identity

$$\Phi = (X_1, X_2) \equiv \{(x_1, x_2) | x_1 \in X_1, x_2 \in X_2\}.$$

Both sides contain the same elements.

We now consider the complex interval spaces on a screen. The following theorem describes the structure of such a screen.

**Theorem 4.25:** Let $\{R, \{o\}, +, \cdot, /, \leq\}$ be a completely and linearly ordered division ringoid with the neutral elements $o$ and $e$, and let $\{S, \leq\}$ be a symmetric screen of $\{R, +, \cdot, \leq\}$. Consider the complex division ringoid $\{CR, \{(o, o)\}, +, \cdot, /, \leq\}$ as well as the semimorphisms $\square: \boldsymbol{PCR} \to \overline{\boldsymbol{ICR}}$ and $\Diamond: \overline{\boldsymbol{ICR}} \to \overline{\boldsymbol{ICS}}$. Then $\{\boldsymbol{ICS}, \bar{N}, \diamondplus, \diamonddot, \diamondslash, \leq, \subseteq\}$ with $\bar{N} := \{A \in \boldsymbol{ICS} | (o, o) \in A\}$ is a completely and weakly ordered division ringoid with respect to $\leq$. The neutral elements are $[(o, o), (o, o)]$ and $[(e, o), (e, o)]$. With respect to $\subseteq$, $\boldsymbol{ICS}$ is an inclusion-isotonally ordered monotone upper screen division ringoid of $\boldsymbol{ICR}$.

*Proof*: We know by Theorem 3.9 that $\boldsymbol{CS}$ is a symmetric screen of $\boldsymbol{CR}$. By Theorem 4.4, $\{\boldsymbol{ICR}, \boxplus, \boxdot\}$ is a ringoid, and by Lemma 4.19(a), all elements of $\boldsymbol{ICS}$ that satisfy (D5) are point intervals, i.e., of the form $X = [(x, y), (x, y)]$. Each such interval, in particular, satisfies (D5a) in $\boldsymbol{ICS}$:

(D5a) $[(x, y), (x, y)] \diamondplus [(e, o), (e, o)] = [(o, o), (o, o)]$.

This implies that $y = o$. Therefore, $X = [(x, o), (x, o)]$. The set of all such intervals is a linearly ordered subset of $\{\boldsymbol{ICS}, \leq\}$ which is order isomorphic to the set $\{S, \leq\}$. Therefore, by Lemma 4.19(*b*)–(*d*), $\{\boldsymbol{ICS}, \bar{N}, \diamondplus, \diamonddot, \diamondslash, \leq, \subseteq\}$ is a completely and weakly ordered division ringoid with respect to $\leq$. By Lemmas 3.4 and 3.5, it is an inclusion-isotonally ordered monotone upper screen division ringoid of $\boldsymbol{ICR}$. ■

The operations $\diamondast$ in $\boldsymbol{ICS}$ are not directly implementable on a computer. Therefore, we express them in terms of implementable formulas. Consider once more the isomorphism $\tau: \boldsymbol{CIR} \to \boldsymbol{ICR}$. It is expressed by the formula

$$\bigwedge_{\Phi, \Psi \in \boldsymbol{CIR}} \tau\Phi \boxast \tau\Psi = \tau(\Phi \circledast \Psi), \qquad * \in \{+, \cdot\}. \tag{1}$$

Using the rounding $\Diamond: \overline{\boldsymbol{IR}} \to \overline{\boldsymbol{IS}}$ and using Theorem 3.11, a rounding $\Diamond: \boldsymbol{CIR} \to \boldsymbol{CIS}$ and operations in $\boldsymbol{CIS}$ are defined by

$$\bigwedge_{\Phi \in \boldsymbol{CIR}} \Diamond \Phi := (\Diamond X_1, \Diamond X_2),$$

$$\bigwedge_{\Phi, \Psi \in \boldsymbol{CIS}} \Phi \diamondast \Psi := \Diamond(\Phi \circledast \Psi), \qquad * \in \{+, \cdot\}.$$

Now we show in complete analogy to the matrix case, which we considered above, that the operations in ***ICS*** and ***CIS*** are isomorphic. Denoting $(Z_1, Z_2) := \Phi \circledast \Psi$, we consider the expression $\tau(\Phi \diamond\!\!\!\ast\, \Psi)$, and we obtain

$$\begin{aligned}\tau(\Phi \diamond\!\!\!\ast\, \Psi) &= \tau(\Diamond(Z_1, Z_2)) = \tau(\Diamond Z_1, \Diamond Z_2)\\ &= \tau(\inf(U(Z_1) \cap \boldsymbol{IS}), \inf(U(Z_2) \cap \boldsymbol{IS}))\\ &= \tau\{\inf(U(Z_1) \cap \boldsymbol{IS}, U(Z_2) \cap \boldsymbol{IS})\}.\end{aligned}$$

Since the inclusion and intersection in ***CIS*** are defined componentwise, we may exchange the upper bound and the pair brackets to obtain

$$\tau(\Phi \diamond\!\!\!\ast\, \Psi) = \tau\{\inf U\{(Z_1, Z_2) \cap \boldsymbol{CIS}\}\}.$$

If we now effect the mapping $\tau$, we obtain

$$\begin{aligned}\tau(\Phi \diamond\!\!\!\ast\, \Psi) &= \inf U\{\tau(\Phi \circledast \Psi) \cap \boldsymbol{ICS}\}\\ &\underset{(1)}{=} \inf U\{\tau\Phi \boxast \tau\Psi) \cap \boldsymbol{ICS}\}\\ &\underset{\text{Theorem 4.18(c)}}{=} \Diamond(\tau\Phi \boxast \tau\Psi).\end{aligned}$$

By the definition of the operations $\diamond\!\!\!\ast$ in ***ICS***, this finally leads to

$$\tau(\Phi \diamond\!\!\!\ast\, \Psi) = \tau\Phi \diamond\!\!\!\ast\, \tau\Psi. \tag{2}$$

Thus the mapping $\tau$ represents an isomorphism between the ringoids $\{\boldsymbol{CIS}, \diamond\!\!\!+, \diamond\!\!\!\cdot, \leq, \subseteq\}$ defined by Theorem 3.11 and $\{\boldsymbol{ICS}, \diamond\!\!\!+, \diamond\!\!\!\cdot, \leq, \subseteq\}$ studied in Theorem 4.25.

The isomorphism reduces the operations in ***ICS***, which are not implementable in practice, to the operations in ***CIS*** that have the properties

$$\begin{aligned}&\Phi \diamond\!\!\!+\, \Psi = (X_1 \diamond\!\!\!+\, Y_1, X_2 \diamond\!\!\!+\, Y_2),\\ &\Phi \diamond\!\!\!\cdot\, \Psi = (\Diamond(X_1 \boxdot Y_1 \boxminus X_2 \boxdot Y_2), \Diamond(X_1 \boxdot Y_2 \boxplus X_2 \boxdot Y_1)).\end{aligned}$$

These operations are implementable on a computer. Addition can be performed componentwise in terms of addition in ***IS***. Multiplication can be performed using Table 1 and one of the algorithms for scalar products with optimal accuracy. See Chapter 6.

The result of these considerations concerning complex intervals is illustrated in Fig. 27. The latter is completely analogous to Fig. 24. The sets $R$, $D$, $S$ may be interpreted as in Fig. 24. We emphasize, however, that the isomorphism shown in Fig. 27 does not include division.

For reasons of simplicity, calculations on a computer are often performed by the vertical method in the ringoids ***CID*** and ***CIS*** defined in Theorem 2.6. Multiplication in these ringoids leads to upper bounds for multiplication defined in the same sets by means of Theorem 3.11. This can be seen by

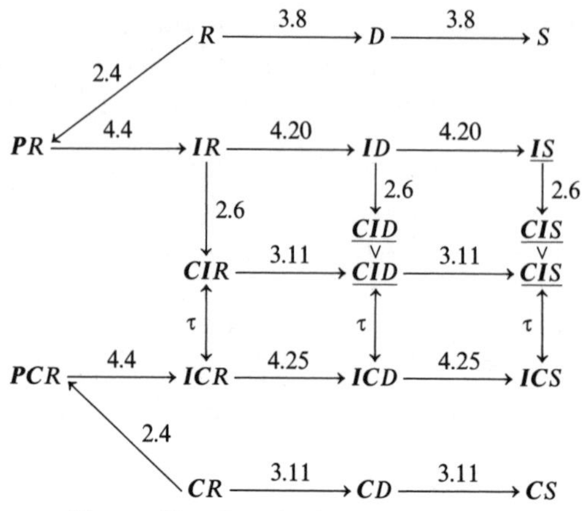


FIGURE 27. Complex interval arithmetic.

means of the formula

$$\begin{aligned}\Phi \mathbin{\Diamond\!\!\!\!\cdot\;} \Psi &= (\Diamond(X_1 \boxdot Y_1 \boxminus X_2 \boxdot Y_2), \Diamond(X_1 \boxdot Y_2 \boxplus X_2 \boxdot Y_1)) \\ &\subseteq (X_1 \mathbin{\Diamond\!\!\!\!\cdot\;} Y_1 \mathbin{\Diamond\!\!\!\!-\;} X_2 \mathbin{\Diamond\!\!\!\!\cdot\;} Y_2, X_1 \mathbin{\Diamond\!\!\!\!\cdot\;} Y_2 \mathbin{\Diamond\!\!\!\!+\;} X_2 \mathbin{\Diamond\!\!\!\!\cdot\;} Y_1).\end{aligned}$$

The last inequality is a consequence of (R3) for the rounding $\Diamond: \overline{\boldsymbol{IR}} \to \overline{\boldsymbol{IS}}$ and the inclusion-isotony of addition and subtraction in $\boldsymbol{IS}$.

## 8. COMPLEX INTERVAL MATRICES AND VECTORS

We now consider the case of complex interval matrices. Again, let $\{R, +, \cdot, \leq\}$ be a completely and linearly ordered ringoid, $\{\boldsymbol{C}R, +, \cdot, \leq\}$ its complexification (Theorem 2.6), and $\{\boldsymbol{IC}R, \boxplus, \boxdot, \leq, \subseteq\}$ the completely and weakly ordered ringoid of intervals over $\boldsymbol{C}R$ (Theorem 4.4). By Theorem 2.5, the set of $n \times n$ matrices $\{M_n\boldsymbol{C}R, +, \cdot, \leq\}$ also forms a completely and weakly ordered ringoid.

We consider the ringoid $\{\boldsymbol{P}M_n\boldsymbol{C}R, +, \cdot\}$ corresponding to the power set of $M_n\boldsymbol{C}R$ (Theorem 2.4). In the subset of intervals $\boldsymbol{I}M_n\boldsymbol{C}R \subset \boldsymbol{P}M_n\boldsymbol{C}R$ we define operations $\circledast$, $* \in \{+, \cdot\}$, by employing the semimorphism

$$\square : \boldsymbol{P}M_n\boldsymbol{C}R \to \overline{\boldsymbol{I}M_n\boldsymbol{C}R} := \boldsymbol{I}M_n\boldsymbol{C}R \cup \{\varnothing\}$$

$$\bigwedge_{\boldsymbol{A},\boldsymbol{B} \in \boldsymbol{I}M_n\boldsymbol{C}R} \boldsymbol{A} \circledast \boldsymbol{B} := \square(\boldsymbol{A} * \boldsymbol{B}) = [\inf(\boldsymbol{A} * \boldsymbol{B}), \sup(\boldsymbol{A} * \boldsymbol{B})].$$

Then using Theorem 4.4, we see that $\{IM_nCR, \boxplus, \boxdot, \leq\}$ is also a completely and weakly ordered ringoid.

Independently of the set $IM_nCR$, we now consider the set $M_nICR$, i.e., the set of $n \times n$ matrices the components of which are intervals over $CR$. Employing the operations in the ringoid $\{ICR, \boxplus, \boxdot, \leq, \subseteq\}$, we define operations and an order relation in $M_nICR$ by means of Theorem 2.5. Then $\{M_nICR, \oplus, \odot, \leq\}$ is seen to be a completely and weakly ordered ringoid.

We consider the question of whether there exists any relation between the completely and weakly ordered ringoids $\{M_nICR, \oplus, \odot, \leq\}$ and $\{IM_nCR, \boxplus, \boxdot, \leq\}$. The assertion of Lemma 4.12 is that both of these ringoids are isomorphic with respect to the algebraic and the order structure if the following formula relating the operations in $PCR$ and $ICR$ holds.

$$\bigwedge_{A_i, B_i \in ICR} \left( \sum_{\nu=1}^{n} A_\nu \boxdot B_\nu \subseteq \square \left( \sum_{\nu=1}^{n} A_\nu \cdot B_\nu \right) \right). \tag{1}$$

Here $\square : PCR \to \overline{ICR} := ICR \cup \{\varnothing\}$ denotes the monotone upwardly directed rounding. We are now going to show that this fōrmula is valid in the present setting.

Because of the isomorphism between the ringoids $CIR$ and $ICR$, which we noted above, we simply identify certain elements of $CIR$ and $ICR$ within the following proof. These elements are the ones that correspond to each other through the mapping

$$\tau : CIR \to ICR \qquad \text{with} \quad \tau([x_1, x_2], [y_1, y_2]) = [(x_1, y_1), (x_2, y_2)].$$

Let $A_i, B_i \in ICR$. Using the isomorphism just referred to, $A_i \equiv (A_{i1}, A_{i2})$ and $B_i \equiv (B_{i1}, B_{i2})$, where $A_{i1}, A_{i2}, B_{i1}, B_{i2} \in IR$. Then

$$x \in \sum_{\nu=1}^{n} A_i \boxdot B_i \Rightarrow \bigwedge_{i=1(1)n} \bigvee_{x_i \in A_i \boxdot B_i} x = \sum_{i=1}^{n} x_i, \qquad \text{with}$$

$$A_i \boxdot B_i \equiv (A_{i1} \boxdot B_{i1} \boxminus A_{i2} \boxdot B_{i2}, A_{i1} \boxdot B_{i2} \boxplus A_{i2} \boxdot B_{i1}).$$

If we now also separate $x_i = (x_{i1}, x_{i2})$ into real and imaginary parts, we obtain

$$\bigwedge_{i=1(1)n} (x_{i1} \in A_{i1} \boxdot B_{i1} \boxminus A_{i2} \boxdot B_{i2} \wedge x_{i2} \in A_{i1} \boxdot B_{i2} \boxplus A_{i2} \boxdot B_{i1})$$

and

$$x = \left( \sum_{i=1}^{n} x_{i1}, \sum_{i=1}^{n} x_{i2} \right).$$

Because of Corollary 4.10, we now get with $A_{ij} = [a_{ij}^1, a_{ij}^2]$, $B_{ij} = [b_{ij}^1, b_{ij}^2]$, $i = 1(1)n, j = 1,2$ for all $i = 1(1)n$ that

$$\min_{r,s=1,2} (a_{i1}^r \cdot b_{i1}^s) - \max_{r,s=1,2} (a_{i2}^r \cdot b_{i2}^s)$$

$$\leq x_{i1} \leq \max_{r,s=1,2} (a_{i1}^r \cdot b_{i1}^s) - \min_{r,s=1,2} (a_{i2}^r \cdot b_{i2}^s)$$

$$\wedge \min_{r,s=1,2} (a_{i1}^r \cdot b_{i2}^s) + \min_{r,s=1,2} (a_{i2}^r \cdot b_{i1}^s)$$

$$\leq x_{i2} \leq \max_{r,s=1,2} (a_{i1}^r \cdot b_{i2}^s) + \max_{r,s=1,2} (a_{i2}^r \cdot b_{i1}^s)$$

$$\underset{(\mathrm{OD1})_R}{\Rightarrow} \sum_{i=1}^{n} \left( \min_{r,s=1,2} (a_{i1}^r \cdot b_{i1}^s) - \max_{r,s=1,2} (a_{i2}^r \cdot b_{i2}^s) \right)$$

$$\leq \sum_{i=1}^{n} x_{i1} \leq \sum_{i=1}^{n} \left( \max_{r,s=1,2} (a_{i1}^r \cdot b_{i1}^s) - \min_{r,s=1,2} (a_{i2}^r \cdot b_{i2}^s) \right)$$

$$\wedge \sum_{i=1}^{n} \left( \min_{r,s=1,2} (a_{i1}^r \cdot b_{i2}^s) + \min_{r,s=1,2} (a_{i2}^r \cdot b_{i1}^s) \right)$$

$$\leq \sum_{i=1}^{n} x_{i2} \leq \sum_{i=1}^{n} \left( \max_{r,s=1,2} (a_{i1}^r \cdot b_{i2}^s) + \max_{r,s=1,2} (a_{i2}^r \cdot b_{i1}^s) \right)$$

and

$$x = \left( \sum_{i=1}^{n} x_{i1}, \sum_{i=1}^{n} x_{i2} \right),$$

i.e., the bounds that we have just established for $x$ consist of points in the set represented by the sum appearing in the expression

$$\square \left( \sum_{i=1}^{n} A_i \cdot B_i \right).$$

This proves (1).

Then by Lemma 4.12, the two completely and weakly ordered ringoids $\{IM_nCR, \boxplus, \boxdot, \leq\}$ and $(M_nICR, \oplus, \odot, \leq\}$ are isomorphic with respect to algebraic and order structure. Figure 28 illustrates this isomorphism by means of a schematic. Through this isomorphism the operations $\boxast$, $* \in \{+,-,\cdot\}$, in $IM_nCR$, which are not implementable in terms of their definition, are reduced to the operations in $M_nICR$. Because of the isomorphism $\tau: ICR \to CIR$, these latter operations are identical to those in the ringoid $M_nCIR$. These operations in turn are defined and can easily be implemented in terms of those in $CIR$.

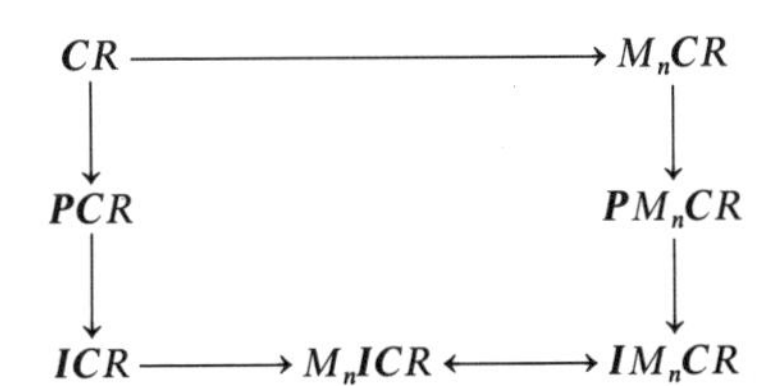


FIGURE 28. Illustration of the isomorphism $M_nICR \leftrightarrow IM_nCR$.

We now consider complex interval matrices on a screen, and we begin with the following theorem that characterizes such a structure.

**Theorem 4.26:** Let $\{R, +, \cdot, \leq\}$ be a completely and linearly ordered ringoid with the neutral elements $o$ and $e$, and let $\{S, \leq\}$ be a symmetric screen of $\{R, +, \cdot, \leq\}$. Consider the complex ringoids $\{CR, +, \cdot, \leq\}$ and $\{M_nCR, +, \cdot, \leq\}$ and the semimorphisms $\square: \boldsymbol{PM_nCR} \to \overline{\boldsymbol{IM_nCR}}$ and $\Diamond: \overline{\boldsymbol{IM_nCR}} \to \overline{\boldsymbol{IM_nCS}}$. Then $\{\boldsymbol{IM_nCS}, \diamondplus, \diamonddot, \leq, \subseteq\}$ is a completely and weakly ordered ringoid with respect to $\leq$. With respect to $\subseteq$, $\boldsymbol{IM_nCS}$ is an inclusion-isotonally ordered monotone upper screen ringoid of $\boldsymbol{IM_nCR}$.

*Proof*: The proof of this theorem is completely analogous to the proofs of Theorems 4.21 and 4.25. Therefore, we only indicate the main steps. By Lemma 4.19(a), all elements of $\boldsymbol{IM_nCS}$ that fulfill (D5) are point intervals of the form $\boldsymbol{X} = [X, X]$ with $X \in M_nCS$. Using (D5a) in $\boldsymbol{IM_nCS}$, we see that all imaginary parts of the components of $X$ vanish and that $X$ is a diagonal matrix. By (D5c) we obtain for all $\boldsymbol{A} \in \boldsymbol{IM_nCS}$ that $\boldsymbol{X} \diamonddot \boldsymbol{A} = \boldsymbol{A} \diamonddot \boldsymbol{X}$. In this relation choose $\boldsymbol{A}$ to be a point interval with all components of both bounds equal to $(e, o)$. Then we obtain that all diagonal elements of $\boldsymbol{X}$ are equal.

The set of all point intervals, the bounds of which are diagonal matrices with constant diagonal elements and vanishing imaginary parts, is a linearly ordered subset of $\{\boldsymbol{IM_nCS}, \leq\}$, which is order isomorphic to $\{S, \leq\}$. The balance of the proof now follows by Lemma 4.19(b)–(d). ■

Now in a manner similar to the discussion following Theorems 4.21 and 4.25, the usual isomorphisms can be studied. Figure 29 gives the resulting diagram.

The case of complex vectoids can be presented as the following corollary of Theorem 4.23.

**Corollary 4.27:** Let $\{R, +, \cdot, \leq\}$ be a completely and linearly ordered ringoid and $\{S, \leq\}$ a symmetric screen of $\{R, +, \cdot, \leq\}$. Consider the complex ringoids $\{CR, +, \cdot, \leq\}$, $\{M_nCR, +, \cdot, \leq\}$ and the completely and weakly ordered vectoids $\{V_nCR, CR, \leq\}$, $\{M_nCR, CR, \leq\}$ and $\{V_nCR, M_nCR, \leq\}$ as well as the semimorphisms $\square: \boldsymbol{PV_nCR} \to \overline{\boldsymbol{IV_nCR}}$, $\Diamond: \overline{\boldsymbol{IV_nCR}} \to \overline{\boldsymbol{IV_nCS}}$ and

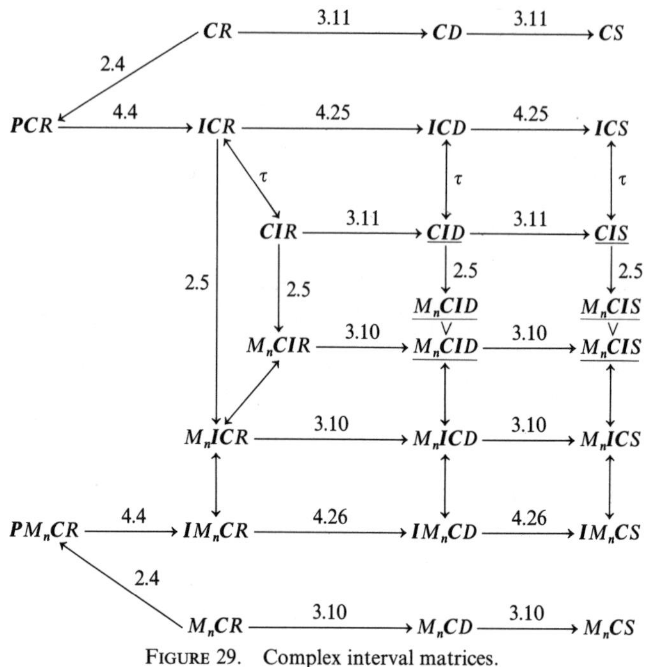


FIGURE 29. Complex interval matrices.

$\square: PM_nCR \to \overline{IM_nCR}$, $\lozenge: \overline{IM_nCR} \to \overline{IM_nCS}$. Then the structures $\{IV_nCS, ICS, \leq, \subseteq\}$, $\{IM_nCS, ICS, \leq, \subseteq\}$, and $\{IV_nCS, IM_nCS, \leq, \subseteq\}$ are completely and weakly ordered vectoids with respect to $\leq$. $\{IM_nCS, ICS\}$ is multiplicative. With respect to $\subseteq$, all of these vectoids are inclusion-isotonally ordered and monotone upper screen vectoids of $\{IV_nCR, ICR\}$, $\{IM_nCR, ICR\}$ and $\{IV_nCR, IM_nCR\}$, respectively. ■

The operations in the interval vectoids occurring in this corollary are not implementable by using their definition. Under our assumptions, however, we may apply the isomorphisms expressed by Theorem 4.15, and in addition, we make use of the isomorphisms $ICR \leftrightarrow CIR$ and $IM_nCR \leftrightarrow M_nCIR$. Summarizing, we obtain the following isomorphisms

$$\begin{aligned}
\{IV_nCR, ICR, \leq, \subseteq\} &\leftrightarrow \{V_nICR, ICR, \leq, \subseteq\} \\
&\leftrightarrow \{V_nCIR, CIR, \leq, \subseteq\}, \\
\{IM_nCR, ICR, \leq, \subseteq\} &\leftrightarrow \{M_nICR, ICR, \leq, \subseteq\} \\
&\leftrightarrow \{M_nCIR, CIR, \leq, \subseteq\}, \\
\{IV_nCR, IM_nCR, \leq, \subseteq\} &\leftrightarrow \{V_nICR, M_nICR, \leq, \subseteq\} \\
&\leftrightarrow \{V_nCIR, M_nCIR, \leq, \subseteq\}.
\end{aligned}$$

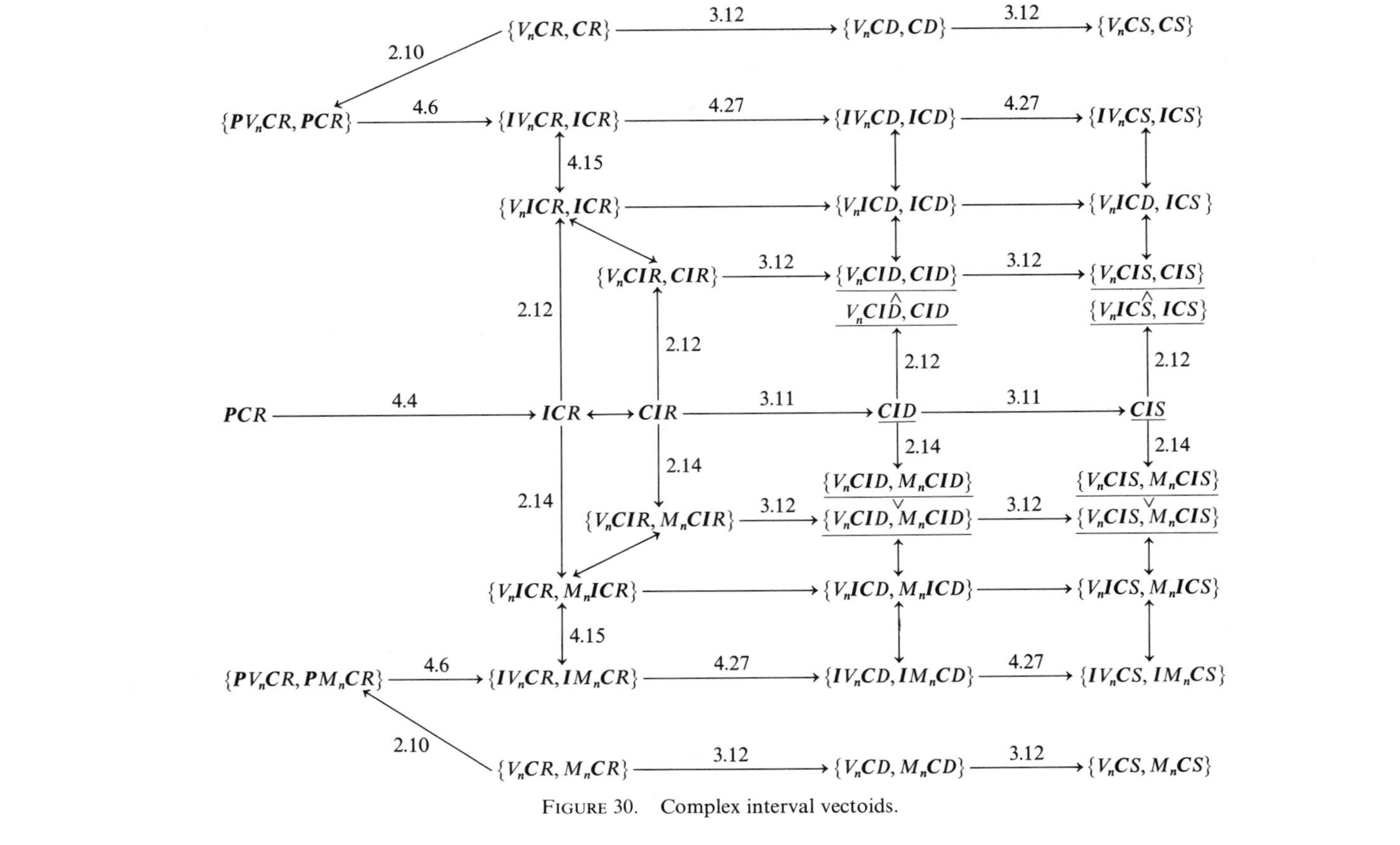


FIGURE 30. Complex interval vectoids.

Now it can be shown in complete analogy to the considerations following Theorems 4.21 and 4.25 that these isomorphisms also induce isomorphisms between the corresponding spaces over $D$ and $S$; for instance

$$\begin{aligned}\{IV_nCS, ICS, \leq, \subseteq\} &\leftrightarrow \{V_nICS, ICS, \leq, \subseteq\}\\ &\leftrightarrow \{V_nCIS, CIS, \leq, \subseteq\},\\ \{IM_nCS, ICS, \leq, \subseteq\} &\leftrightarrow \{M_nICS, ICS, \leq, \subseteq\}\\ &\leftrightarrow \{M_nCIS, CIS, \leq, \subseteq\},\\ \{IV_nCS, IM_nCS, \leq, \subseteq\} &\leftrightarrow \{V_nICS, M_nICS, \leq, \subseteq\}\\ &\leftrightarrow \{V_nCIS, M_nCIS, \leq, \subseteq\}.\end{aligned}$$

Figure 30 summarizes all these results. As in the similar figures displayed before, double arrows indicate isomorphisms, horizontal arrows denote semimorphisms, and vertical arrows indicate a vertical definition of the arithmetic operations. All operations defined by semimorphisms within vectoids ending with $CID$ or $CIS$ can be implemented on a computer by one of the algorithms for scalar products with optimal accuracy. See Chapter 6. Upper bounds for these operations can be obtained if within the same sets, operations are also defined by the vertical method. See Fig. 30.

# Part 2 / IMPLEMENTATION OF ARITHMETIC ON COMPUTERS

# Chapter 5 / FLOATING-POINT ARITHMETIC

**Summary:** Thus far our considerations have proceeded under the assumption that the set $R$ in Fig. 1 is a linearly ordered ringoid. We are now going to be more specific and assume that $R$ is the linearly ordered field of real numbers and that the subsets $D$ and $S$ are special screens, which are called floating-point systems. In Section 1 we briefly review the definition of the real numbers and their representations by $b$-adic expansions. Section 2 deals with floating-point numbers and floating-point systems. We also discuss several basic roundings of the real numbers into a floating-point system, and we derive error bounds for these roundings.

In Section 3 we consider floating-point operations defined by semimorphisms, and we derive error bounds for these operations. Then we consider the operations defined in the other sets listed under $S$ and $D$ in Fig. 1, and we derive error formulas and bounds for these operations also. All this is done under the assumption that the operations are defined by semimorphisms.

In the case of the scalar product and matrix multiplication, we also derive error formulas and bounds for the vertical definition of the operations. The error formulas are simpler if the operations are defined by semimorphisms, and the error bounds are smaller by a factor of $n$ compared to the bounds obtained by the vertical definition of the operations. These two properties are reproduced in the error analysis of many algorithms.

In Section 4 we consider extended floating-point systems in which a unique additive inverse exists for every element. We show that this uniqueness property simplifies the proofs that many of the structures listed in Fig. 1 are ringoids.

## 1. DEFINITION AND PROPERTIES OF THE REAL NUMBERS

Two methods of defining the real numbers are most commonly used. These may be called the constructive method and the axiomatic method.

The constructive method begins with the definition of the natural numbers by the so-called Peano axioms. Then the whole numbers (integers), the rational numbers, the real numbers, and the complex numbers are defined. The whole numbers are defined as pairs of natural numbers, and the difference of any two whole numbers is defined and is itself a whole number. The rational numbers are defined as pairs of whole numbers, and the quotient of any two rational numbers is defined and is itself a rational number (except for division by zero). The complex numbers are defined as pairs of real numbers, and the square root of any complex number is defined and is a complex number. The passage from the rational numbers to the real numbers cannot be performed by considering pairs. Different methods of defining the real numbers have been developed and all are equivalent. The two best-known procedures for defining real numbers use Dedekind cuts in the set of rational numbers or fundamental sequences of rational numbers in the sense of G. Cantor. Each of these procedures requires infinitely many rational numbers to define an irrational number.

We now consider the axiomatic method of defining the real numbers.

**Definition 5.1:** A set $R$, in which an addition $+$, a multiplication $\cdot$, and an order relation $\leq$ are defined, is called a conditionally complete, linearly ordered field if the following properties hold:

I. $\{R, +, \cdot\}$ is a field.

II. $\{R, \leq\}$ is a conditionally complete, linearly ordered set.

III. The following compatibility properties hold between the algebraic and the order structure in $R$:

(a) $$\bigwedge_{a,b,c \in R} (a \leq b \Rightarrow a + c \leq b + c),$$

(b) $$\bigwedge_{a,b,c \in R} (a \leq b \wedge c \geq 0 \Rightarrow a \cdot c \leq b \cdot c). \quad \blacksquare$$

Because of the importance of this definition in the whole of analysis, we enumerate the axioms of the concepts I and II in more detail:

I. (F11) $$\bigwedge_{a,b,c \in R} (a + b) + c = a + (b + c)$$

(F21) $$\bigwedge_{a,b,c \in R} (a \cdot b) \cdot c = a \cdot (b \cdot c)$$

(associativity of addition and of multiplication),

(F12) $\bigwedge_{a,b \in R} a + b = b + a$ 

(F22) $\bigwedge_{a,b \in R} a \cdot b = b \cdot a$ (commutativity of addition and of multiplication),

(F13) $\bigwedge_{a \in R} \bigvee_{0 \in R} a + 0 = a$

(F23) $\bigvee_{1 \in R \setminus \{0\}} \bigwedge_{a \in R} a \cdot 1 = a$ (existence of neutral elements),

(F14) $\bigwedge_{a \in R} \bigvee_{-a \in R} a + (-a) = 0$

(F24) $\bigwedge_{a \in R \setminus \{0\}} \bigvee_{a^{-1} \in R} a \cdot a^{-1} = 1$ (existence of inverse elements),

(F) $\bigwedge_{a,b,c \in R} a \cdot (b + c) = a \cdot b + a \cdot c$ (distributivity).

II. (O1) $\bigwedge_{a \in R} a \leq a$ (reflexivity),

(O2) $\bigwedge_{a,b,c \in R} (a \leq b \wedge b \leq c \Rightarrow a \leq c)$ (transitivity),

(O3) $\bigwedge_{a,b \in R} (a \leq b \wedge b \leq a \Rightarrow a = b)$ (antisymmetry),

(O4) $\bigwedge_{a,b \in R} (a \leq b \vee b \leq a)$ (linearity),

(O5*) Every nonempty bounded subset of $R$ has an infimum and a supremum.

These few properties are the basis for the huge construct that is real analysis. All further properties of the real numbers are derived from them. In this connection the following theorem is of fundamental interest.

**Theorem 5.2:** Two conditionally complete, linearly ordered fields $R$ and $R'$ are isomorphic. ■

For the proof of this theorem, see, for instance, McShane and Botts [51].

Since corresponding elements of isomorphic structures can be identified with each other, Theorem 5.2 asserts that there exists at most one conditionally complete, linearly ordered field. Such a structure does in fact exist since the constructive method mentioned above actually produces one. In this sense the real numbers defined by that method represent the only realization of a conditionally complete, linearly ordered field.

Theorem 5.2 also implies that the real numbers cannot be extended by additional elements without giving up or weakening one of the above properties. By partly abandoning the order structure, the complex numbers may be obtained. Sometimes the two elements $-\infty$ and $+\infty$ with the property $-\infty < a < +\infty$, for all $a \in \boldsymbol{R}$, are adjoined to the real numbers. The following rules are assumed for this extended set:

(a) $a + \infty = +\infty, \quad a - \infty = -\infty, \quad a/(+\infty) = a/(-\infty) = 0,$
(b) $a > 0 \Rightarrow a \cdot (+\infty) = +\infty, \quad a \cdot (-\infty) = -\infty, \quad a/0 = +\infty,$
(c) $a < 0 \Rightarrow a \cdot (+\infty) = -\infty, \quad a \cdot (-\infty) = +\infty, \quad a/0 = -\infty,$

and sometimes even

(d) $0 \cdot (+\infty) = (+\infty) \cdot 0 = 0 \cdot (-\infty) = (-\infty) \cdot 0 = 0.$

With these rules $\{\boldsymbol{R} \cup \{-\infty\} \cup \{+\infty\}, \leq\}$ is a complete lattice. However, it no longer is a linearly ordered field. For instance, $a + \infty = b + \infty$ even if $a < b$.

We now give the definition of the absolute value.

**Definition 5.3:** A mapping $|\,|:\boldsymbol{R} \to \boldsymbol{R}_+ := \{x \in \boldsymbol{R} | x \geq 0\}$ is called the *absolute value* if

$$\bigwedge_{a \in \boldsymbol{R}} |a| := \sup(a, -a) = \begin{cases} a & \text{for } a \geq 0 \\ 0 & \text{for } a = 0 \\ -a & \text{for } a \leq 0. \end{cases} \quad \blacksquare$$

Using the absolute value, a distance $\rho(a, b) := |a - b|$ can be defined, which has all properties of a metric. With the latter, the concepts of the limit of a sequence and the limit of a series can be defined.

**Definition 5.4:** A sequence[†] $\{a_n\} \in \boldsymbol{R}^{\boldsymbol{N}} := \{\varphi | \varphi : \boldsymbol{N} \to \boldsymbol{R}\}$ of elements $a_n \in \boldsymbol{R}$ is called *convergent* to a number $a \in \boldsymbol{R}$, in which event we write $\lim_{n \to \infty} a_n = a$ if

$$\bigwedge_{\varepsilon > 0} \bigvee_{N(\varepsilon) \in \boldsymbol{N}} \bigwedge_{n \geq N(\varepsilon)} |a_n - a| < \varepsilon.$$

Otherwise the sequence is called *divergent*.

Using a sequence $\{a_n\}$, we may define an associated sequence $\{s_n\}$, where

$$s_n := \sum_{\nu=1}^{n} a_\nu .$$

If $\{s_n\}$ is convergent to a limit $s$, we say that the *infinite series*

$$\sum_{\nu=1}^{\infty} a_\nu = a_1 + a_2 + a_3 + \cdots$$

[†] $\boldsymbol{N}$ denotes the set of natural numbers.

is convergent and write

$$s = \sum_{\nu=1}^{\infty} a_\nu .$$

Here $s$ is called the *sum* of the series. The numbers $s_n$ are called the *partial sums*. If $\{s_n\}$ is divergent, the series is called divergent. ■

According to Definition 5.1, the real numbers are defined as a set of elements, which have the properties I–III. The constructive method demonstrates the existence of such elements. With this existence, the development of most of analysis can then be performed. Indeed, this is a common approach that provides an elegant but abstract body of mathematics.

However, the applicability of analysis or even arithmetic to other areas (or even to mathematics itself) from such an abstract base is quite limited. Bridging this limitation requires the development of additional mathematical concepts and an associated body of theory. This additional development must furnish more concrete form for the real numbers, which lie at the basis of analysis. We ought at least to furnish a setting in which the operations defined among the real numbers can be implemented, in some sense.

To accomplish this, we study representations of the real numbers. This study will allow us to conceptualize the reals more quantitatively. We may then think of numbers as objects with values, we may execute the operations among the numbers to find the results of such operations, etc. We begin our study of representations with the following well-known theorem on *b-adic systems*. This theorem may be shown to follow from the axioms I–III, see [55].

**Theorem 5.5:** Every real number $x$ is uniquely represented by a $b$-adic expansion of the form

$$x = * d_n d_{n-1} \ldots d_1 d_0 . d_{-1} d_{-2} d_{-3} \ldots = * \sum_{\nu=n}^{-\infty} d_\nu b^\nu \tag{1}$$

with $* \in \{+, -\}$ and $b \in \mathbf{N}$, $b > 1$. Here the $d_i$, $i = n(-1)-\infty$, are integers that satisfy the inequalities

$$0 \le d_i \le b - 1 \qquad \text{for all} \quad i = n(-1)-\infty, \tag{2}$$

$$d_i \le b - 2 \qquad \text{for infinitely many} \quad i. \tag{3}$$

Every $b$-adic expansion (1)–(3) represents exactly one real number.

If we approximate (1) by the finite sum

$$s_m := * \sum_{\nu=n}^{-m} d_\nu b^\nu,$$

the following error bound holds:

$$0 \leq |x - s_m| < b^{-m}. \quad \blacksquare \tag{4}$$

In (1) $b$ is called the *base* of the number system. The $d_i$, $i = n(-1)-\infty$, are called the *digits*.

Theorem 5.5 asserts that there exists a one-to-one correspondence between the real numbers and the $b$-adic expansions of the form (1)–(3). Condition (3) needs some interpretation. Without it the representation of a real number by a $b$-adic expansion (1) need not be unique. Consider, for instance, the two decimal expansions with the partial sums

$$s_n = 3 \cdot 10^{-1} + 0 \cdot 10^{-2} + 0 \cdot 10^{-3} + \cdots + 0 \cdot 10^{-n}$$

and

$$s_n' = 2 \cdot 10^{-1} + 9 \cdot 10^{-2} + 9 \cdot 10^{-3} + \cdots + 9 \cdot 10^{-n}.$$

Both sequences $\{s_n\}$ and $\{s_n'\}$ converge[†] to and therefore represent the same real numbers $s = 0.3$.

Although $b$-adic expansions are familiar objects, and especially so within the decimal system, it will be useful to review several additional facts concerning them. If the digits of a $b$-adic expansion are all zero for $i < -m$, it is called finite. The zeros in question can be omitted.

$$x = *d_n d_{n-1} \ldots d_1 d_0 . d_{-1} d_{-2} \ldots d_{-m} = * \sum_{\nu=n}^{-m} d_\nu b^\nu, \qquad * \in \{+, -\}. \tag{5}$$

The significance of $b$-adic expansions lies in the fact that all real numbers, in particular the irrational numbers, can be approximated by finite expansions. If we terminate a $b$-adic expansion with the digit $d_{-m}$, then according to (4) the resulting error is less than $b^{-m}$. By taking $m$ large enough, the error can be made arbitrarily small.

Multiplication by $b^m$ transforms each finite $b$-adic expansion (5) into a whole number. Thus each such expansion represents a rational number. On the other hand, not every rational number has a finite $b$-adic expansion. For instance, in the decimal system, we have $\frac{1}{3} = 0.333\ldots$. We say that the decimal expansion of $\frac{1}{3}$ has a period length 1. In general in a $b$-adic system the rational numbers are characterized by the fact that they have periodic $b$-adic expansions.

If we define a $b$-adic expansion by selecting its digits randomly out of the set $\{0, 1, 2, \ldots, b-1\}$, the chance of obtaining a periodic expansion is negligible. This suggests that *most* of these expansions represent irrational

[†] A monotone increasing sequence of real numbers that is bounded above converges to its supremum.

numbers. In what follows we approximate real numbers by finite $b$-adic expansions, i.e., by a subset of the rational numbers. In the decimal system, for instance, the number $\frac{1}{3}$ is not a member of this approximating subset.

It is possible that the one-to-one correspondence between the real numbers and the $b$-adic expansions permits the definition of the real numbers through their $b$-adic expansions. Two $b$-adic expansions would then be called equal if they are identical. We can also easily decide which of two numbers is the smaller. Consider the two expansions

$$c = c_p c_{p-1} \dots c_1 c_0 . c_{-1} c_{-2} c_{-3} \dots$$

and

$$d = d_q d_{q-1} \dots d_1 d_0 . d_{-1} d_{-2} d_{-3} \dots,$$

and let $n$ be the largest index for which $c_n \neq d_n$. Without loss of generality we may assume $c_n < d_n$, i.e., $c_n + 1 \leq d_n$. Then $c < d$ since

$$c = \sum_{\nu=p}^{-\infty} c_\nu b^\nu = \sum_{\nu=p}^{n} c_\nu b^\nu + \sum_{\nu=n-1}^{-\infty} c_\nu b^\nu < \sum_{\nu=p}^{n} c_\nu b^\nu + b^n \leq \sum_{\nu=q}^{n} d_\nu b^\nu \leq d.$$

Negative $b$-adic expansions are treated correspondingly.

The definition of the operations for $b$-adic expansions, however, is an involved one. An infinite $b$-adic expansion cannot easily be added or multiplied in a simple way. The operations for infinite $b$-adic expansions must be defined in terms of a sequence of approximations.

For finite $b$-adic expansions of the form (5), the operations of addition, subtraction, multiplication, and division can easily be executed by well-known rules (at least in the case $b = 10$). These algorithms can easily be derived by using the representation (5). They reduce the operations with real numbers to operations with single digits. In order to prescribe these algorithms for addition, subtraction, and multiplication, tables for the addition, subtraction, and multiplication of all possible combinations of the digits 0, 1, 2, 3, ..., $b - 1$ suffice. Division can be executed as a repeated subtraction. Since these algorithms are so well known from the computations with decimal numbers, we refrain from deriving them here.

However, direct operation with infinite $b$-adic expansions is not possible. Operations are defined by means of successive approximations. Since the error of such approximations can, in principle, be made arbitrarily small, they can be used as legitimate replacements for the operations for real numbers. We describe this process in some further detail. Once again let

$$c = c_p c_{p-1} \dots c_1 c_0 . c_{-1} c_{-2} c_{-3} \dots \quad \text{and} \quad d = d_q d_{q-1} \dots d_1 d_0 . d_{-1} d_{-2} d_{-3} \dots$$

be the $b$-adic expansions of two real numbers. We denote the $n$th partial sums by

$$\gamma_n = c_p c_{p-1} \dots c_1 c_0 . c_{-1} c_{-2} \dots c_{-n},$$
$$\delta_n = d_q d_{q-1} \dots d_1 d_0 . d_{-1} d_{-2} \dots d_{-n}.$$

In order to define $s := c * d, * \in \{+, -, \cdot, /\}$, in terms of the approximation $\gamma_n$ and $\delta_n$, we form the sequence $\{\sigma_n\}$ with

$$\sigma_n := \gamma_n * \delta_n, \qquad * \in \{+, -, \cdot, /\}.$$

If in the case of division, $\lim_{n\to\infty} \delta_n = d \neq 0$, the sequence $\{\sigma_n\}$ exists for sufficiently large $n$. The quantity $\sigma_n$ can be calculated by the well-known algorithms for finite $b$-adic expansions. Appealing to a well-known theorem of analysis, we have

$$\lim_{n\to\infty} \sigma_n = \lim_{n\to\infty} (\gamma_n * \delta_n) = \lim_{n\to\infty} \gamma_n * \lim_{n\to\infty} \delta_n = c * d, \qquad * \in \{+, -, \cdot, /\},$$

and therefore $s = \lim_{n\to\infty} \sigma_n$.

However, the sequence $\{\sigma_n\}$ is not identical with the sequence $\{s_n\}$ of partial sums of $s$. We illustrate this with two examples.

1. $\frac{1}{29} + \frac{2}{29} = \frac{3}{29}$ or $0.03448 \dots + 0.06896 \dots = 0.10344 \dots$.

| $n$ | $\sigma_n$ | $s_n$ |
|---|---|---|
| $-1$ | 0.0 | 0.1 |
| $-2$ | 0.09 | 0.10 |
| $-3$ | 0.102 | 0.103 |
| $-4$ | 0.1033 | 0.1034 |
| ⋮ | ⋮ | ⋮ |

2. $\frac{1}{3} \cdot \frac{1}{3} = \frac{1}{9}$ or $0.33333 \dots \cdot 0.33333 \dots = 0.11111 \dots$.

| $n$ | $\sigma_n$ | $s_n$ |
|---|---|---|
| $-1$ | 0.09 | 0.1 |
| $-2$ | 0.1089 | 0.11 |
| $-3$ | 0.110889 | 0.111 |
| $-4$ | 0.11108889 | 0.1111 |
| ⋮ | ⋮ | ⋮ |

Note that this example also illustrates the fact that double precision arithmetic does not guarantee single precision accuracy. Nevertheless, the real number $s$ is well-defined as the limit of the sequence $\{\sigma_n\}$. The irrational numbers can be defined as Cauchy sequences of rational numbers (G. Cantor). Two sequences are equivalent if they have the same limit. In this sense the two sequences $\{\sigma_n\}$ and $\{s_n\}$ belong to the same equivalence class that defines $s$.

## 2. FLOATING-POINT NUMBERS AND ROUNDINGS

We begin our discussion by recalling that according to Theorem 5.5, every real number $x$ can be uniquely represented by a $b$-adic expansion of the form

$$x = *d_n d_{n-1} \ldots d_1 d_0 . d_{-1} d_{-2} \ldots = * \sum_{\nu=n}^{-\infty} d_\nu b^\nu \tag{1}$$

with $* \in \{+, -\}$, $b \in \mathbf{N}$, $b > 1$, and

$$0 \leq d_i \leq b - 1 \qquad \text{for all} \quad i = n(-1) - \infty, \tag{2}$$

$$d_i \leq b - 2 \qquad \text{for infinitely many} \quad i. \tag{3}$$

We may further assume that $d_n \neq 0$.

In (1) the ($b$-ary) point may be shifted to any other position if we compensate for this shifting by a multiplication with a corresponding power of $b$. If the point is shifted immediately to the left of the first nonzero digit $d_n$, the resulting expression is referred to as the *normalized b-adic representation* of the number $x$. Zero is the only real number that has no such representation. Conditions (2) and (3) assure the unicity of the normalized $b$-adic representation of all other real numbers. We employ the symbol $\boldsymbol{R}_b$ to denote all these numbers, including zero:

$$\begin{aligned} \boldsymbol{R}_b := \{0\} \cup \{x = *m \cdot b^e \mid\, & * \in \{+, -\}, b \in \mathbf{N}, b > 1, e \in \mathbf{Z},^\dagger \\ & m = \sum_{i=1}^{\infty} x[i] b^{-i}, \\ & x[i] \in \{0, 1, \ldots, b-1\}, x[1] \neq 0, \\ & x[i] \leq b - 2 \text{ for infinitely many } i\}. \end{aligned} \tag{4}$$

Here $b$ is called the *base* of the representation, $*$ the *sign of* $x$ (sgn($x$)), $m$ the *mantissa* of $x$ (mant($x$)), and $e$ the *exponent* of $x$ (exp($x$)).

In general the elements of $\boldsymbol{R}_b$ cannot be represented on a computer. Only truncated versions of these elements can be so represented. The following definition distinguishes two special subsets of $\boldsymbol{R}_b$, which we use in the development to follow.

† $\boldsymbol{Z}$ denotes the set of integers.

**Definition 5.6:** A real number is called a *normalized floating-point number* or simply a *floating-point number* if it is an element of one of the following sets $S_{b,l}$ or $S = S(b, l, e1, e2)$. These sets are called *floating-point systems*:

1. $S_{b,l} := \{x \in \mathbf{R}_b \mid m = \sum_{i=1}^{l} x[i] b^{-i}\}$.
2. $S = S(b, l, e1, e2) := \{x \in S_{b,l} \mid e1 \le e \le e2, e1, e2 \in \mathbf{Z}\}$.

In order to have a unique representation of zero available in $S$, we assume additionally that $\operatorname{sgn}(0) = +$, $\operatorname{mant}(0) = 0.00 \ldots 0$ ($l$ zeros after the ($b$-ary) point), and $\exp(0) = e1$. ■

In $S_{b,l}$ the mantissa is restricted to a finite length $l$. $S(b, l, e1, e2)$ is the subset of $S_{b,l}$, where the exponent is bounded by $e1$ and $e2$. We have

$$S(b, l, e1, e2) \subset S_{b,l} \subset \mathbf{R}_b = \mathbf{R}.$$

Both sets $S_{b,l}$ and $S = S(b, l, e1, e2)$ have the properties

(S3) $\quad 0, 1 \in S \wedge \bigwedge_{x \in S} -x \in S$

and

$$b^{-1} \le |m| < 1. \tag{5}$$

$S_{b,l}$ is a countable set, and its cardinality is that of the rational numbers. Every element of $S(b, l, e1, e2)$ represents a rational number. The representation is unique. The number of elements in $S(b, l, e1, e2)$ is $2(b-1) \cdot b^{l-1}(e2 - e1 + 1) + 1$.

The greatest floating-point number in $S(b, l, e1, e2)$ is

$$B := +0.(b-1)(b-1) \ldots (b-1) \cdot b^{e2}.$$

The least number in $S$ is $-B$. The nonzero floating-point numbers in $S$ that are of least absolute value are

$$-0.100 \ldots 0 \cdot b^{e1} \qquad \text{and} \qquad +0.100 \ldots 0 \cdot b^{e1}. \tag{6}$$

The floating-point numbers in $S$ are not uniformly distributed between the numbers in (6) and $-B$ (resp. $+B$). The density decreases with increasing exponent. To illustrate this, we use an example of a floating-point system of 33 elements. This example is from [14] and corresponds to $b = 2$, $l = 3$, $e1 = -1$, and $e2 = 2$. See Fig. 31.

Both sets $S_{b,l}$ and $S(b, l, e1, e2)$ are discrete. $S_{b,l}$ is a screen of $\mathbf{R}$.

To continue, we introduce the following notation:

$$\mathbf{R}_b^* := \mathbf{R}_b \cup \{-\infty\} \cup \{+\infty\} = \mathbf{R}^* := \mathbf{R} \cup \{-\infty\} \cup \{+\infty\},$$
$$S_{b,l}^* := S_{b,l} \cup \{-\infty\} \cup \{+\infty\},$$
$$S^* := S^*(b, l, e1, e2) := S(b, l, e1, e2) \cup \{-\infty\} \cup \{+\infty\},$$
$$\bar{S} := [-B, +B] \subset \mathbf{R}.$$

| e | $2^e$ | $m_1$ | $m_2$ | $m_3$ | $m_4$ |
|---|---|---|---|---|---|
| -1 | 1/2 | 0.100 | 0.101 | 0.110 | 0.111 |
| 0 | 1 | 0.100 | 0.101 | 0.110 | 0.111 |
| 1 | 2 | 0.100 | 0.101 | 0.110 | 0.111 |
| 2 | 4 | 0.100 | 0.101 | 0.110 | 0.111 |

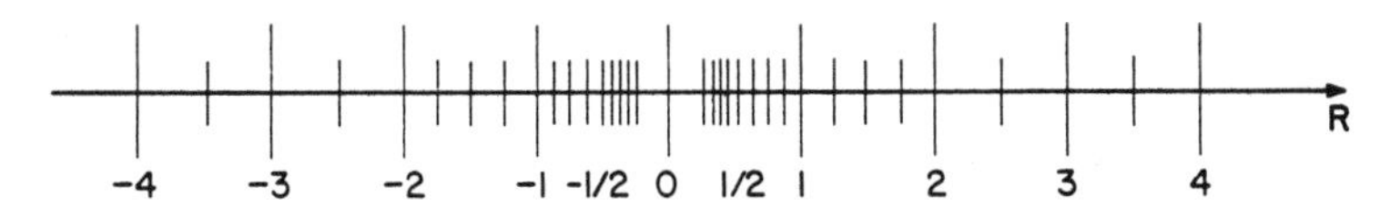


FIGURE 31. A simple floating-point system.

Then $S^*(b, l, e1, e2)$ is a screen of $S_{b,l}$ and $\boldsymbol{R}_b$ as well as of $S^*_{b,l}$ and $\boldsymbol{R}^*_b$. Further, $S(b, l, e1, e2)$ is a screen of $\bar{S}$. All of these screens are symmetric, i.e., they have the property (S3).

We are now going to consider roundings from $\boldsymbol{R}^*$ into $S^*(b, l, e1, e2)$. Similar roundings could also be defined from $\boldsymbol{R}^*$ into $S^*_{b,l}$, from $\boldsymbol{R}$ into $S_{b,l}$, or from $\bar{S}$ into $S(b, l, e1, e2)$.

A rounding $\square : \boldsymbol{R}^* \to S^*$ is defined by the property

(R1) $\bigwedge_{x \in S^*} \square x = x.$

In the following development we are especially concerned with the monotone directed roundings $\triangledown$ and $\triangle$, as well as the monotone and antisymmetric roundings. We shall consider the following roundings:

$\bigwedge_{x \in \boldsymbol{R}^*} \triangledown x := \max\{y \in S^* \mid y \leq x\},$ monotone downwardly directed rounding,

$\bigwedge_{x \in \boldsymbol{R}^*} \triangle x := \min\{y \in S^* \mid x \leq y\},$ monotone upwardly directed rounding,

$\bigwedge_{x \geq 0} \square_b x := \triangledown x \wedge \bigwedge_{x < 0} \square_b x = -\square_b(-x),$ monotone rounding toward zero,

$\bigwedge_{x \geq 0} \square_0 x := \triangle x \wedge \bigwedge_{x < 0} \square_0 x = -\square_0(-x),$ montone rounding away from zero.

Using the notation

$$\bigwedge_{x \geq 0} s_\mu(x) := \triangledown x + (\triangle x - \triangledown x) \cdot \mu/b, \qquad \mu = 1(1)b - 1,$$

we define the following roundings $\square_\mu : \boldsymbol{R}^* \to S^*$, $\mu = 1(1)b-1$, which are in common use:

$$\bigwedge_{x \in [0, b^{e1-1})} \square_\mu x := 0,$$

$$\bigwedge_{x \geq b^{e1-1}} \square_\mu x := \begin{cases} \nabla x & \text{for} \quad x \in [\nabla x, s_\mu(x)) \\ \Delta x & \text{for} \quad x \in [s_\mu(x), \Delta x], \end{cases}$$

$$\bigwedge_{x<0} \square_\mu x := -\square_\mu(-x).$$

If $b$ is an even number, $\bigcirc := \square_{b/2}$ denotes the rounding to the nearest floating-point number. Figure 32 contains a diagram of this rounding in the case of the floating-point system $S(2, 3, -1, 2)$ illustrated above (see Fig. 31).

The roundings listed above have many familiar properties. For instance, we have

$$\nabla x = -\Delta(-x) \wedge \Delta x = -\nabla(-x)$$
$$\square_b x = \operatorname{sgn}(x) \cdot \nabla|x| \wedge \square_0 x = \operatorname{sgn}(x) \cdot \Delta|x|.$$

All of these roundings are monotone (R2). The roundings $\square_\mu$, $\mu = 0(1)b$, are also antisymmetric (R4).

We recall that in a linearly ordered set every monotone rounding and therefore, in particular, all the roundings $\square_\mu$, $\mu = 0(1)b$, can be expressed in terms of the monotone downwardly or upwardly directed rounding $\nabla$ or $\Delta$. (See Section 3.5.) Since $\Delta$ can be expressed in terms of $\nabla$, we give an explicit description of the rounding $\nabla : \boldsymbol{R}^* \to S^* = S^*(b, l, e1, e2)$ only. In

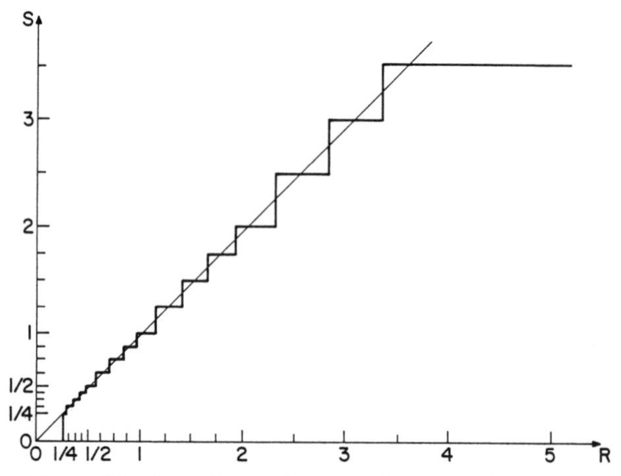


FIGURE 32. Rounding to the nearest floating-point number.

this description we use the abbreviation $B := 0.(b-1)(b-1)\ldots(b-1)\cdot b^{e2}$. We obtain

$$\nabla x = \begin{cases} +\infty & \text{for } x = +\infty \\ +B & \text{for } +B \le x < +\infty \\ +0.x[1]x[2]\ldots x[l]\cdot b^{e} & \text{for } b^{e1-1} \le x < +B \\ 0.00\ldots 0\cdot b^{e1} & \text{for } 0 \le x < b^{e1-1} \\ -0 = 0.100\ldots 0\cdot b^{e1} & \text{for } -b^{e1-1} \le x < 0 \\ -0.x[1]x[2]\ldots x[l]\cdot b^{e} & \text{for } -B \le x < -b^{e1-1} \\ & \quad \wedge \bigwedge_{i\ge 1} x[l+i] = 0 \\ -0.100\ldots 0\cdot b^{l+1} & \text{for } -B < x < -b^{e1-1} \\ & \quad \wedge \bigwedge_{i=1(1)l} x[i] \\ & \quad = b-1 \\ & \quad \wedge \bigvee_{i\ge 1} x[l+i] \ne 0 \\ -(0.x[1]x[2]\ldots x[l] + b^{-l})\cdot b^{e} & \text{for } -B < x < -b^{e1-1} \\ & \quad \wedge \bigvee_{i\in\{1,2,\ldots,l\}} x[i] \\ & \quad \ne b-1 \\ & \quad \wedge \bigvee_{i\ge 1} x[l+i] \ne 0 \\ -\infty & \text{for } -\infty \le x < -B. \end{cases}$$

Using the function $[x]$ (the greatest integer less than or equal to $x$), the description of $\nabla x$ can be shortened. In the following description of $\nabla x$, we assume that $x$ is represented in the form $x = m \cdot b^{e} \in \boldsymbol{R}_b$ whenever $x \ne +\infty$ and $x \ne -\infty$.

$$\nabla x = \begin{cases} +\infty & \text{for } x = +\infty \\ +B & \text{for } +B \le x < +\infty \\ [m\cdot b^{l}]\cdot b^{-l}\cdot b^{e} & \text{for } b^{e1-1} \le |x| < +B \\ 0 = 0.00\ldots 0\cdot b^{e1} & \text{for } 0 \le x < b^{e1-1} \\ -0.100\ldots 0\cdot b^{e1} & \text{for } -b^{e1-1} \le x < 0 \\ -\infty & \text{for } -\infty \le x < -B. \end{cases}$$

Thus the implementation of the rounding $\nabla$ on a computer is simplified if the function $[x]$ is available. In the algorithms for the implementation of computer arithmetic, which we describe in Chapter 6, we shall use this prescription of $\triangledown x$.

The roundings of $\boldsymbol{R}^* \to S^*(b, l, e1, e2)$, which we have been discussing, are not the only ones that are used in computer arithmetic or in numerical analysis. Typically for error analyses of numerical processes, it is usually assumed that the rounding has the property

$$\bigwedge_{x \in \boldsymbol{R}} \square x = x(1 - \varepsilon) \qquad \text{with} \quad |\varepsilon| \leq \varepsilon^* = \text{const.} \tag{7}$$

Figure 33 illustrates a rounding with this property. The graph corresponding to $\square$ has to stay within a cone around the identity mapping. It is not necessarily a monotone nor even an antisymmetric function.

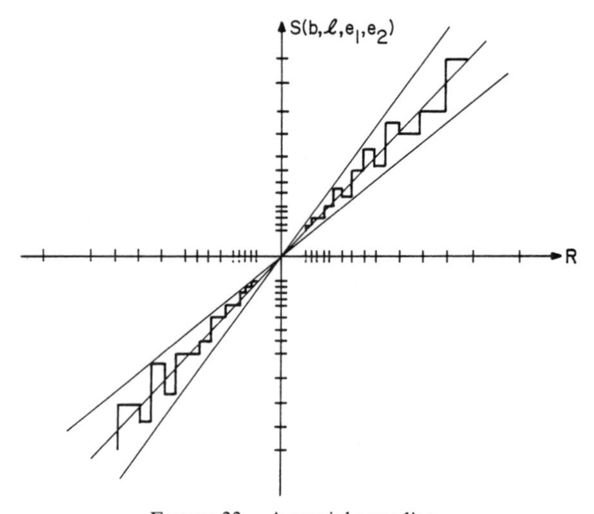


FIGURE 33. A special rounding.

Every rounding corresponds to a step function that crosses the graph of the identity mapping for all values in $S(b, l, e1, e2)$. This implies, in particular, that within an interval around zero the rounding cannot be described by Eq. (7). Compare Figs. 32 and 33.

Although Eq. (7) does not necessarily describe a monotone mapping, we show in the following theorem that within the interval $b^{e1-1} \leq |x| \leq B$, every monotone rounding has this property (7).

**Theorem 5.7:** Let $S^* = S^*(b, l, e1, e2)$ be a floating-point system and $\square : \boldsymbol{R}^* \to S^*$ any monotone rounding. Further, let $\delta(x) := x - \square x$ be the rounding error and $\varepsilon_1 := \delta(x)/x$ and $\varepsilon_2 := \delta(x)/\square x$ the relative rounding errors. Then

$$\bigwedge_{x \in \mathbf{R}} (b^{e1-1} \le |x| \le B \Rightarrow \Box x = x(1-\varepsilon_1) \quad \text{with} \quad |\varepsilon_1| < \varepsilon^*$$

$$\wedge\, x = \Box x(1-\varepsilon_2) \quad \text{with} \quad |\varepsilon_2| < \varepsilon^*$$

$$\wedge\, |x - \Box x| < \varepsilon^* |x| \wedge |x - \Box x| < \varepsilon^* |\Box x|).$$

Here $\varepsilon^*$ is independent of $x$, and

$$\varepsilon^* = \begin{cases} \frac{1}{2} b^{1-l} & \text{for} \quad \Box = \bigcirc \\ b^{1-l} & \text{for} \quad \Box \neq \bigcirc. \end{cases}$$

Here $\bigcirc$ denotes the rounding to the nearest floating-point number of $S^*$, and $B = 0.(b-1)(b-1)\ldots(b-1) \cdot b^{e2}$.

*Proof*: We consider the formula that includes $\varepsilon_1$. The formula with $\varepsilon_2$ can be dealt with analogously. We have $x = \Box x + \delta(x) = \Box x + \varepsilon_1 x \Rightarrow \Box x = x(1-\varepsilon_1)$. With $x = m \cdot b^e$ we obtain using (5) that $b^{-1} \cdot b^e \le |x| < b^e$. Since $|\delta(x)| \le b^{-l} \cdot b^e$, we get

$$|\varepsilon_1| = |\delta(x)|/|x| \le (b^{-l} \cdot b^e)/(b^{-1} \cdot b^e) = b^{1-l}.$$

It is easy to see that actually $|\varepsilon_1| < b^{1-l}$: *If* $x = b^{-1} \cdot b^e$, then $x$ is a screen-point, and because of (R1), $\delta(x) = 0$. If $\delta(x) = b^{-l} \cdot b^e$, then $|x| > b^{-1} \cdot b^e$. The case $\Box = \bigcirc$ can be proved analogously. ■

Theorem 5.7 furnishes error bounds for a rounding only for $x \in [b^{e1-1}, B]$. For $|x| < b^{e1-1}$, the relative error $\varepsilon_1$ can—depending on the definition of the rounding function—be identically unity or even tend to infinity. For $|x| > B$, the relative error $\varepsilon_1$ tends to unity as $x$ goes to infinity.

Figure 34 shows the relative error of the rounding illustrated in Fig. 32. For $b^{e1-1} \le |x| \le B$, we have $|\varepsilon_1| < \varepsilon^* = \frac{1}{2} \cdot b^{1-l}$. Within this region the relative error can only be made smaller by enlarging the number $l$ of digits in the mantissa.

Figure 34 also shows that it is difficult (resp. impossible) to approximate real numbers well on the computer when they lie to the left (resp. right) of this region. For $|x| < b^{e1-1}$, the relative error is unity. For $|x| \ge B$, it tends to unity as $x$ goes to infinity. The only way to decrease the size of these outer regions is to decrease $e1$ and to increase $e2$. This requires enlarging the number of digits used for the representation of the exponent.

We summarize these results in the following two rules:

(a) In order to approximate better within the region $b^{e1-1} \le |x| \le B$, the number of digits used for the representation of the mantissa has to be enlarged.

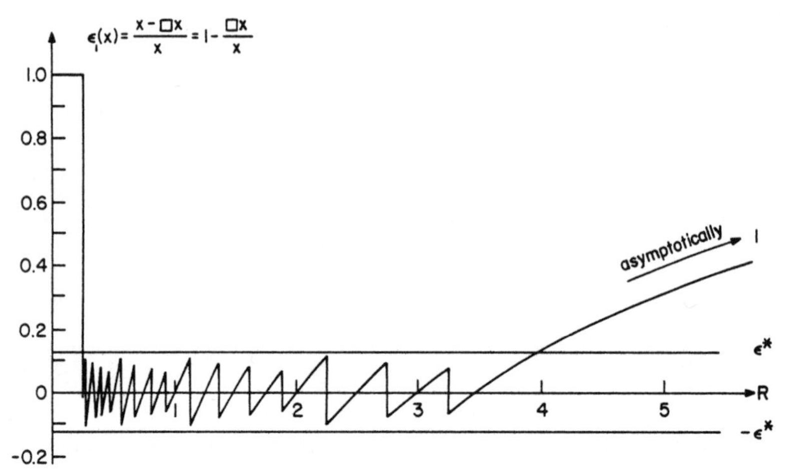


FIGURE 34. Relative rounding error.

(b) In order to enlarge the region $b^{e1-1} \leq |x| \leq B$ of good approximation, the numbers of digits used for the representation of the exponent has to be enlarged.

If a number $|x| \in (0, b^{e1-1})$ occurs in a virtual sense on a computer, one speaks of *exponent underflow* or simply of *underflow*. Correspondingly, one speaks of *exponent overflow* or simply of *overflow* if $|x| > B$. With these concepts, the condition $b^{e1-1} \leq |x| \leq B$, which appears in Theorem 5.7, is usually expressed by saying: *If no overflow and no underflow occurs, then....*

## 3. FLOATING-POINT OPERATIONS

We now turn to consideration of arithmetic operations in floating-point systems followed by a treatment of errors for such operations.

In Chapters 3 and 4 we established that the arithmetic in the subsets under $D$ and $S$ and in all rows of Fig. 1 can be defined by semimorphisms. We recall that if $M$ is any ringoid or vectoid and $N \subseteq M$ a symmetric lower or upper screen (or both) of $M$, then a semimorphism $\square$ is defined by the following formulas:

(R1) $\bigwedge_{x \in N} \square x = x,$

(R2) $\bigwedge_{x,y \in M} (x \leq y \Rightarrow \square x \leq \square y),$

(R4) $\bigwedge_{x \in M} (\square(-x) = -(\square x)),$

(RG) $\bigwedge_{x,y \in N} (x \boxast y := \square(x * y)).$

In general the operations defined by (RG) are not associative, nor are addition and multiplication distributive. We show this by means of a few simple examples. Consider the floating-point system $S = S(10, 1, -1, 1)$. Let $x, y, z \in S$.

1. Consider the operation $x \overset{\triangle}{+} y := \triangle(x + y)$. For $x := 0.6, y := 0.5, z := 0.4$, we obtain

$$(x \overset{\triangle}{+} y) \overset{\triangle}{+} z = (\triangle 1.1) \overset{\triangle}{+} 0.4 = 0.2 \cdot 10 \overset{\triangle}{+} 0.4 = \triangle(0.24 \cdot 10) = 0.3 \cdot 10.$$
$$x \overset{\triangle}{+} (y \overset{\triangle}{+} z) = 0.6 \overset{\triangle}{+} 0.9 = \triangle 1.5 = 0.2 \cdot 10,$$

i.e., $(x \overset{\triangle}{+} y) \overset{\triangle}{+} z \neq x \overset{\triangle}{+} (y \overset{\triangle}{+} z)$.

2. Consider the operation $x \overset{\triangledown}{\cdot} y := \triangledown(x \cdot y)$. For $x := 0.3, y := 0.4, z = 0.4$, we obtain

$$(x \overset{\triangledown}{\cdot} y) \overset{\triangledown}{\cdot} z = (\triangledown 0.12) \overset{\triangledown}{\cdot} 0.4 = 0.1 \overset{\triangledown}{\cdot} 0.4 = 0.4 \cdot 10^{-1},$$
$$x \overset{\triangledown}{\cdot} (y \overset{\triangledown}{\cdot} z) = 0.3 \overset{\triangledown}{\cdot} (\triangledown 0.16) = 0.3 \overset{\triangledown}{\cdot} 0.1 = 0.3 \cdot 10^{-1},$$

i.e., $(x \overset{\triangledown}{\cdot} y) \overset{\triangledown}{\cdot} z \neq x \overset{\triangledown}{\cdot} (y \overset{\triangledown}{\cdot} z)$.

The same effect can be exhibited for nondirected roundings. For instance, let $\square : \boldsymbol{R}^* \to S^*$ be the rounding to the nearest number of the screen with the property that the midpoint between two neighboring screen numbers is rounded downwardly. Then

3. With $x := 0.7, y := 0.7, z := 0.9$, we obtain for the operation $x \boxplus y := \square(x + y)$

$$(x \boxplus y) \boxplus z = \square(1.4) \boxplus 0.9 = 0.1 \cdot 10 \boxplus 0.9 = \square 0.19 \cdot 10 = 0.2 \cdot 10,$$
$$x \boxplus (y \boxplus z) = 0.7 \boxplus (\square 1.6) = 0.7 \boxplus 0.2 \cdot 10 = \square 0.27 \cdot 10 = 0.3 \cdot 10,$$

i.e., $(x \boxplus y) \boxplus z \neq x \boxplus (y \boxplus z)$.

4. With $x := 0.3, y := 0.4, z := 0.4$, we obtain for the operations $x \boxdot y := \square(x \cdot y)$

$$(x \boxdot y) \boxdot z = (\square 0.12) \boxdot 0.4 = 0.1 \boxdot 0.4 = \square 0.04 = 0.4 \cdot 10^{-1},$$
$$x \boxdot (y \boxdot z) = 0.3 \boxdot (\square 0.16) = 0.3 \boxdot 0.2 = \square 0.06 = 0.6 \cdot 10^{-1},$$

i.e., $(x \boxdot y) \boxdot z \neq x \boxdot (y \boxdot z)$. The distributive law doesn't hold either:

5. For $x := 0.3, y := 0.7, z := 0.9$, we obtain

$$x \boxdot (y \boxplus z) = 0.3 \boxdot (\square 1.6) = 0.3 \boxdot 0.2 \cdot 10 = 0.6,$$
$$x \boxdot y \boxplus x \boxdot z = (\square 0.21) \boxplus (\square 0.27) = 0.2 \boxplus 0.3 = 0.5,$$

i.e, $x \boxdot (y \boxplus x) \neq x \boxdot y \boxplus x \boxdot z$.

If the operations in the subset $S^*$ of $\boldsymbol{R}^*$ are defined by (RG), employing a monotone rounding, then error bounds may be obtained. These are given in the following theorem whose proof follows directly from Theorem 5.7.

**Theorem 5.8:** Let $S^* = S^*(b, l, e1, e2)$ be a floating-point system, $\square : \boldsymbol{R}^* \to S^*$ be a monotone rounding, and let operations in $S^*$ be defined by

$$\text{(RG)} \quad \bigwedge_{x,y \in S^*} x \boxast y := \square(x * y) \qquad \text{for all} \quad * \in \{+, -, \cdot, /\}.$$

If $\delta := x * y - x \boxast y$ denotes the absolute error and $\varepsilon_1 := \delta/(x * y)$ and $\varepsilon_2 := \delta/(x \boxast y)$ relative errors, then for all operations $* \in \{+, -, \cdot, /\}$, the following error relations hold:

$$\bigwedge_{x,y \in S^*} (b^{e1-1} \leq |x * y| \leq B \Rightarrow x \boxast y = (x * y)(1 - \varepsilon_1) \qquad \text{with} \quad |\varepsilon_1| < \varepsilon^*$$

$$\wedge\, x * y = (x \boxast y)(1 - \varepsilon_2) \qquad \text{with} \quad |\varepsilon_2| < \varepsilon^*$$

$$\wedge\, |x * y - (x \boxast y)| < \varepsilon^* |x * y|$$

$$\wedge\, |x * y - (x \boxast y)| < \varepsilon^* |x \boxast y|).$$

Here $\varepsilon^*$ and $B$ are the constants defined in Theorem 5.7. ■

In Theorems 5.7 and 5.8, as well as several of the following theorems, the condition on the left-hand side of the $\Rightarrow$ sign means that neither underflow nor overflow occurs. The error estimates in these theorems are valid in this situation.

Note that the relations on the right-hand side of the $\Rightarrow$ sign in Theorem 5.8 are precise quantitative statements about the errors in arithmetic. Contrast this with the usual error estimates in numerical analysis. Those estimates have the same form as the bounds here, but are only qualitative statements.

Theorem 5.8 is valid for all roundings $\square \in \{\bigtriangledown, \bigtriangleup, \square_\mu, \mu = 0(1)b\}$ and in particular for all semimorphisms.

Then as a result of the horizontal definition of the arithmetic in rows II, III, VII, VIII, and IX of Fig. 1, similar error relations can also be derived. As an example, we consider the case of matrix operations in row III. The results are summarized in the following theorem. (See also Theorem 3.10.)

**Theorem 5.9:** Let $S^* = S^*(b, l, e1, e2)$ be a floating-point system, $\square : \boldsymbol{R}^* \to S^*$ a monotone rounding, and $M_n\boldsymbol{R}$ the set of $n \times n$ matrices over $\boldsymbol{R}$. We define a rounding $\square : M_n\boldsymbol{R}^* \to M_nS^*$ by

$$\bigwedge_{X = (x_{ij}) \in M_n\boldsymbol{R}^*} \square X := (\square x_{ij})$$

and operations $\boxast$ by

$$\text{(RG)} \quad \bigwedge_{X,Y \in M_nS^*} X \boxast Y := \square(X * Y), \qquad * \in \{+, \cdot\}.$$

Then

$$\bigwedge_{X=(x_{ij})\in M_nR}\Big(\bigwedge_{i,j} b^{e1-1}\le |x_{ij}|\le B \Rightarrow \square X=(x_{ij}(1-\varepsilon_{ij}^1)) \quad \text{with } |\varepsilon_{ij}^1|<\varepsilon^*$$
$$\wedge\, X=(\square x_{ij}(1-\varepsilon_{ij}^2)) \quad \text{with } |\varepsilon_{ij}^2|<\varepsilon^*$$
$$\wedge\, |X-\square X|<\varepsilon^*|X|$$
$$\wedge\, |X-\square X|<\varepsilon^*|\square X|\Big).$$

With $Z := (z_{ij}) := X * Y$, $* \in \{+, \cdot\}$, we obtain the following error relations for the operations:

$$\bigwedge_{X,Y\in M_nS}\Big(\bigwedge_{i,j} b^{e1-1}\le |z_{ij}|\le B \Rightarrow X \boxast Y=(z_{ij}(1-\varepsilon_{ij}^1)) \quad \text{with } |\varepsilon_{ij}^1|<\varepsilon^*$$
$$\wedge\, X*Y=(\square z_{ij}(1-\varepsilon_{ij}^2)) \quad \text{with } |\varepsilon_{ij}^2|<\varepsilon^*$$
$$\wedge\, |X*Y-X\boxast Y|<\varepsilon^*|X*Y|$$
$$\wedge\, |X*Y-X\boxast Y|<\varepsilon^*|X\boxast Y|\Big).$$

Here all absolute values are to be taken componentwise. $\varepsilon^*$ and $B$ are defined as in Theorem 5.7. ■

As before, we may note now that Theorem 5.9 holds for all roundings $\square \in \{\nabla, \Delta, \square_\mu, \mu = 0(1)b\}$ and in particular for all semimorphisms. It is clear that corresponding formulas hold for the matrix-vector multiplication. Corresponding theorems can also be derived for all complex operations and for the complex matrix and vector operations. All these results lead to error relations that are especially simple by comparison with error estimates derived on the basis of the vertical definition of arithmetic. In order to illustrate this, we derive the corresponding error relations for matrix multiplication when the latter is defined by the vertical method.

The essence of the point to be made is illustrated by a comparison of the error relations for the scalar product defined by the horizontal and the vertical methods. Actually, the error relations for matrix multiplication derived in Theorem 5.9 deal with such scalar products if the former are written componentwise. If $x = (x_i)$ and $y = (y_i)$ with $x_i, y_i \in S^*, i = 1(1)n$, the horizontal method simply leads to the error relations

$$\square\Big(\sum_{i=1}^n x_i y_i\Big) = (1-\varepsilon_1)\sum_{i=1}^n x_i y_i \qquad \text{with} \quad |\varepsilon_1|<\varepsilon^*, \tag{1}$$

$$\sum_{i=1}^n x_i y_i = (1-\varepsilon_2)\Big(\square \sum_{i=1}^n x_i y_i\Big) \qquad \text{with} \quad |\varepsilon_2|<\varepsilon^*, \tag{2}$$

and to the error bounds

$$\left|\sum_{i=1}^{n} x_i y_i - \square \sum_{i=1}^{n} x_i y_i\right| < \varepsilon^* \left|\sum_{i=1}^{n} x_i y_i\right|, \tag{3}$$

$$\left|\sum_{i=1}^{n} x_i y_i - \square \sum_{i=1}^{n} x_i y_i\right| < \varepsilon^* \left|\square \sum_{i=1}^{n} x_i y_i\right|, \tag{4}$$

provided that neither underflow nor overflow occurs. These relations hold for all monotone roundings and in particular for all $\square \in \{\nabla, \Delta, \square_\mu, \mu = 0(1)b\}$. $\varepsilon^*$ is defined as in Theorem 5.7.

If the floating-point matrix product is defined by the vertical method, the scalar products are defined by the formula

$$\boxed{\sum_{i=1}^{n}} x_i \boxdot y_i = (x_1 \boxdot y_1) \boxplus (x_2 \boxdot y_2) \boxplus \cdots \boxplus (x_n \boxdot y_n).$$

If we apply Theorem 5.8 to this relation, we obtain the following expression in the case $n = 4$:

$$\begin{aligned}
(x_1 \boxdot y_1) &\boxplus (x_2 \boxdot y_2) \boxplus (x_3 \boxdot y_3) \boxplus (x_4 \boxdot y_4) \\
&= (((x_1 y_1(1 - \varepsilon_1) + x_2 y_2(1 - \varepsilon_2))(1 - \varepsilon_5) \\
&\quad + x_3 y_3(1 - \varepsilon_3))(1 - \varepsilon_6) + x_4 y_4(1 - \varepsilon_4))(1 - \varepsilon_7) \\
&= x_1 y_1(1 - \varepsilon_1)(1 - \varepsilon_5)(1 - \varepsilon_6)(1 - \varepsilon_7) \\
&\quad + x_2 y_2(1 - \varepsilon_2)(1 - \varepsilon_5)(1 - \varepsilon_6)(1 - \varepsilon_7) \\
&\quad + x_3 y_3(1 - \varepsilon_3)(1 - \varepsilon_6)(1 - \varepsilon_7) \\
&\quad + x_4 y_4(1 - \varepsilon_4)(1 - \varepsilon_7).
\end{aligned} \tag{5}$$

Here $|\varepsilon_i| < \varepsilon^*$ for all $i = 1(1)7$.

The many epsilons occurring in this expression make it more complicated than the simple formulas (1) and (2), which occur in Theorem 5.8. We may expect that an error analysis of numerical algorithms is more troublesome if it is based on (5) instead of (1) and (2). Moreover, in general (1) and (2) are more accurate than (5). We now estimate the error in (5) as a means of simplifying it. This requires estimating products of the form

$$\prod_{i=1}^{n} (1 - \varepsilon_i) \qquad \text{with} \quad |\varepsilon_i| < \varepsilon^*.$$

If we apply Bernoulli's inequality [53],

$$1 + nz \leq (1 + z)^n \leq 1/(1 - nz), \qquad n \geq 2, \qquad -1 < z < 1/n,$$

we obtain

$$1 - n\varepsilon^* \leq (1 - \varepsilon^*)^n < \prod_{i=1}^{n} (1 - \varepsilon_i) < (1 + \varepsilon^*)^n \leq \frac{1}{1 - n\varepsilon^*}.$$

Using this inequality and appealing to (5), we obtain

$$\sum_{i=1}^{n} x_i y_i (1 - n\varepsilon^*) < \boxed{\sum_{i=1}^{n}} x_i \boxdot y_i < \sum_{i=1}^{n} x_i y_i \frac{1}{1 - n\varepsilon^*}.$$

This leads to the error bounds

$$\left| \sum_{i=1}^{n} x_i y_i - \boxed{\sum_{i=1}^{n}} x_i \boxdot y_i \right| < n\varepsilon^* \left| \sum_{i=1}^{n} x_i y_i \right| \tag{6}$$

and

$$\left| \sum_{i=1}^{n} x_i y_i - \boxed{\sum_{i=1}^{n}} x_i \boxdot y_i \right| < n\varepsilon^* \left| \boxed{\sum_{i=1}^{n}} x_i y_i \right| \tag{7}$$

whenever $n\varepsilon^* < 1$.

These bounds for the relative error are $n$ times as large as those we obtained for the horizontal method. For matrix multiplication defined by the vertical method, error bounds may also be obtained by this process. Letting $Z := (z_{ij}) := X \cdot Y$, these bounds are

$$\bigwedge_{X,Y \in M_n S^*} \left( \bigwedge_{i,j} b^{e1-1} \leq |z_{ij}| \leq B \Rightarrow |X \cdot Y - X \boxdot Y| \leq n\varepsilon^* |X \cdot Y| \right.$$
$$\left. \wedge |X \cdot Y - X \boxdot Y| \leq n\varepsilon^* |X \boxdot Y| \right)$$

if $n\varepsilon^* < 1$.

As before, these bounds are also $n$ times as large as those for the horizontal method. The more accurate formulas (5) are more complicated than the corresponding formulas for the horizontal method (Theorem 5.9). These two properties are reproduced in the error analysis of many algorithms such as the Gauss algorithm for systems of linear equations.

The error relations established so far can also be extended to the interval rows of Fig. 1. Of course, such an extension would be mainly of theoretical interest since a basic principle of interval mathematics is that the computer controls the rounding error automatically. Nevertheless, as an example of interval arithmetic error relations, we give Theorem 5.10 below. Since for all interval rows in Fig. 1, arithmetic is defined by the horizontal method, the validity of this theorem follows easily by employing Theorems 5.7 and 5.8.

In order to state Theorem 5.10, we require a few concepts of interval mathematics. See [3]. The distance between the two intervals $A = [a_1, a_2]$,

$B = [b_1, b_2] \in \boldsymbol{IR}$ is defined by

$$q(A,B) := \max\{|a_1 - b_1|, |a_2 - b_2|\},$$

while the absolute value of the interval $A$ is given by

$$|A| := q(A,[0,0]) = \max(|a_1|,|a_2|) = \max_{a\in A} |a|.$$

If $\boldsymbol{A} = (A_{ij}), \boldsymbol{B} = (B_{ij}) \in M_n\boldsymbol{IR}$ are matrices with interval components, the distance matrix and the absolute value matrix are defined by

$$q(\boldsymbol{A},\boldsymbol{B}) := (q(A_{ij}, B_{ij})), \qquad |\boldsymbol{A}| := (|A_{ij}|).$$

**Theorem 5.10:** Let $S^* = S^*(b, l, e1, e2)$ be a floating-point system, and let $\Diamond : \overline{\boldsymbol{IR}^*} \to \overline{\boldsymbol{IS}^*}$ be the monotone upwardly directed rounding. Then

$$\bigwedge_{X=[x_1,x_2]\in \boldsymbol{IR}^*} (b^{e1-1} \le |x_1|,|x_2| \le B$$
$$\Rightarrow \Diamond X = [\nabla x_1, \triangle x_2] = [x_1(1-\varepsilon_1), x_2(1-\varepsilon_2)] \quad \text{with} \quad |\varepsilon_1|,|\varepsilon_2| < \overline{\varepsilon} = b^{1-l}$$
$$\wedge\ X = [x_1, x_2] = [\nabla x_1(1-\varepsilon_1), \triangle x_2(1-\varepsilon_2)] \quad \text{with} \quad |\varepsilon_1|,|\varepsilon_2| < \overline{\varepsilon} = b^{1-l}$$
$$\wedge\ q(X,\Diamond X) < \overline{\varepsilon}|X| \ \wedge\ q(X,\Diamond X) < \overline{\varepsilon}|\Diamond X|).$$

If $X, Y \in \boldsymbol{IS}^*$ and $Z := [z_1, z_2] := X \boxast Y, * \in \{+, \cdot, /\}$, then

$$\bigwedge_{X,Y\in \boldsymbol{IS}^*} (b^{e1-1} \le |z_1|,|z_2| \le B$$
$$\Rightarrow X \circledast Y := \Diamond\ (X \boxast Y) = [z_1(1-\varepsilon_1), z_2(1-\varepsilon_2)] \quad \text{with} \quad |\varepsilon_1|,|\varepsilon_2| < \overline{\varepsilon}$$
$$\wedge\ X \boxast Y = [\nabla z_1(1-\varepsilon_1), \triangle z_2(1-\varepsilon_2)] \quad \text{with} \quad |\varepsilon_1|,|\varepsilon_2| < \overline{\varepsilon}$$
$$\wedge\, q(X \boxast Y, X \circledast Y) < \overline{\varepsilon}|X \boxast Y| \wedge q(X \boxast Y, X \circledast Y) < \overline{\varepsilon}|X \circledast Y|).$$

If, moreover, a rounding $\Diamond : M_n\boldsymbol{IR}^* \to M_n\boldsymbol{IS}^*$ and operations $\circledast$ in $M_n\boldsymbol{IS}^*$ are defined by

$$\bigwedge_{\boldsymbol{X}=(X_{ij})\in M_n\boldsymbol{IR}^*} \Diamond \boldsymbol{X} := (\Diamond X_{ij}),$$

$$\text{(RG)} \qquad \bigwedge_{\boldsymbol{X},\boldsymbol{Y}\in M_n\boldsymbol{IS}^*} \boldsymbol{X} \circledast \boldsymbol{Y} := \quad (\boldsymbol{X} \boxast \boldsymbol{Y}), \qquad * \in \{+, \cdot\},$$

then

$$\bigwedge_{X=([x_{ij}^1,x_{ij}^2])\in M_nIR^*}\left(\bigwedge_{i,j} b^{e1-1} \le |x_{ij}^1|, |x_{ij}^2| \le B\right.$$

$$\Rightarrow \Diamond X = ([x_{ij}^1(1-\varepsilon_{ij}^1), x_{ij}^2(1-\varepsilon_{ij}^2)]) \quad \text{with} \quad |\varepsilon_{ij}^1|, |\varepsilon_{ij}^2| < \bar{\varepsilon} = b^{1-l}$$

$$\wedge X = \Diamond([x_{ij}^1(1-\varepsilon_{ij}^1), x_{ij}^2(1-\varepsilon_{ij}^2)]) \quad \text{with} \quad |\varepsilon_{ij}^1|, |\varepsilon_{ij}^2| < \bar{\varepsilon} = b^{1-l}$$

$$\left.\wedge\, q(X, \Diamond X) < \bar{\varepsilon}|X| \wedge q(X, \Diamond X) < \bar{\varepsilon}|\Diamond X|\right).$$

If $X$, $Y \in M_nIS^*$, and $Z := ([z_{ij}^1, z_{ij}^2]) := X \boxast Y$, $* \in \{+, \cdot\}$, then

$$\bigwedge_{X,Y\in M_nIS^*}\left(\bigwedge_{i,j} b^{e1-1} \le |z_{ij}^1|, |z_{ij}^2| \le B\right.$$

$$\Rightarrow X \circledast Y = ([z_{ij}^1(1-\varepsilon_{ij}^1), z_{ij}^2(1-\varepsilon_{ij}^2)]) \quad \text{with} \quad |\varepsilon_{ij}^1|, |\varepsilon_{ij}^2| < \bar{\varepsilon}$$

$$\wedge X \boxast Y = \Diamond([z_{ij}^1(1-\varepsilon_{ij}^1), z_{ij}^2(1-\varepsilon_{ij}^2)]) \quad \text{with} \quad |\varepsilon_{ij}^1|, |\varepsilon_{ij}^2| < \bar{\varepsilon}$$

$$\left.\wedge\, q(X \boxast Y, X \circledast Y) < \bar{\varepsilon}|X \boxast Y| \wedge q(X \boxast Y, X \circledast Y) < \bar{\varepsilon}|X \circledast Y|\right).$$

*Proof*: The proof is simple. We only indicate it in the case $IR$. The matrix properties then follow componentwise.

$$\begin{aligned} q(X, \Diamond X) &= \max\{|x_1 - \nabla x_1|, |x_2 - \Delta x_2|\} \\ &= \max\{|\varepsilon_1 x_1|, |\varepsilon_2 x_2|\} < \bar{\varepsilon}|X|, \\ q(X, \Diamond X) &= \max\{|\nabla x_1(1-\varepsilon_1) - \nabla x_1|, |\Delta x_2(1-\varepsilon_2) - \Delta x_2|\} \\ &= \max\{|\varepsilon_1 \nabla x_1|, |\varepsilon_2 \Delta x_2|\} < \bar{\varepsilon}|\Diamond X|. \end{aligned}$$

Using (RG), (R1), and (R2), we observe even more importantly than these error bounds that *semimorphisms provide maximal accuracy* in the sense that there is no element of the screen $N$ between the results of the operation $*$ and of its approximation $\boxast$.

## 4. EXTENDED FLOATING-POINT ARITHMETIC

We now consider the question of the additive inverse on screens. This will lead to the concept of extended floating-point systems and several theoretical results concerning ringoids. We begin by noting that the property, Theorem 2.2(t), assures us that in a linearly ordered ringoid $R$ we have

$$\bigwedge_{a\in R} a + (-a) = o,$$

i.e., for all $a \in R$ the element $-a$ is inverse to $a$. For applications, the most important linearly ordered ringoids are those on a screen of the real numbers

**R**. Conditions under which addition on a screen ringoid of the real numbers has a unique inverse are furnished by the following theorem.

**Theorem 5.11:** Let $S \subseteq \boldsymbol{R}$ be a symmetric screen of $\boldsymbol{R}$ and $\square : \boldsymbol{R} \to S$ be a semimorphism. If $\varepsilon > 0$ denotes the least distance of neighboring screenpoints of $S$, then for all $a \in S$ the element $b = -a$ is the unique inverse of $a$ if and only if

$$\square^{-1} 0 \subseteq (-\varepsilon, \varepsilon).^{\dagger} \tag{1}$$

*Proof*: Since $S$ is a symmetric screen, it contains $-a$ if it contains $a$. The definition of addition in $S$ implies that $-a$ is inverse to $a$. There remains to show uniqueness.

(a) (1) *is sufficient*: Let us assume that $b \neq -a$ is another inverse of $a$. Then $a \boxplus b = 0$ and $a + b \in \square^{-1} 0$. By (1), therefore,

$$-\varepsilon < a + b < \varepsilon. \tag{2}$$

Since $-b \neq a$, we obtain, on the other hand, $|a - (-b)| = |a + b| \geq \varepsilon$, using the hypothesis concerning $\varepsilon$. This contradicts (2), demonstrating the sufficiency.

(b) (1) *is necessary*: Since $\square$ is a rounding, we have $0 \in \square^{-1} 0$. Since $\square$ is a monotone, $\square^{-1} 0$ is connected, and because of the antisymmetry of $\square$, the set $\square^{-1} 0$ is symmetric with respect to zero. Suppose that (1) were false. Then $\square^{-1} 0 \supset (-\varepsilon, \varepsilon)$. If we now consider two elements $a, b \in S$ with distance $\varepsilon$, we obtain $|a - b| = \varepsilon$ and therefore $a + (-b) \in \square^{-1} 0$ or $a \boxplus (-b) = 0$, i.e., $-b$ is inverse to $a$ and $-b \neq a$. ■

Note incidentally that the hypothesis $\varepsilon > 0$ in Theorem 5.11 assures us that the screen is discrete.

We recall that $\{S, \boxplus, \boxdot\}$ is a ringoid (cf. Theorem 3.8) and that in general it is the four properties (D5a)–(D5d) that furnish the uniqueness of the minus operator. However, under the assumptions of the theorem, the uniqueness of the minus operator in $S$ is a consequence of (D5a) alone.

A simpler sufficient condition than (1) for the existence of a unique additive inverse on a symmetric screen of the real numbers is $\square^{-1} 0 = 0$. This condition, however, precludes some customary and convenient underflow treatment.

If $\boldsymbol{R}$, $S$, $\square$, and $\varepsilon$ are defined as in Theorem 5.11, then condition (1) is automatically fulfilled if $\varepsilon$ itself is a screenpoint. This follows from the monotonicity of the rounding since assuming a zero value outside the interval $(-\varepsilon, \varepsilon)$ would then be a contradiction. Therefore, in any such screen that contains $\varepsilon$ as an element, there exists a unique additive inverse for all $a \in S$.

† $(-\varepsilon, \varepsilon)$ denotes the open interval in $\boldsymbol{R}$ between the bounds $-\varepsilon$ and $\varepsilon$.

Such screens often occur in practice. For instance, the whole numbers or the so-called fixed-point numbers as subsets of the real numbers have this property. The floating-point numbers also are sometimes extended into a screen with this property. This can be accomplished if in the case of $e = e1$, nonnormalized mantissas are admitted in addition to the usual floating-point numbers of $S(b, l, e1, e2)$ (see Definition 5.6). In this case $\varepsilon$ itself becomes an element of the screen, and we obtain a unique additive inverse. This extended set of floating-point numbers is specified by the following definition.

**Definition 5.12:** Let $S(b, l, e1, e2)$ be a floating-point system, and let

$$F = F(b, l, e1) := \left\{ x = *mb^{e1} \,|\, * \in \{+, -\}, \right.$$

$$\left. m = \sum_{i=1}^{l} x[i]b^{-i}, x[i] \in \{0, 1, \ldots, b-1\} \right\}.$$

Then the set

$$E = E(b, l, e1, e2) := S(b, l, e1, e2) \cup F(b, l, e1)$$

is called an *extended floating-point system.* ■

In extended floating-point systems, every element has a unique additive inverse. This makes it possible to define a minus operator and subtraction solely through use of addition in a manner similar to the way these concepts are defined in an additive group. More generally, therefore, let us consider ringoids with unique additive inverse. We may ask under which circumstances the minus operator and subtraction, defined by means of the additive inverse, on the one hand, and intrinsically for the ringoid, on the other, are identical. The following theorem shows that this actually is the case under quite natural hypotheses.

**Theorem 5.13:** Let $\{R, +, \cdot\}$ be a ringoid, and $\ominus : R \to R$ be an operator with the property that for all $x \in R$, the element $y = \ominus x$ is the unique additive inverse of $x$, i.e.,

(a) $\bigwedge_{x \in R} x + \ominus x = o$ (existence of a unique inverse).

Further, let $\ominus$ have the properties

(b) $\bigwedge_{x,y \in R} \ominus(xy) = (\ominus x)y = x(\ominus y)$ (multiplicative),

(c) $\bigwedge_{x,y \in R} \ominus(x + y) = (\ominus x) + \ominus y)$ (additive).

Then $\ominus$ is identical with the minus operator of the ringoid $\{R, +, \cdot\}$, i.e., $\bigwedge_{x \in R} \ominus x = -x$.

*Proof*: We prove the theorem by showing that the following three properties hold:

1. $\ominus(\ominus x) = x$.
2. $\ominus x = (\ominus e) \cdot x$.
3. $\ominus e$ fulfills (D5a–(D5d) in $R$.

1: Applying (a) to $\ominus x$ and then to $x$, we obtain

$$\ominus x + (\ominus(\ominus x)) = o \wedge x + (\ominus x) = \ominus x + x = o \Rightarrow x = \ominus(\ominus x),$$

because of the uniqueness of the inverse.

2: (b) $\Rightarrow_{x=e} \ominus y = (\ominus e) \cdot y$.

3: (D5b): $(\ominus e)(\ominus e) =_{(b)} \ominus(\ominus e) =_1 \cdot e$.

(D5a) follows from (a) for $x = e$, (D5c) from (b), and (D5d) from (c). ■

Ringoids in which an additive inverse exists and is unique, such as the case of an extended floating-point system, carry several theoretical advantages. The reason for this, we shall see, stems from the fact that the uniqueness of the element $-e$ is already a consequence of (D5a) alone. Using this property, we may easily simplify the proof of several theorems previously stated, for instance, that

(a) the power set of a ringoid is a ringoid (see Theorem 2.4),
(b) the set of matrices over a ringoid is a ringoid (see Theorem 2.5),
(c) the complexification of a ringoid leads to a ringoid (Theorem 2.6),
(d) the intervals over a ringoid form a ringoid (Theorem 4.4).

That semimorphisms of these structures also generate ringoids is the subject of the following generalization of Theorem 5.11.

**Theorem 5.14:** Let $\{R, +, \cdot\}$ be a ringoid, $\{R, \leq\}$ a complete lattice, $\{S, \leq\}$ a symmetric lower or upper screen (or both) of $\{R, \leq\}$, and $\square : R \to S$ a semimorphism. Further, let $\{M, \leq\}$ be an ordered set and $q: R \times R \to M$ a function with the property

(Q1) $\bigwedge_{a,b,c \in R} q(a + c, b + c) = q(a, b)$.

Further, let the screen $S$ and the rounding $\square$ have the properties

(Q2) $\bigwedge_{a \in R} (\square a = o \Rightarrow q(a, o) < \varepsilon)$,

(Q3) $\bigwedge_{a,b \in S} (a \neq b \Rightarrow q(a, b) \geq \varepsilon)$.

Then $-e$ is the only element in $S$ that fulfills (D5a), and $\{S, \boxplus, \boxdot\}$ is a ringoid.

*Proof*: Lemma 3.4 provides the properties (D1)–(D5) in $S$. There remains to prove (D6).

(D6): Let $x \neq -e$ be any element of $S$ with the property $x \boxplus e = o$. Then

$$\square(x + e) = o \underset{(Q2)}{\Rightarrow} q(x + e, o) < \varepsilon. \tag{3}$$

On the other hand, we obtain

$$x \neq -e \underset{(Q3)}{\Rightarrow} q(x, -e) \geq \varepsilon \underset{(Q1)}{\Rightarrow} q(x + e, o) \geq \varepsilon. \tag{4}$$

(4) is a contradiction to (3), i.e., there exists no element $x \neq -e$ in $S$ that satisfies (D5a). $-e$ therefore is the only element with this property, and $\{S, \boxplus, \boxdot\}$ is a ringoid. ■

Typically, the function $q$ defined in Theorem 5.14 will be a metric, and the ordered set $\{M, \leq\}$ taken to be the set $\{\mathbf{R}_+, \leq\}$. However, we stress the fact that in the proof of Theorem 5.14, use is made only of the properties (Q1)–(Q3) and the fact that $\{M, \leq\}$ is an ordered set. Then in applications of the theorem, use is permitted of generalized concepts of metric systems, wherein the metric takes values in an ordered or an ordered linear space.

If $q$ is a metric, then $\varepsilon > 0$ by (Q2) and (Q3) replaces the hypothesis of Theorem 5.11 that the screen is discrete. (Q2) is a generalization of (1). The property (Q1) often holds in an additive metric space.

We now briefly consider some simple applications of Theorem 5.14. Let $R$ be chosen to be $\mathbf{R}$, the set of real numbers, and let $S$ be an extended floating-point system with the minimal distance among screenpoints $\varepsilon > 0$.

First, let us consider the ringoid of matrices $\{M_n\mathbf{R}, +, \cdot\}$ and any semimorphism $\square : M_n\mathbf{R} \to M_nS$ defined by

$$\bigwedge_{A = (a_{ij}) \in M_n\mathbf{R}} \square A = (\square a_{ij}),$$

where $\square : \mathbf{R} \to S$ is a monotone and antisymmetric rounding. If $A = (a_{ij})$, $B = (b_{ij}) \in M_n\mathbf{R}$, let $q$ be the function $q(A, B) := (|a_{ij} - b_{ij}|)$. $q$ has the properties (Q1), (Q2), and (Q3) as well. Therefore, $-E$ is the only element in $M_nS$ that fulfills (D5a), and $\{M_nS, \boxplus, \boxdot\}$ is a ringoid.

As a second application, we consider the ringoid $\{\mathbf{C}, +, \cdot\}$ and any semimorphism $\square : \mathbf{C} \to \mathbf{C}S$ defined by

$$\bigwedge_{\alpha = (a_1, a_2) \in \mathbf{C}} \square \alpha := (\square a_1, \square a_2).$$

If $\alpha = (a_1, a_2)$, $\beta = (b_1, b_2) \in \mathbf{C}$, then the function $q(\alpha, \beta) := (|a_1 - b_1|, |a_2 - b_2|)$ has the properties (Q1)–(Q3). Therefore, $-(1, 0) = (-1, 0)$ is the only element in $\mathbf{C}S$ that fulfills (D5a), and $\{\mathbf{C}S, \boxplus, \boxdot\}$ is a ringoid.

For our last application, consider $\{IR, \boxplus, \boxdot\}$, the ringoid of intervals over $R$. We consider the semimorphism $\lozenge: IR \to IS$ defined by the monotone upwardly directed rounding. With $A = [a_1, a_2]$, $B = [b_1, b_2] \in IR$, the function

$$q(A, B) := \max\{|a_1 - b_1|, |a_2 - b_2|\}$$

obviously has the property (Q1). For all $A \in IR$ the rounding $\lozenge$ has the property

$$\lozenge A = \inf(U(A) \cap IS).$$

If $\lozenge A = [0, 0]$, then $A = [0, 0]$ and $q(A, [0, 0]) = 0 < \varepsilon$, which means that (Q2) holds. If $A = [a_1, a_2]$, $B = [b_1, b_2] \in IS$, and $A \neq B$, then $a_1 \neq b_1$ and/or $a_2 \neq b_2$. Therefore,

$$q(A, B) = \max\{|a_1 - b_1|, |a_2 - b_2|\} \geq \varepsilon,$$

i.e., (Q3) holds. Therefore, $-[1, 1]$ is the only element in $IS$ that fulfills (D5a), and $\{IS, \diamondplus, \diamonddot\}$ is a ringoid.

These considerations can easily be extended to interval matrices, complex matrices, complex intervals, and so on if the generalized metric is defined componentwise.

# Chapter 6 / IMPLEMENTATION OF FLOATING-POINT ARITHMETIC ON A COMPUTER

**Summary:** In this chapter we deal with the implementation of arithmetic on a computer, and in particular by means of a floating-point screen $S$. This floating-point implementation will be described for all rows displayed in Fig. 1. We split the implementation into three stages or levels with the details of level 2 based on level 1 and the details of level 3 based in turn on level 2.

Level 1 supplies basic routines such as integer arithmetic, shifting routines, test for zero, the comparison relations, and the function $[x] :=$ entire $(x)$, which determines the greatest integer less than or equal to $x$. All arithmetic routines may be built up out of these basic level 1 routines. Level 1 routines are treated in Section 1. We suppose that these basic functions are more or less made available by the hardware designer. However, since more primitive processors supply only an addition routine, we give some hints how the other basic operations can then be obtained.

In subsequent sections we treat the level 2 routines. These include different floating-point operations for the first row of Fig. 1. These operations are defined by formula (RG) for all roundings of the set $\{\nabla, \Delta, \Box_\mu, \mu = 0(1)b\}$. We show that this process can be split up into several independent routines, which among others consist of approximation of the arithmetic operations, normalization routines, execution of the different roundings, and so on. We discuss most of these routines for two different kinds of accumulators, which we call the *long* and the *short* accumulator. Finally, we discuss and implement two algorithms that calculate arbitrary sums with maximum accuracy. We view them as level 2 operations. They are fundamental for the optimal computation of scalar products, the latter being used in the arithmetic of many of the rows of Fig. 1.

The last section of this chapter contains a brief discussion of the level 3 routines. We simply summarize the definition of the operations in the different rows of Fig. 1 and point out that they all can be performed by using the level 2 operations. These ideas were already extensively studied in preceding chapters. We don't derive algorithms for these operations because they are simple and offer no difficulties in principle provided the level 2 routines are available and the operations are clearly defined.

## 1. THE LEVEL 1 ROUTINES

In this section we list several basic routines that suffice for building up any higher-order arithmetic. We assume that these routines are available in the computer, possibly in the hardware. Since some simpler processors may not provide all the basic routines, we offer several hints that will be useful for obtaining them.

In principle, we assume that the following basic routines are available on a computer:

(1) Test of whether a number $x$ is zero or not.
(2) Comparison relations $=, \leq, \geq, <, >$.

These relations may be realized by

(a) Comparison of the difference $x - y$ of the operands $x$ and $y$ with zero.

(b) Comparison of the operands by the following steps:

($\alpha$) comparison of the signs;
($\beta$) comparison of the digits, successively.

(3) Shifting operations. We presume that the computer is binary-encoded and that shifting operations for one binary digit to the left and the right are available.

If the arithmetic uses a base $b > 2$, further shifting operations for one base $b$ digit to the left and right are useful.

If the computer is byte-oriented, then shifting operations of one byte to the left and to the right should be available.

In the following algorithms, the shifting operations are represented by a multiplication or a division by the base $b$.

(4) Integer addition and subtraction. If necessary, subtraction can be reduced to addition by means of a change of sign.

(5) Integer multiplication. If multiplication is not explicitly available, the following techniques can be used: Let the operands be $x$ and $y$. Let

$$y := y[m]b^m + y[m-1]b^{m-1} + \cdots + y[1]b^1 + y[0]b^0.$$

Then

$$z := x \cdot y := x \cdot y[0] + x \cdot y[1] \cdot b + x \cdot y[2] \cdot b^2 + \cdots + x \cdot y[m] \cdot b^m,$$

or in recursive terminology,

$$z := 0,$$
$$z := z + x \cdot y[i] \cdot b^i \qquad \text{for} \quad i = 0(1)m.$$

Here the intermediate products $x \cdot y[i]$ may be obtained by various methods. For example,

($\alpha$) Add $x$ a total of $y[i]$ times.
($\beta$) Look up the products $x \cdot y[i]$ in a table that has been prepared.

The final additions of the products $x \cdot y[i]$ are performed after appropriate shifting.

(6) Integer division $x/y$. Division is a reversion of multiplication. This leads directly to the algorithms for division. The latter are defined in terms of continued subtraction. Similar to multiplication, division can be performed in two different ways.

($\alpha$) Continued subtraction of $y$ itself shifted appropriately as determined by $x$.

($\beta$) Subtraction of proper multiples of $x$ (also shifted appropriately) found in a prepared table.

Either method can be carried out in the form of a combined addition–subtraction process. If proper positioning is assumed, this means the following: Subtract $y$ from $x$ until what remains becomes negative. The number of subtractions minus one is the first digit of the quotient. Then shift $y$ by one position to the right (or what remained by one position to the left). Now add $y$ until what remains becomes positive. The second digit of the quotient is $b$ minus the number of these additions. Now again shift $y$ by one position to the right (or what remains by one position to the left), and so on.

Another and a very useful level 1 routine is the following.

(7) The function $[x] :=$ entire($x$), which determines the greatest integer less than or equal to $x$.

## 2. INTRODUCTORY REMARKS ABOUT THE LEVEL 2 ROUTINES

The level 2 routines are designed as ingredients for programming the operations for the product sets and interval sets displayed in Fig. 1. It turns out that the level 2 routines include, in particular, all operations of row 1 of Fig. 1 for the different roundings. Among others, this includes those routines that, in a narrower interpretation, are often called floating-point arithmetic. We develop these algorithms in the following sections. To some extent the description of these algorithms depends on the computer being used and on the representation of numbers in that computer.

We use the so-called *sign–magnitude representation*, at least for the mantissas. This representation is employed in common practice, and it appears to be the most natural representation on computers as well. The following algorithms are described for an arbitrary base $b > 1$. However, the base 10 is a natural base at least for the representation of mantissas. The base 10 is used in common practice, and it avoids many unnecessary anomalies and confusion that may occur when other bases are used in computers. In particular, errors arising from the conversion of numbers to and from the decimal system are avoided.

Since the exponent in the floating-point representation is an integer, a representation for the exponent, which uses a so-called characteristic and/or a $b$- or $(b - 1)$-complement representation of negative numbers, may be used.

The architecture of the computer that we use to describe the algorithms is that of the classical computer, circa 1960. The floating-point numbers (sign, mantissa, and exponent together) are stored in *words*. The accumulator is able to accommodate twice as many digits as are in the floating-point mantissa plus a few additional. This computer is a very convenient tool with which to express the essential steps of the algorithms.

For a different computer, for instance, one for which the storage is organized bytewise or in which the accumulator consists only of 8 or 16 bits, a few additional considerations of minor difficulty are necessary. A convenient way of implementing the arithmetic that we describe consists in a simulation of our ideal computer on the actual computer.

Now we derive the algorithms for the implementation of arithmetic for the first row of Fig. 1. According to our theory, the only way to define arithmetic for this row is by means of semimorphisms. Thus we derive the algorithms for the operations defined by the formula

$$\text{(RG)} \quad \bigwedge_{x,y \in S} x \boxast y := \square(x * y), \qquad * \in \{+, -, \cdot, /\}.$$

We implement (RG) for the roundings $\square_\mu$, $\mu = 0(1)b$, and for purposes of application to interval arithmetic, we implement it as well for $\nabla$ and $\Delta$.

At first sight it seems to be doubtful that formula (RG) can be implemented on computers at all. In order to determine the approximation $x \boxast y$, the exact but unknown result $x * y$ seems to be required in (RG). If, for instance, in the case of addition in a decimal floating-point system, $x$ is of the magnitude $10^{50}$ and $y$ of the magnitude $10^{-50}$, about 100 decimal digits in the accumulator would be necessary in order to represent $x + y$. Even if the largest computers had such long accumulators, it would hardly be necessary to employ so many digits. We will show by means of the following algorithms for $* \in \{+, -, \cdot, /\}$ that whenever $x * y$ is not representable on the computer, it is sufficient to replace it by an appropriate and representable value $x \tilde{*} y$. The latter will have the property $\square(x * y) = \square(x \tilde{*} y)$ for all roundings $\square \in \{\nabla, \Delta, \square_\mu, \mu = 0(1)b\}$. Then $x \tilde{*} y$ can be used to define $x \boxast y$ by means of the relations

$$\bigwedge_{x,y \in S} x \boxast y = \square(x * y) = \square(x \tilde{*} y).$$

The algorithms that implement this relation can, in principle, be separated into the following five steps:

1. Decomposition of $x$ and $y$, i.e., separation of $x$ and $y$ into mantissa and exponent. If floating-point numbers are not stored in a common word, this step is vacuous.
2. Determination of $x \tilde{*} y$. It may be that $x \tilde{*} y = x * y$.
3. Normalization of $x \tilde{*} y$. If the result of 2 is already normalized, this step can be skipped.
4. Rounding of $x \tilde{*} y$ to determine $x \boxast y = \square(x \tilde{*} y) = \square(x * y)$.
5. Composition, i.e., assembling of the mantissa and exponent of the result into a floating-point number. If the floating-point numbers are not stored in words, this step is vacuous.

Figure 35 shows a graphical representation of these five steps in the form of a flow diagram. Labels are introduced between the single steps in order to denote the segments of the detailed flow diagrams, which we discuss in subsequent sections.

We shall see that division can be executed in a manner that eliminates the need for normalization. Since we deal with monotone roundings only, the

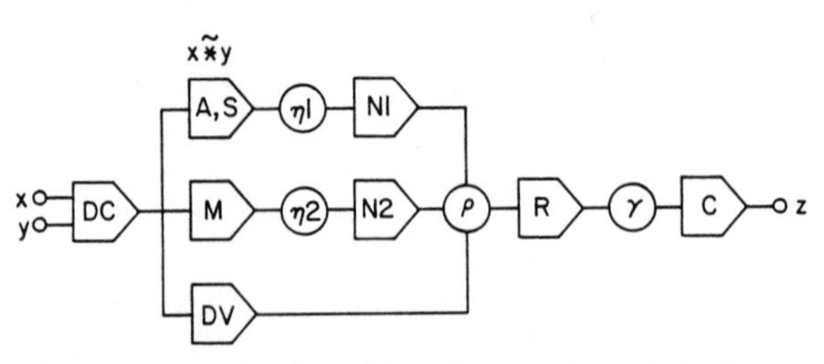


FIGURE 35. Flow diagram for the arithmetic operations. DC: decomposition; A, S: addition and subtraction; M: multiplication; DV: division; N1, N2: normalization; R: rounding; C: Composition.

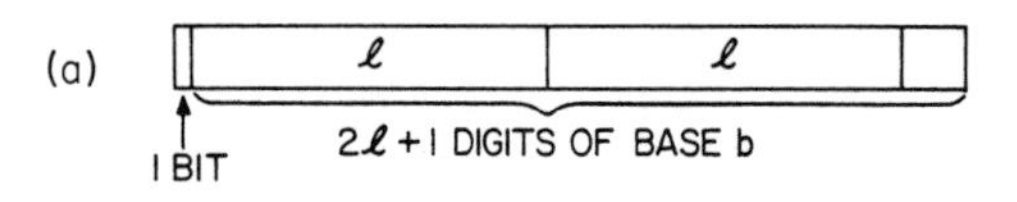


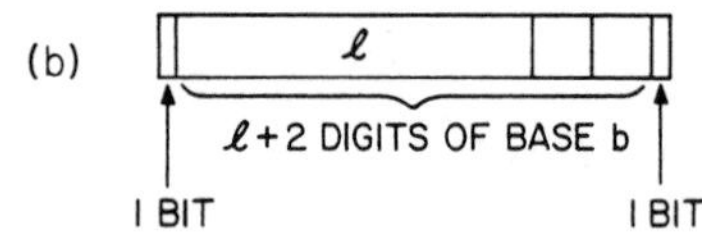


FIGURE 36. (a) Long accumulator; (b) short accumulator.

normalization has to be performed before the rounding since otherwise the monotonicity is destroyed.

In the following sections we briefly discuss the main features of the steps of the algorithms that are enumerated in Fig. 35. We summarize the results by using flow diagrams.

At first we develop algorithms for the floating-point operations using a so-called *long accumulator*. By this we understand a computer register with one digit, which may be a binary digit, in front of the point and $2l + 1$ digits of base $b$ after the point.† See Fig. 36a. Since many computers do not have such a long accumulator, we shall discuss algorithms for the execution of the floating-point operations using a so-called *short accumulator* later on. See Fig. 36b. By this we understand a computer register with one digit, which can be a binary digit, in front of the point and $l + 2$ digits of base $b$ plus one binary digit after the point. The algorithms show that further reduction of the length of the accumulator is not possible if the operations are defined by

† By point we mean the *b-ary point*, i.e., decimal point for $b = 10$.

(RG) for all roundings $\square \in \{\nabla, \triangle, \square_\mu, \mu = 0(1)b\}$. In the following algorithms, all operator symbols $+, -, \cdot, /, =, \leq, \geq, <, >$ are well defined since they are used in the sense of integer arithmetic, as given by the level 1 routines.

The question whether the algorithms using the short or the long accumulator are faster has no simple answer. If the accumulator of appropriate length and the algorithms are implemented in hardware, both algorithms may be nearly equal in speed. If the accumulator is available in hardware while the algorithms are to be implemented in software, the algorithms that use the long accumulator are likely to be faster because they are simpler. If for a computer at hand the accumulator as well as the storage word are relatively short (for instance, 8 or 16 bits) and if the accumulator has to be simulated, then the algorithms using the short accumulator are probably faster.

Key to the whole implementation is to take care that the formulas (RG), as well as the rounding properties (R1)–(R4), which are operative, are strictly realized. This means that these formulas have to be valid for all $x, y \in S$ (resp. $\boldsymbol{R}$) and not only for some or most of such $x, y$. Even an interval around zero, however small, may not be excluded from this requirement.

With these provisos, a principal result of the following sections is that the whole implementation can be separated into five *independent* steps as indicated in Fig. 35 and its context. This means, in particular, that the provisional result, $x \mathbin{\tilde{*}} y$ for all $* \in \{+, -, \cdot, /\}$, may be determined independently of the rounding function. The latter is to be applied so that

$$\bigwedge_{\square \in \{\nabla, \triangle, \square_\mu, \ \mu = 0(1)b\}} \ \bigwedge_{x, y \in S} \square(x * y) = \square\,(x \mathbin{\tilde{*}} y).$$

Consequently, in the flow diagram of Fig. 35, the rounding $R$ may be any of the roundings $\square \in \{\nabla, \triangle, \square_\mu, \mu = 0(1)b\}$, and the entire algorithm delivers the result defined by (RG) and for this particular rounding.

In the following flow diagrams, we use the usual conventions: rectangles denote statements; circles, labels; and figures with six edges, conditions. The flow diagrams are otherwise self-explanatory.

## 3. ADDITION AND SUBTRACTION

We suppose that the long accumulator is being used. Without loss of generality, we may assume that $ex \geq ey$. We distinguish two cases.

Case 1. $ex - ey \geq l + 2$. In this case $y$ is too small in absolute value to influence the first $l$ digits of the sum $x + y$. In case of the roundings $\square_\mu$, $\mu = 1(1)b - 1$, we therefore simply obtain $mz := mx$. However, an arbitrarily small $y$ can change the mantissa of $x$ in the case of the roundings $\nabla$, $\Delta$, $\square_0$, or $\square_b$. To handle these cases, we set

$$ez := ex$$

and

$$mz := \begin{cases} mx & \text{if} \quad my = 0, \\ b^{-(l+2)}(mx \cdot b^{l+2} - 1)^{\dagger} & \text{if} \quad my < 0, \\ b^{-(l+2)}(mx \cdot b^{l+2} + 1) & \text{if} \quad my > 0. \end{cases}$$

Case 2. $ex - ey \leq l + 1$. In this case we divide $my$ by $b^{ex-ey}$, i.e., we shift $my$ to the right by $ex - ey$ digits of base $b$. Then we compute the correct sum $mz := mx + my \cdot b^{(ex-ey)}$.

$2l + 1$ digits of base $b$ suffice for the representation of this sum in all cases. The binary digit in front of the point, which functions in case of a mantissa overflow, is not strictly necessary. It is convenient to use it in order to avoid complicated shiftings, which slow down the addition.

Figure 37 displays a flow diagram for the addition algorithm that we just sketched. The first statement in this diagram represents the decomposition of $x$ and $y$. Upon completion of this addition algorithm, the result is to be normalized and then rounded.

By way of comment on the condition $ex - ey \geq l + 2$ displayed in the flowchart in Fig. 37, we interpolate an example in which $ex - ey = l + 1$ and which shows that in the case of a rounding $\square_\mu$, $1 \leq \mu \leq b - 1$, addition of a digit in the $(l + 2)$nd place can change all digits of $mx$. To see this, let us consider a floating-point system with a mantissa of three digits. In the following three examples we assume that the operands $x$ and $y$ are exact and are already appropriately positioned as indicated.

| | | | | | | | |
|---|---|---|---|---|---|---|---|
| $x$ | 0. | 1 | 0 | 0 | 0 | 0 | $\cdot b^3$ |
| $y$ | $-0.$ | 0 | 0 | 0 | 0 | $b - \mu + 1$ | $\cdot b^3$ |
| $x + y$ | 0. | 0 | $b - 1$ | $b - 1$ | $b - 1$ | $\mu - 1$ | $\cdot b^3$ |
| $x + y$ normalized | 0. | $b - 1$ | $b - 1$ | $b - 1$ | $\mu - 1$ | | $\cdot b^2$ |
| $\square_\mu(x + y)$ | 0. | $b - 1$ | $b - 1$ | $b - 1$ | | | $\cdot b^2$ |

$^{\dagger}$ This expression is a symbolic description of subtracting unity in the $(l + 2)$nd digit of $mx$; the latter has $2l + 1$ digits when it resides in the accumulator, where these operations are being performed. This convention will be used hereafter.

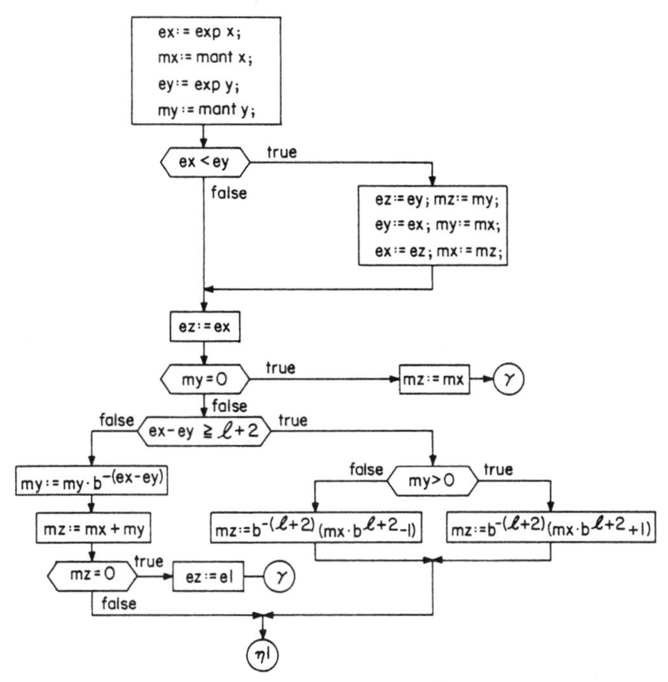


FIGURE 37. Execution of the addition $x \mp y$.

The following example shows that a corresponding addition in the $(l + 3)$rd place, however, does not change the mantissa of $x$.

| | | | | | | | | |
|---|---|---|---|---|---|---|---|---|
| $x$ | 0. | 1 | 0 | 0 | 0 | 0 | 0 | $\cdot b^3$ |
| $y$ | $-0.$ | 0 | 0 | 0 | 0 | 0 | $b - \mu + 1$ | $\cdot b^3$ |
| $x + y$ | 0. | 0 | $b - 1$ | $b - 1$ | $b - 1$ | $b - 1$ | $\mu - 1$ | $\cdot b^3$ |
| $x + y$ normalized | 0. | $b - 1$ | $b - 1$ | $b - 1$ | $b - 1$ | $\mu - 1$ | | $\cdot b^2$ |
| $\square_\mu(x + y)$ | 0. | 1 | 0 | 0 | | | | $\cdot b^3$ |

These examples are sensible only if $b - \mu + 1$ is a nonzero digit of the number system that is used, i.e., if $1 \leq \quad - \mu + 1 \leq b - 1$. From this inequality we obtain $2 \leq \mu \leq b - 1$, and therefore $b \geq 3$. We comment on the binary case $b = 2$ below. If $b \geq 3$, the rounding $\square_1$ is certainly of only minor interest.

Now we consider the case $b = 2$. The two examples studied above are not relevant in this case. The rounding $\square_1$ corresponds to the standard rounding to the nearest number of the screen. We show again by a simple example that an addition can change all digits of $mx$ if the exponents differ by $l+1$:

| | | | | | | | | | | |
|---|---|---|---|---|---|---|---|---|---|---|
| $x$ | 0. | 1 | 0 | 0 | 0 | 0 | 0 | 0 | $\cdot b^3$ |
| $y$ | $-0.$ | 0 | 0 | 0 | 0 | 1 | 0 | 1 | $\cdot b^3$ |
| $x+y$ | 0. | 0 | 1 | 1 | 1 | 0 | 1 | 1 | $\cdot b^3$ |
| $x+y$ normalized | 0. | 1 | 1 | 1 | 0 | 1 | 1 | | $\cdot b^2$ |
| $\square_1(x+y)$ | 0. | 1 | 1 | 1 | | | | | $\cdot b^2$ |

An addition, however, does not change the mantissa of $mx$, if the difference in the exponents is $l+2$. We illustrate this by a further example:

| | | | | | | | | | |
|---|---|---|---|---|---|---|---|---|---|
| $x$ | 0. | 1 | 0 | 0 | 0 | 0 | 0 | 0 | $\cdot b^3$ |
| $y$ | $-0.$ | 0 | 0 | 0 | 0 | 0 | 1 | 1 | $\cdot b^3$ |
| $x+y$ | 0. | 0 | 1 | 1 | 1 | 1 | 0 | 1 | $\cdot b^3$ |
| $x+y$ normalized | 0. | 1 | 1 | 1 | 1 | 0 | 1 | | $\cdot b^2$ |
| $\square_1(x+y)$ | 0. | 1 | 0 | 0 | | | | | $\cdot b^3$ |

In Case 2, these examples show that if $b \geq 3$ and $\mu \geq 2$, as well as if $b=2$ and $\mu=1$, the situation $ex - ey = l+1$ requires that nondegenerate operations be performed on the mantissa. This state of affairs occurs in practice, for instance, when $b = 2, 10$, or 16 and for rounding to the nearest number of the screen ($\mu = b/2$).

In normalized floating-point representation in the binary system, the first digit after the point is always unity. Therefore, in principle this digit can be eliminated. Thus in the binary system, the long accumulator can be reduced to a length of $2l$.

We illustrate the addition algorithm by means of several examples. Let $l = 4$.

1. We consider the case $ex - ey \geq l+2$.

(a) $x = +0.d_1d_2d_3d_4 \cdot b^3$, $d_4 \neq b-1$, and $y > 0$. Then

$$x \mp y = 0.d_1d_2d_3d_401 \cdot b^3$$

and

$$\nabla(x+y) = \nabla(x \mp y) = 0.d_1d_2d_3d_4 \cdot b^3,$$
$$\Delta(x+y) = \Delta(x \mp y) = 0.d_1d_2d_3(d_4+1) \cdot b^3,$$
$$\square_\mu(x+y) = \square_\mu(x \mp y) = 0.d_1d_2d_3d_4 \cdot b^3 \quad \text{for} \quad 1 \leq \mu \leq b-1.$$

(b) $x = -0.d_1d_2d_3d_4 \cdot b^3, d_4 \neq b-1$, and $y < 0$. Then

$$x \mp y = -0.d_1d_2d_3d_401 \cdot b^3$$

and

$$\begin{aligned}
&\triangledown(x+y) = \triangledown(x \tilde{\mp} y) = -0.d_1 d_2 d_3 (d_4 + 1) \cdot b^3, \\
&\triangle(x+y) = \triangle(x \tilde{\mp} y) = -0.d_1 d_2 d_3 d_4 \cdot b^3, \\
&\square_\mu(x+y) = \square_\mu(x \tilde{\mp} y) = -0.d_1 d_2 d_3 d_4 \cdot b^3 \qquad \text{for } 1 \le \mu \le b-1.
\end{aligned}$$

(c) $x = +0.1000 \cdot b^3$ and $y < 0$. Then

$$x \mp y = 0.0(b-1)(b-1)(b-1)|(b-1)(b-1) \cdot b^3$$

and normalized

$$\begin{aligned}
x \tilde{\mp} y &= 0.(b-1)(b-1)(b-1)(b-1)(b-1) \cdot b^2, \\
\triangledown(x+y) = \triangledown(x \tilde{\mp} y) &= 0.(b-1)(b-1)(b-1)(b-1) \cdot b^2, \\
\triangle(x+y) = \quad (x \tilde{\mp} y) &= 0.1000 \cdot b^3, \\
\square_\mu(x+y) = \square_\mu(x \tilde{\mp} y) &= 0.1000 \cdot b^3 \qquad \text{for} \quad 1 \le \mu \le b-1.
\end{aligned}$$

(d) $x = -0.1000 \cdot b^3$ and $y > 0$. Then for the normalized sum we have $x \tilde{\mp} y = -0.(b-1)(b-1)(b-1)(b-1)(b-1) \cdot b^2$ and $\triangledown\ (x+y) = \triangledown(x+y) = -0.1000 \cdot b^3$, $\triangle(x+y) = \triangle(x \tilde{\mp} y) = -0.(b-1)(b-1)(b-1)(b-1) \cdot b^2$, $\square_\mu(x+y) = \square_\mu(x \tilde{\mp} y) = -0.1000 \cdot b^3$ for $1 \le \mu \le b-1$.

2. Now we consider the case $ex - ey \le l + 1$ and in particular for $b = 10$.

(a) $x = 0.1000 \cdot 10^6$ and $y = -0.5001 \cdot 10^1$. Then $x + y = 0.0999|9499|9 \cdot 10^6$ and normalized $x + y = 0.9999|4999 \cdot 10^5$, $\triangledown(x+y) = 0.9999 \cdot 10^5$, $\triangle(x+y) = 0.1000 \cdot 10^6$, $\square_\mu(x+y) = 0.9999 \cdot 10^5$ for $5 \le \mu \le 9$.

(b) $x = 0.1000 \cdot 10^6$ and $y = -0.5000 \cdot 10^1$. Then

$$\begin{aligned}
x + y &= 0.0999|9500|0 \cdot 10^6, \\
\triangledown(x+y) &= 0.9999 \cdot 10^5, \\
\triangle(x+y) &= 0.1000 \cdot 10^6 \\
\square_\mu(x+y) &= 0.1000 \cdot 10^6 \text{ for } \mu = 5.
\end{aligned}$$

In the last two examples the operands are identical except for the last digit of $y$. If the $(2l+1)$st digit of the accumulator was in fact not present, the corresponding sums $x + y$ would be identical. Since these sums differ, these two examples show that it is really necessary to have $2l + 1$ digits available in the accumulator if the addition is executed as prescribed and the result is required to be correct for all the roundings $\triangledown, \triangle$, and $\square_5$.

## 4. NORMALIZATION

A normalization consists of appropriate right or left shifts of the mantissa and a corresponding correction of the exponent.

The addition algorithm described in Fig. 37 can result in a mantissa with the property $|mz| \geq 1$. The inequality

$$|mx| + |my| \cdot b^{-(ex-ey)} < 1 + 1,$$

however, shows that a shift of at most one digit to the right may be required. It also shows that a single bit is sufficient in order to record such an overflow of the mantissa in case of an addition since the leading digit can only be unity. If this bit is one after an addition, then necessarily the $2l$th as well as the $(2l + 1)$st digits in the accumulator are zero. Thus the result after the right shift is still represented correctly to all digits.

After an addition, however, a left shift of more than one digit may be necessary. Fig. 38 shows the flow diagram for the execution of the normalization after an addition or subtraction.

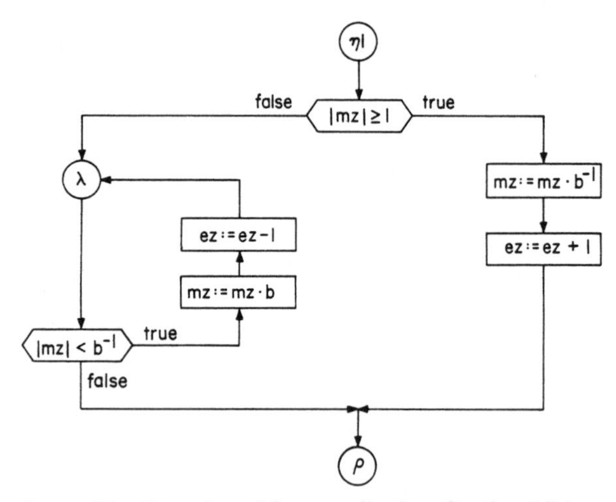


FIGURE 38. Execution of the normalization after the addition.

The algorithm for multiplication will be discussed in Section 6. However, we may remark here that the normalization procedure for multiplication is simpler. To see this, note that since $b^{-1} \leq |mx| < 1$ and $b^{-1} \leq |my| < 1$, we obtain $b^{-2} \leq |mx| \cdot |my| < 1$. Then no right shift and at most one left shift may be necessary after a multiplication. Fig. 39 shows the flow diagram for the execution of the normalization after a multiplication.

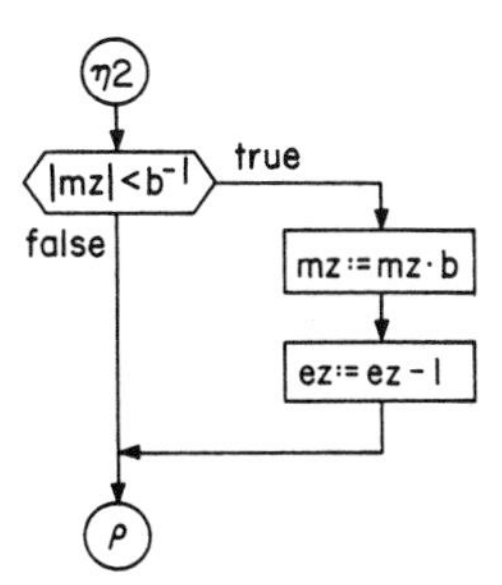


FIGURE 39. Execution of the normalization after a multiplication.

## 5. ROUNDING

After normalization the mantissa, which in general has a length of $2l + 1$ digits, is to be rounded to $l$ digits. This rounding can alter the mantissa so that $|mz| = 1$. In this case a right shift by one digit and a corresponding correction of the exponent becomes necessary.

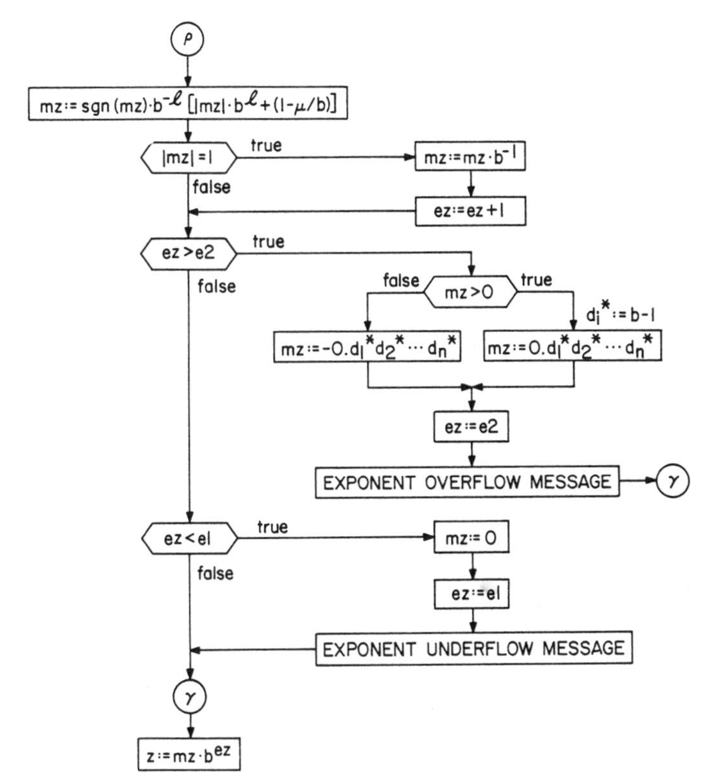


FIGURE 40. Execution of the rounding $\square_\mu$, $1 \leq \mu \leq b - 1$.

Figure 40 shows the flow diagram for the execution of the roundings $\square_\mu$, $1 \leq \mu \leq b - 1$. It is easy to see that the algorithm of Fig. 40 is applicable for the case $\mu = b$.

Figure 41 gives the flow diagram for the rounding $\bigtriangledown$. The last statement in each of Figs. 40 and 41 denotes the composition of $mz$ and $ez$ to form $z$. The rounding $\square_b$ is defined in terms of $\bigtriangledown$ by means of the formula

$$\bigwedge_{z \in \mathbf{R}} \square_b z = \operatorname{sgn}(z) \cdot \bigtriangledown |z|.$$

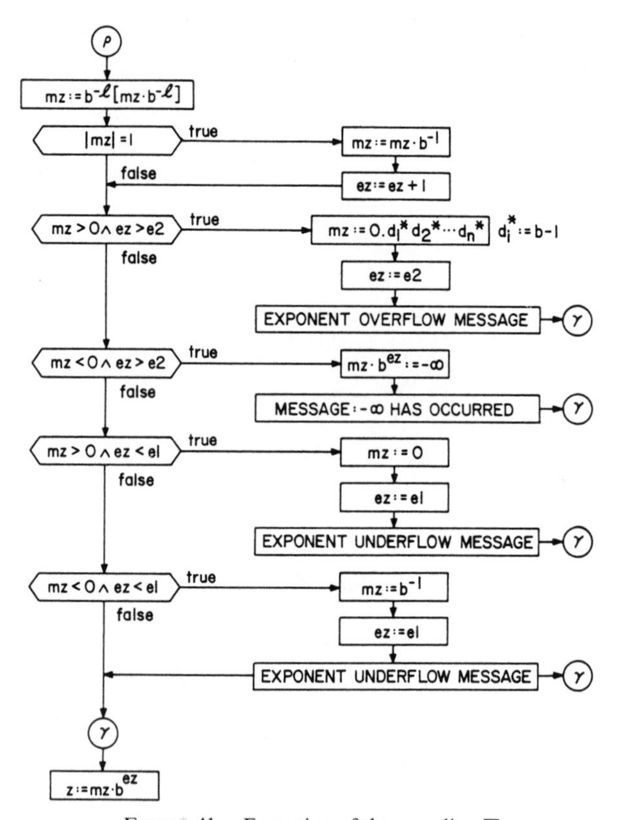


FIGURE 41. Execution of the rounding $\bigtriangledown$.

A flow diagram for the rounding $\bigtriangleup$ can be given in complete analogy to the one for $\bigtriangledown$. $\bigtriangleup$ also can be produced in terms of $\bigtriangledown$ by using the formula

$$\bigwedge_{z \in \mathbf{R}} \bigtriangleup z = -\bigtriangledown(-z).$$

A critical step in the algorithm for the rounding $\triangledown$ is the statement $mz := b^{-1}[mz \cdot b^1]$. If we denote this operation by $\triangledown_* mz$, then each rounding $\square_\mu$, $0 \le \mu \le b$, can easily be expressed in terms of $\triangledown_*$ if the first statement in Fig. 40 is replaced by Fig. 42.

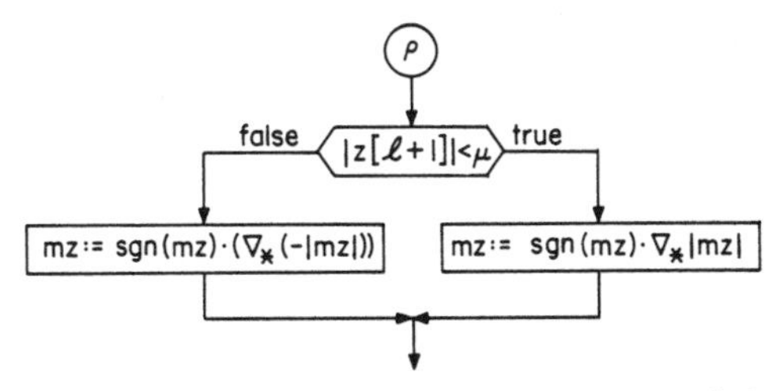


FIGURE 42. Execution of the rounding $\square_\mu$, $0 \le \mu \le b$, by $\triangledown_*$.

All the algorithms given in this section exhibit the central role of the function $[x]: \boldsymbol{R} \to \boldsymbol{Z}$, which determines the greatest integer less or equal to $x$. If this function is available, all roundings in the set $\{\triangledown, \triangle, \square_\mu, \mu = 0(1)b\}$ can easily be realized on the computer. If it is not available, the more literal definition of $\triangledown$ given in Section 2 of Chapter 5 has to be implemented.

Finally, we remark once more that the algorithms for all operations $* \in \{+, -, \cdot, /\}$ show that the intermediate result $x \mathbin{\tilde{*}} y$ can be chosen so that for all roundings $\square \in \{\triangledown, \triangle, \square_\mu, \mu = 0(1)b\}$, the property

$$\bigwedge_{x,y \in S} \square(x * y) = \square(x \mathbin{\tilde{*}} y)$$

holds. This assures us that for the rounding step $R$ in the execution of arithmetic operations as in Fig. 35, any of the roundings $\triangledown, \triangle, \square_\mu, \mu = 0(1)$b, can be employed. The result obtained is the one defined by (RG) and that rounding. No other part of the entire algorithm in Fig. 35 need be changed.

## 6. MULTIPLICATION

If $x = 0$ or $y = 0$ or $x = y = 0$, then $x \cdot y = 0$ and $\square(x \cdot y) = 0$. Otherwise we determine $ez := ex + ey$ (the possibility that $ez$ lies outside the range $[e1, e2]$ is considered in Section 9) and $mz := mx \cdot my$. This multiplication can be executed correctly within the long accumulator. Because $b^{-2} \le |mx| \cdot |my| < 1$, a normalization consisting of at most one left shift may be necessary. Figure 43 displays the flow diagram for multiplication. The corresponding normalization algorithm was already described in Section 4.

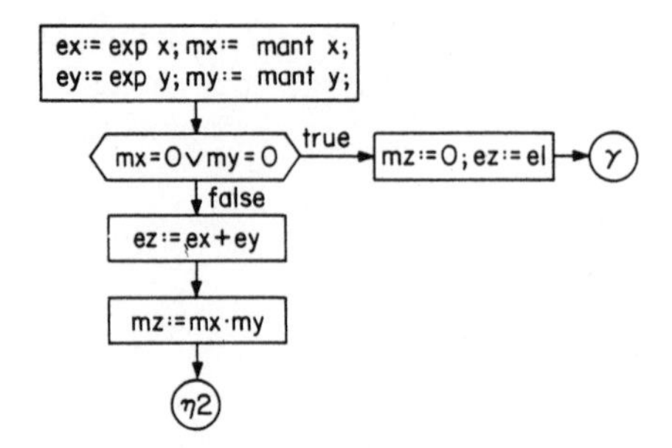


FIGURE 43. Execution of multiplication.

## 7. DIVISION

We determine the quotient $x \tilde{/} y$. If $y = 0$, an error message is to be given. If $x = 0$ and $y \neq 0$, then $x/y = 0$ and $\square(x/y) = 0$. If $x \neq 0$ and $y \neq 0$, we determine $ez := ex - ey$. We read $mx$ into the first $l$ digits of the accumulator and set all the remaining digits of the accumulator to zero. If $|mx| \geq |my|$, we shift the contents of the accumulator to the right by one digit and increase $ez$ by unity. Now we divide the contents of the accumulator by $my$. It is sufficient to determine the first $l + 1$ digits of the quotient. We denote this

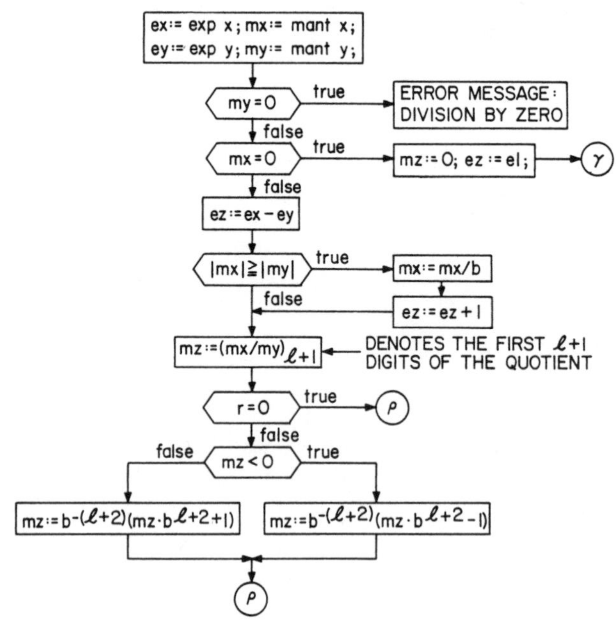


FIGURE 44. Execution of the division $x \tilde{/} y$.

quotient by $(mx/my)_{l+1}$. It is already normalized. Only the $(l+1)$st digit is needed for the execution of the roundings $\square_\mu$, $1 \leq \mu \leq b-1$. However in the case of one of the roundings $\triangledown$, $\triangle$, $\square_0$, and $\square_b$, even if the $(l+1)$st digit is zero, an arbitrarily small remainder $r$ can influence the mantissa (even its leading digit). Here the contents of the accumulator after the first $l+1$ digits denote the remainder $r$.

For proper treatment of all of these cases, we proceed as follows:

1. $mz := (mx/my)_{l+1}$, i.e., compute the first $l+1$ digits of $mz$.
2. If $r = 0$, there is no further treatment of $mz$.
3. If $r \neq 0$ and $mz > 0$, $mz := b^{-(l+2)}(mz \cdot b^{l+2} + 1)$.
4. If $r \neq 0$ and $mz < 0$, $mz := b^{-(l+2)}(mz \cdot b^{l+2} - 1)$.

Figure 44 shows the flow chart for the execution of division.

## 8. ALGORITHMS USING THE SHORT ACCUMULATOR

In the case $ex - ey \leq l + 1$ of the algorithm for the addition shown in Fig. 37, the mantissa $my$ may be shifted to the right as many as $l+1$ digits. Thus the ensuing addition, as described above, requires an accumulator of $2l+1$ digits (after the point). The multiplication algorithm commences with a correct calculation of the product of the mantissas $mx \cdot my$. As with addition, this operation can be conveniently accommodated within an accumulator of $2l+1$ digits (after the point). However, many computers do not provide so long an accumulator. In such a case extra care must be taken so that the operations are executed appropriately, i.e., that formula (RG) holds.

Figure 45 displays a flow chart for the execution of addition by use of the *short accumulator*. This accumulator consists of $l+2$ digits of base $b$ followed by an additional bit as well as preceded by an additional bit in front of the point. The latter bit is used for treatment of mantissa overflow in the case of addition. See Fig. 36. The contents of the accumulator which follow the point are denoted by

$$.d_1 d_2 \ldots d_l d_{l+1} d_{l+2} d.$$

Here the $d_i$ are digits of base $b$, while $d$ is a binary digit. To permit a simple description of the algorithms corresponding to the flowcharts in Figs. 45 and 46, we take the $(l+3)$rd digit of the accumulators to be a full digit of base $b$. It could, however, be replaced by a single binary digit or a flag as indicated here.

The branches $ex - ey \geq l + 2$ in Figs. 37 and 45 are the same. The complementary branch $ex - ey \leq l + 1$ in Fig. 45 is redivided by the condition

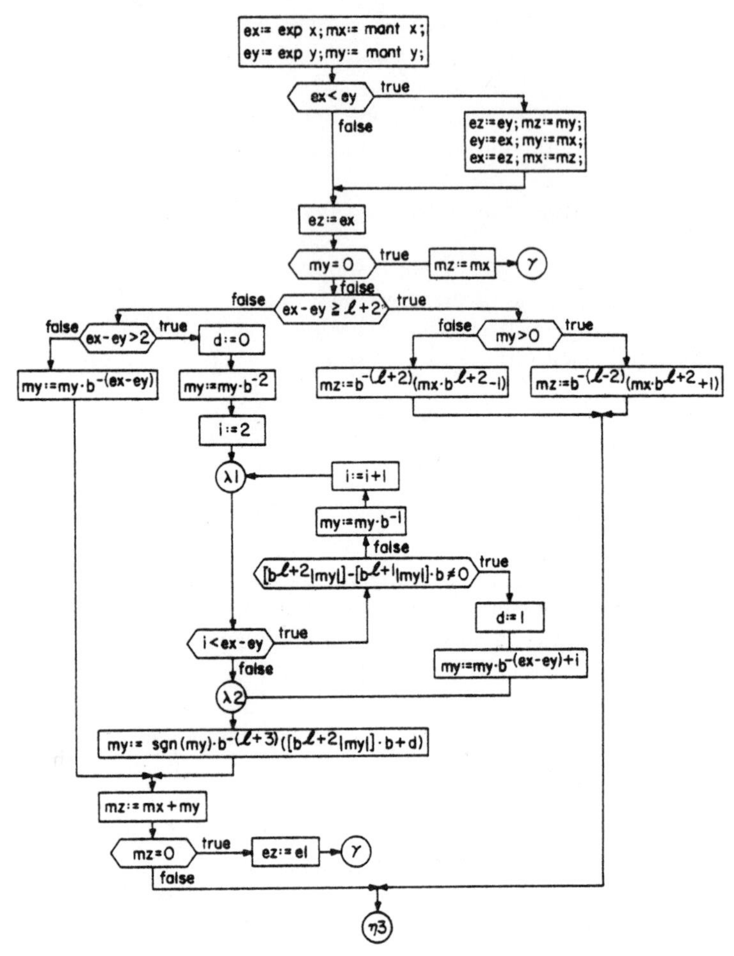


FIGURE 45. Execution of addition with the short accumulator.

$ex - ey > 2$. If $ex - ey \leq 2$, addition with the short accumulator proceeds as before. The condition $ex - ey > 2 \wedge ex - ey \leq l + 1$ specifies that branch of the algorithm of Fig. 45 in which addition is executed differently than in the algorithm of Fig. 37. In this case, $|mz| = |mx + my| \geq \big||mx| - |my|\big| > b^{-1} - b^{-2} = (b - 1)b^{-2} \geq b^{-2}$, i.e., the first nonzero digit of $mz$ occurs within the first or second place after the point. Then in the case of the rounding $\square_\mu$, $1 \leq \mu \leq b - 1$, the digit of $mz$, which keys the rounding, is not more remote than the $(l + 2)$nd place to the right of the point. However, in the case of the rounding $\bigtriangledown$, $\bigtriangleup$, $\square_0$, or $\square_b$, a digit beyond the $(l + 2)$nd can influence

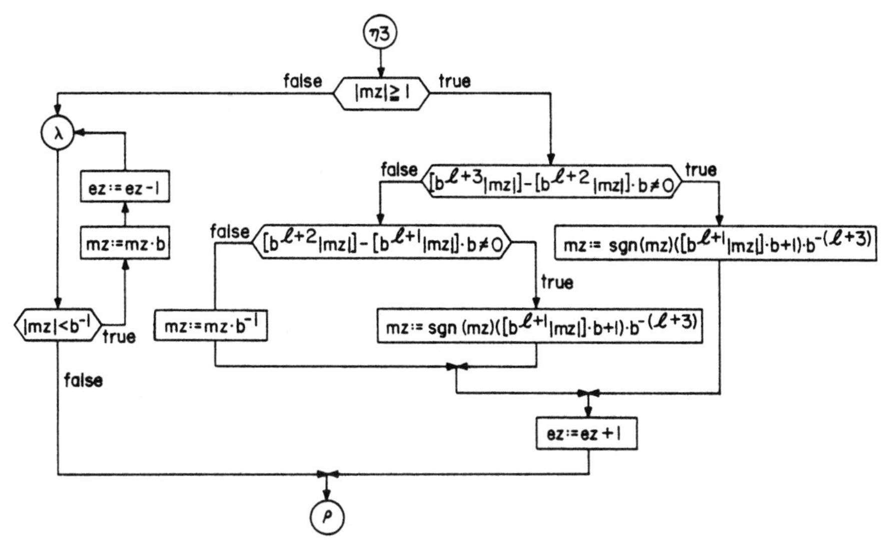


Figure 46. Execution of normalization with the short accumulator.

the result. Then during any shifting of $my$ to the right, note must be taken whether digits shifted beyond the $(l+2)$nd place are zero or not. This information is recorded in the $(l+3)$rd digit $d$. Thus $d$ need only be a binary digit.

As before, the addition $mz := mx + my$ may cause an overflow in the mantissa. This would require a right shift within the normalization algorithm. The $(l+3)$rd digit may be altered as a result of this right shift. Figure 46 displays the algorithm for the execution of the normalization using the short accumulator.

To simplify the description of the algorithm in Figs. 45 and 46, we assume that the $(l+3)$rd digit after the point is a full digit of base $b$. In fact as we have already remarked, it suffices to treat this digit as a binary digit.

We now illustrate the algorithms of Figs. 45 and 46 with several examples. We take $l = 4$ and the decimal system $b = 10$. The provisional and possibly normalized and unrounded result is denoted by $x \mathbin{\tilde{\mp}} y$. We begin with the following case.

(a) $\operatorname{sgn}(x) = \operatorname{sgn}(y)$

1. $x = 0.9999 \cdot b^3, \qquad y = 0.1001 \cdot b^0,$
$x + y = 1.0000001 \cdot b^3, \qquad x \mathbin{\tilde{\mp}} y = 0.1000001 \cdot b^4,$
$\triangledown(x+y) = \triangledown(x \mathbin{\tilde{\mp}} y) = 0.1000 \cdot b^4,$
$\triangle(x+y) = \triangle(x \mathbin{\tilde{\mp}} y) = 0.1001 \cdot b^4,$
$\square_\mu(x+y) = \square_\mu(x \mathbin{\tilde{\mp}} y) = 0.1000 \cdot b^4 \qquad \text{for} \quad 1 \le \mu \le 9.$

2. $x = 0.9998 \cdot b^3, \qquad y = 0.2070 \cdot b^0,$
   $x + y = 1.0000070 \cdot b^3, \qquad x \,\tilde{+}\, y = 0.1000001 \cdot b^4,$
   $\bigtriangledown(x + y) = \bigtriangledown(x \,\tilde{+}\, y) = 0.1000 \cdot b^4,$
   $\bigtriangleup(x + y) = \bigtriangleup(x \,\tilde{+}\, y) = 0.1001 \cdot b^4,$
   $\square_\mu(x + y) = \square_\mu(x \,\tilde{+}\, y) = 0.1000 \cdot b^4 \quad \text{for} \quad 1 \le \mu \le 9.$
3. $x = 0.8234 \cdot b^5, \qquad y = 0.5012 \cdot b^1,$
   $x + y = 0.82345012 \cdot b^5, \qquad x \,\tilde{+}\, y = 0.8234501 \cdot b^5,$
   $\bigtriangledown(x + y) = \bigtriangledown(x \,\tilde{+}\, y) = 0.8234 \cdot b^5,$
   $\bigtriangleup(x + y) = \bigtriangleup(x \,\tilde{+}\, y) = 0.8235 \cdot b^5,$
   $\square_\mu(x + y) = \square_\mu(x \,\tilde{+}\, y) = 0.8235 \cdot b^5 \quad \text{for} \quad 1 \le \mu \le 5,$
   $\square_\mu(x + y) = \square_\mu(x \,\tilde{+}\, y) = 0.8234 \cdot b^5 \quad \text{for} \quad 6 \le \mu \le 9.$
4. $x = -0.8234 \cdot b^5, \qquad y = -0.5021 \cdot b^1,$
   $x + y = -0.82345021 \cdot b^5, \qquad x \,\tilde{+}\, y = -0.8234501 \cdot b^5,$
   $\bigtriangledown(x + y) = \bigtriangledown(x \,\tilde{+}\, y) = -0.8235 \cdot b^5,$
   $\bigtriangleup(x + y) = \bigtriangleup(x \,\tilde{+}\, y) = -0.8234 \cdot b^5,$
   $\square_\mu(x + y) = \square_\mu(x \,\tilde{+}\, y) = -0.8235 \cdot b^5 \quad \text{for} \quad 1 \le \mu \le 5,$
   $\square_\mu(x + y) = \square_\mu(x \,\tilde{+}\, y) = -0.8234 \cdot b^5 \quad \text{for} \quad 6 \le \mu \le 9.$
5. $x = -0.9998 \cdot b^3, \qquad y = -0.2070 \cdot b^0,$
   $x + y = -1.0000070 \cdot b^3, \qquad x \,\tilde{+}\, y = -0.1000001 \cdot b^4,$
   $\bigtriangledown(x + y) = \bigtriangledown(x \,\tilde{+}\, y) = -0.1001 \cdot b^4,$
   $\bigtriangleup(x + y) = \bigtriangleup(x \,\tilde{+}\, y) = -0.1000 \cdot b^4,$
   $\square_\mu(x + y) = \square_\mu(x \,\tilde{+}\, y) = -0.1000 \cdot b^4 \quad \text{for} \quad 1 \le \mu \le 9.$

(b) Now let $\operatorname{sgn}(x) \neq \operatorname{sgn}(y)$.

6. $x = 0.1000 \cdot b^4, \qquad y = -0.1001 \cdot b^{-1},$
   $x + y = 0.099998999 \cdot b^4, \qquad x \,\tilde{+}\, y = 0.0999989 \cdot b^4,$
   $\bigtriangledown(x + y) = \bigtriangledown(x \,\tilde{+}\, y) = 0.9999 \cdot b^3,$
   $\bigtriangleup(x + y) = \bigtriangleup(x \,\tilde{+}\, y) = 0.1000 \cdot b^4,$
   $\square_\mu(x + y) = \square_\mu(x \,\tilde{+}\, y) = 0.9999 \cdot b^4 \quad \text{for} \quad \mu = 9,$
   $\square_\mu(x + y) = \square_\mu(x \,\tilde{+}\, y) = 0.1000 \cdot b^3 \quad \text{for} \quad 1 \le \mu \le 8.$
7. $x = 0.1000 \cdot b^6, \qquad y = -0.5001 \cdot b^1,$
   $x + y = 0.099994999 \cdot b^6, \qquad x \,\tilde{+}\, y = 0.0999949 \cdot b^6,$
   $\bigtriangledown(x + y) = \bigtriangledown(x \,\tilde{+}\, y) = 0.9999 \cdot b^5,$
   $\bigtriangleup(x + y) = \bigtriangleup(x \,\tilde{+}\, y) = 0.1000 \cdot b^6,$
   $\square_\mu(x + y) = \square_\mu(x \,\tilde{+}\, y) = 0.9999 \cdot b^5 \quad \text{for} \quad 5 \le \mu \le 9,$
   $\square_\mu(x + y) = \square_\mu(x \,\tilde{+}\, y) = 0.1000 \cdot b^6 \quad \text{for} \quad 1 \le \mu \le 4.$
8. $x = 0.1000 \cdot b^6, \qquad y = -0.5000 \cdot b^1,$
   $x + y = 0.099995000 \cdot b^6, \qquad x \,\tilde{+}\, y = 0.0999950 \cdot b^6,$
   $\bigtriangledown(x + y) = \bigtriangledown(x \,\tilde{+}\, y) = 0.9999 \cdot b^5,$
   $\bigtriangleup(x + y) = \bigtriangleup(x \,\tilde{+}\, y) = 0.1000 \cdot b^6,$
   $\square_\mu(x + y) = \square_\mu(x \,\tilde{+}\, y) = 0.1000 \cdot b^6 \quad \text{for} \quad 1 \le \mu \le 5,$
   $\square_\mu(x + y) = \square_\mu(x \,\tilde{+}\, y) = 0.9999 \cdot b^5 \quad \text{for} \quad 6 \le \mu \le 9.$

9. $x = 0.1000 \cdot b^4, \quad y = -0.5412 \cdot b^0,$
$x + y = 0.09994588 \cdot b^4, \quad x \mathbin{\tilde{\mp}} y = 0.0999459 \cdot b^4,$
$\nabla(x + y) = \nabla(x \mathbin{\tilde{\mp}} y) = 0.9994 \cdot b^3,$
$\triangle(x + y) = \triangle(x \mathbin{\tilde{\mp}} y) = 0.9995 \cdot b^3,$
$\square_\mu(x + y) = \square_\mu(x \mathbin{\tilde{\mp}} y) = 0.9995 \cdot b^3 \quad \text{for} \quad 1 \le \mu \le 5,$
$\square_\mu(x + y) = \square_\mu(x \mathbin{\tilde{\mp}} y) = 0.9994 \cdot b^3 \quad \text{for} \quad 6 \le \mu \le 9.$

10. $x = -0.1000 \cdot b^6, \quad y = 0.5001 \cdot b^1,$
$x + y = -0.099994999 \cdot b^6, \quad x \mathbin{\tilde{\mp}} y = -0.0999949 \cdot b^6,$
$\nabla(x + y) = \nabla(x \mathbin{\tilde{\mp}} y) = -0.1000 \cdot b^6,$
$\triangle(x + y) = \triangle(x \mathbin{\tilde{\mp}} y) = -0.9999 \cdot b^5,$
$\square_\mu(x + y) = \square_\mu(x \mathbin{\tilde{\mp}} y) = -0.1000 \cdot b^6 \quad \text{for} \quad 1 \le \mu \le 4$
$\square_\mu(x + y) = \square_\mu(x \mathbin{\tilde{\mp}} y) = -0.9999 \cdot b^5 \quad \text{for} \quad 5 \le \mu \le 9.$

11. $x = -0.8234 \cdot b^4, \quad y = 0.5021 \cdot b^0,$
$x + y = -0.82334979 \cdot b^4, \quad x \mathbin{\tilde{\mp}} y = -0.8233499 \cdot b^4,$
$\nabla(x + y) = \nabla(x \mathbin{\tilde{\mp}} y) = -0.8234 \cdot b^4,$
$\triangle(x + y) = \triangle(x \mathbin{\tilde{\mp}} y) = -0.8233 \cdot b^4,$
$\square_\mu(x + y) = \square_\mu(x \mathbin{\tilde{\mp}} y) = -0.8234 \cdot b^4 \quad \text{for} \quad 1 \le \mu \le 4,$
$\square_\mu(x + y) = \square_\mu(x \mathbin{\tilde{\mp}} y) = -0.8233 \cdot b^4 \quad \text{for} \quad 5 \le \mu \le 9.$

The next example confirms that an accumulator of $l + 1$ digits followed by an additional binary digit $d$ after the point is not capable of delivering correct results as defined by (RG) in all cases.

12. $x = 0.1000 \cdot b^6, \quad y = -0.5001 \cdot b^1,$
$x + y = 0.099994999 \cdot b^6, \quad x \mathbin{\tilde{\mp}} y = 0.099999 \cdot b^6,$
$\nabla(x + y) = \nabla(x \mathbin{\tilde{\mp}} y) = 0.9999 \cdot b^5,$
$\triangle(x + y) = \triangle(x \mathbin{\tilde{\mp}} y) = 0.1000 \cdot b^6, \quad \text{but}$
$\square_\mu(x + y) = 0.9999 \cdot b^5 \ne 0.1000 \cdot b^6$
$= \square_\mu(x \mathbin{\tilde{\mp}} y) \quad \text{for} \quad 5 \le \mu \le 9.$

The preceding study shows that in case of addition and subtraction, the length of the short accumulator (see Fig. 36) is both necessary and sufficient in order to realize formula (RG) for all roundings in the set $\{\nabla, \triangle, \square_\mu, \mu = 0(1)b\}$.

Now we consider the question whether a short accumulator is also sufficient to perform the multiplication and division defined by (RG) for all of these roundings.

In Section 1 we indicated that this is possible for multiplication. Let the mantissas of the operands $x$ and $y$ be

$$mx = 0.x[1]x[2] \ldots x[l] \quad \text{and} \quad my = 0.y[1]y[2] \ldots y[l].$$

Then

$$mx \cdot my = mx \cdot y[1] \cdot b^{-1} + mx \cdot y[2] \cdot b^{-2} + \cdots + mx \cdot y[l] \cdot b^{-l}.$$

Here summands have at most $l + 1$ nonzero digits. Then by using the Bohlender algorithm (see Section 10), this implies that the sum rounded to

$l$ digits for all roundings of the set $\{\nabla, \triangle, \square_\mu, \mu = 0(1)b\}$ can be calculated with an accumulator of $l + 3$ digits of base $b$ plus one binary digit after the point and an additional binary digit in front of the point.

Proceeding in a straightforward manner, we can show that the short accumulator is sufficient to perform the division (defined by (RG)) of two floating-point numbers and any one of the roundings $\{\nabla, \triangle, \square_\mu, \mu = 0(1)b\}$.

## 9. UNDERFLOW AND OVERFLOW TREATMENT

For fixed-point computations, all computed values are less than unity and the problem data themselves must be preprocessed, typically by scaling, so that the computation proceeds coherently. There are analogous requirements for the use of all classical calculating devices such as analog computers, planometers, and so forth. It is easy to forget that the same requirement exists for modern digital computers, even though they employ floating-point operations and an enormous range of representable numbers. Indeed, computations frequently occur during which this representable range is exceeded, even on modern computers. The terms underflow and overflow characterize these occurrences, and we begin now to consider them.

We saw in Chapter 5 that real numbers $x$ with the property

$$b^{e1-1} \leq |x| \leq B := 0\,.\,(b-1)(b-1)\cdots(b-1) \cdot b^{e2} \tag{1}$$

are conveniently mapped into a floating-point system. Any associated error need not be larger than the distance between the two neighboring floating-point numbers of $x$. Whenever

$$0 < |x| < b^{e1-1} \qquad (\text{resp.}\quad |x| > B), \tag{2}$$

we speak of an exponent underflow (resp. an exponent overflow). An attempt to represent a real number of this latter type in the floating-point system results in the loss of nearly all information about the number. If during a computation there is an incidence of a number of the type specified by (2), there should be an automatic call of a well-defined underflow (resp. overflow) routine. Such an occurrence should be made known to the user.

The algorithms discussed in this chapter show that underflow and overflow can occur in the performance of all operations and in particular within the normalization and rounding procedure. On the other hand, certain minor underflows and overflows are rectifiable. These are incidental to the algorithms for the operations $x \mathbin{\tilde{*}} y$ and may be corrected by the normalization and rounding routine. Therefore, it is sensible to check whether a nonrectifiable underflow or overflow occurred only after the rounding as indeed is the case in the algorithms described above. The arithmetic performed on the exponents is an integer arithmetic. In order to be able to handle the rectifiable underflow or overflow this integer arithmetic should be executed with more digits than are used for the representation itself, of the exponent within a floating-point number. This allows the algorithms for the arithmetic operations described above to be continued until a possible nonrectifiable underflow or overflow check is reached.

While there exists no complete recovery process from underflow and overflow on a computer, which may occur during the execution of numerical algorithms, two different recovery procedures are in common use.

The first is a trial and error approach. During the preparation and programming of the problem, care is to be taken so that the entire computation can be performed within the available range of numbers. The special difficulty is that it is hard to judge in advance that all intermediate results stay in this range. It is a convenient and perhaps the most customary practice, therefore, simply to go ahead with the computation without too much analysis concerning the range of numbers that will occur. If an underflow or overflow occurs, the computer stops the computation with an error message. The user then tries a little harder and preprocesses the problem somewhat further, typically by rescaling. After this he starts the computation anew and so on. Several such scaling steps are not uncommon.

The second method exploits the observation that some underflows and overflows are benign with respect to further computation. As an example, let us consider the case of underflow. Suppose that the sum

$$S := \sum_{i=1}^{n} u_i(x_1, x_2, \ldots, x_n)$$

is to be calculated, leading to a result of magnitude $10^{30}$. If the computation of one of the summands $u_\nu$, causes an underflow, viz., $|u_\nu| < 10^{e1-1}$, then $u_\nu$ does not influence the floating-point sum at all and can be neglected. The computer need not interrupt the computation but continues it. On the premise that it is sometimes sensible to continue a computation which has underflowed or overflowed, a well-defined underflow or overflow procedure is to be invoked. It often consists of mapping the numbers $x$, such that $|x| < b^{e1-1}$, to zero, and such that $x \leq -B$ (resp. $B \leq x$) to $-B$ (resp. $B$). Additionally, a detailed underflow or overflow message is given to the user. At the end of the computation the user can then decide whether the underflow or overflow that occurred during his calculation was harmless or not. In the former case the computation is finished. In the latter, a proper scaling of the problem is required followed by recomputation.

Each of these two recovery processes has advantages and disadvantages. The first may cause too many interruptions, especially in cases where the interruptions are unnecessary from the point of view of the second method.

The second method may consume computer time, finishing a computation that is afterwards rejected as incorrect.

## 10. COMPUTATION WITH MAXIMUM ACCURACY OF THE SUM OF *n* FLOATING-POINT NUMBERS

In this section we derive an algorithm for computing the rounded sum of $n$ floating-point numbers

$$\square \sum_{i=1}^{n} x_i \tag{1}$$

for all roundings $\square \in \{\nabla, \triangle, \square_\mu, \mu = 0(1)b\}$. This sum is the exact real summation of the $x_i$, which is rounded in turn to one of the neighboring screen points. It is in this sense a summation process of maximal accuracy. An earlier algorithm with this property was first given in [43]. See [40] also. This algorithm proceeds by sorting the numbers $x_i$ according to their exponents. We discuss this algorithm in Section 11. Which of these two algorithms is the faster may depend on special properties of the computer being used.

In this section we describe an algorithm for computing the sum (1), which avoids sorting. We have already referred to the algorithm as the *Bohlender algorithm*. See [8]. It uses ideas of an earlier algorithm given in [56].

As is customary, it is not necessary to know the precise value of the sum $\sum_{i=1}^{n} x_i$ in order to determine the rounded sum (1). It is sufficient to determine an approximation $\widetilde{\sum_{i=1}^{n} x_i}$ with the property

$$\bigwedge_{(x_i) \in S_{b,l}^n} \; \bigwedge_{\square \in \{\nabla, \triangle, \square_\mu,\, \mu = 0(1)b\}} \square \widetilde{\sum_{i=1}^{n} x_i} = \square \sum_{i=1}^{n} x_i. \tag{2}$$

Here $S_{b,l}^n$ is the $n$-fold product set of the floating-point system $S_{b,l}$ specified in Definition 5.6. We use this set $S_{b,l}$ in this section, allowing arbitrary integer exponents. We do this to avoid complicated underflow and overflow considerations during the derivation of the algorithm. Once the algorithm is derived, we discuss the range of exponents that can occur during its execution.

We recall our notation for a floating-point number $x \in S_{b,l}$:

$$\begin{aligned} x = mx \cdot b^{ex} = 0.x[1]x[2]\cdots x[l] \cdot b^{ex} &= \sum_{i=1}^{l} x[i] \cdot b^{ex-i} \\ &= x[1] \cdot b^{ex-1} + x[2] \cdot b^{ex-2} + \cdots + x[l] \cdot b^{ex-l}. \end{aligned}$$

We now define a binary relation $\prec$ on the set

$$S_b^* := \bigcup_{l=1}^{\infty} S_{b,l}$$

as

$$\bigwedge_{x,y \in S_b^*} (x \prec y :\Leftrightarrow x = 0 \vee y = 0 \vee (ex \leq ey \wedge y \in S_{b,ey-ex})).$$

Here $\prec$ is an antireflexive ordering on the set $S_b^* \backslash \{0\}$. Since $y \in S_{b,ey-ex}$ means that

$$y = y[1]b^{ey-1} + y[2]b^{ey-2} + \cdots + y[ey-ex]b^{ex} = \sum_{i=1}^{ey-ex} y[i] \cdot b^{ey-i},$$

then $x \prec y$ in $S_b^* \backslash \{0\}$ if and only if all digits of $x$ have smaller exponents than all nonzero digits of $y$.

The following lemma connects the relation $\prec$ with addition.

**Lemma 6.1:** Let $S = S_{b,l}$ ($b$ even) be a floating-point system and $\bigcirc : \boldsymbol{R} \to S$ the rounding to the nearest floating-point number. Then for all $x, y \in S$ the following properties hold:

(a) $s := \bigcirc(x + y) \in S \wedge r := (x + y) - s \in S$,

(b) $r \prec s, r \neq 0 \Rightarrow es - er \geq l$,

(c) $\bigwedge_{z \in S} (z \prec x \wedge z \prec y \Rightarrow z \prec r \wedge z \prec s)$,

(d) $\bigwedge_{z \in S} (x \prec z \vee y \prec z \Rightarrow r \prec z)$.

*Proof*: If $x = 0$ or $y = 0$ or $x + y = 0$, then $r = 0$ also, and obviously (a)–(d) hold. Otherwise, we define $d := ex - ey$ and assume without loss of generality that $d \geq 0$. We consider the following cases:

**Case 1:** $d > l \wedge \neg(d = l + 1 \wedge \text{sgn}(x) \neq \text{sgn}(y) \wedge mx = b^{-1})$. Then $s = x \wedge r = y \Rightarrow$ (a)–(d). We illustrate by a simple example that the case in the parentheses contradicts the conclusion that $s = x \wedge r = y$. Let $b = 10$, $l = 3$, $d = l + 1 = 4$ and $x = 0.100 \cdot 10^6$ and $y = -0.513 \cdot 10^2$. Then we obtain $x + y = 0.0999487 \cdot 10^6$, $s = \bigcirc(x + y) = 0.999 \cdot 10^5$, and $r = x + y - s = 0.487 \cdot 10^2$, which shows that $s \neq x$ and $r \neq y$. It is obvious, however, that the case in the parentheses is the only situation consistent with $d > l$, for which the conclusion cannot be drawn.

**Case 2:** $(d \leq l \vee (d = l + 1 \wedge \text{sgn}(x) \neq \text{sgn}(y) \wedge mx = b^{-1}))$. Then

$$x + y \in S_{b,2l} \backslash \{0\}^{\dagger} \Rightarrow x + y = * \sum_{i=1}^{2l} m[i] \cdot b^{e-i}, \tag{3}$$

† We exclude the case $x + y = 0$, the proof having already dealt with it.

with $* \in \{+, -\}, m[i] \in \{0, 1, \ldots, b-1\}, m[1] \neq 0, e \in \mathbf{Z}$. One of the following two situations prevail. Either

$$s = * \sum_{i=1}^{l} m[i] \cdot b^{e-i} \wedge r = * \sum_{i=l+1}^{2l} m[i] \cdot b^{e-i} \tag{4}$$

or

$$s = * \left( \sum_{i=1}^{l} m[i] \cdot b^{-i} + b^{-l} \right) \cdot b^{e} \wedge r = * \left( \sum_{i=l+1}^{2l} m[i] \cdot b^{-i} - b^{-l} \right) \cdot b^{e}. \tag{5}$$

Here $r$ may be denormalized or even zero. In either of the situations (4) or (5), the properties (a) and (b) obviously hold.

There remains to show (c) and (d).

(c): The cases $x = 0 \vee y = 0$ were already considered above. If $z = 0$, (c) holds by definition of $\prec$. If $x \neq 0$, $y \neq 0$, and $z \neq 0$, then

$$z \prec x \wedge z \prec y :\Leftrightarrow (ez \leq ex \wedge x \in S_{b,ex-ez} \wedge ez \leq ey \wedge y \in S_{b,ey-ez}).$$

The case $x + y = 0$ has already been considered. If $x + y \neq 0$, it is obvious that

$$z \prec x \wedge z \prec y \Rightarrow z \prec x + y.$$

Now a consideration of (3)–(5) shows that $z \prec s \wedge z \prec r$.

(d): If $r = 0$, then $r \prec z$. Otherwise, $er \leq \min(ex, ey)$, in which case $r \prec z$ as well. ■

As in Lemma 6.1, let $x$, $y$ be floating-point numbers in $S_{b,l}$, and let $s := \bigcirc(x + y) = x \oplus y$ and $r := (x + y) - s$. In the expressions for $s$ and $r$, the operations $+$ and $-$ denote exact addition and subtraction for real numbers. The essence of Lemma 6.1 is that for any pair $x, y \in S_{b,l}$, both the rounded sum $s$ and the correct error are elements of $S_{b,l}$, i.e., they can be stored as floating-point numbers in the computer. Indeed, by examining the proof of Lemma 6.1, we see that $r$ is produced in the computer through calculation of $s$ itself.

Now consider $n$ floating-point numbers $x_1, x_2, \ldots, x_n \in S_{b,l}$. The most direct way of adding them is the following:

$$\begin{aligned} &s := 0; \\ &\text{for} \quad i := 1(1)n \quad \text{do} \quad s := s \oplus x. \end{aligned} \tag{6}$$

Note that (6) is frequently used even though it does not in general deliver the accuracy required for (1). Alternatively, we may execute (6) and store the errors as well. This leads to the following process:

$$\begin{array}{ll}
s_1 := x_1; & \\
s_2 := s_1 \oplus x_2; & r_1 := (s_1 + x_2) - s_2; \\
s_3 := s_2 \oplus x_3; & r_2 := (s_2 + x_3) - s_3; \\
\vdots & \vdots \\
s_n := s_{n-1} \oplus x_n; & r_{n-1} := (s_{n-1} + x_n) - s_n.
\end{array}$$

Now the values of the errors can be used successively to correct the sum $s_n$, until the requisite accuracy is achieved. The following summation algorithm, which is described in Lemma 6.2, is based on this idea.

**Lemma 6.2:** Let $S = S_{b,l}$, and consider an $n$-tuple of floating-point numbers $x^{(0)} = (x_1^{(0)}, x_2^{(0)}, \ldots, x_n^{(0)}) \in S^n$. Starting with $x^{(0)}$, we determine a sequence $\{x^{(k)}\}_{k=0,1,2,}$ with $x^{(k)} = (x_1^{(k)}, x_2^{(k)}, \ldots, x_n^{(k)}) \in S^n$ recursively as follows:

$$\begin{array}{ll}
s_1^{(k)} := x_1^{(k-1)}; & \\
s_2^{(k)} := s_1^{(k)} \oplus x_2^{(k-1)}; & x_1^{(k)} := (s_1^{(k)} + x_2^{(k-1)}) - s_2^{(k)}; \\
s_3^{(k)} := s_2^{(k)} \oplus x_3^{(k-1)}; & x_2^{(k)} := (s_2^{(k)} + x_3^{(k-1)}) - s_3^{(k)}; \\
\vdots & \vdots \\
s_{n-2}^{(k)} := s_{n-3}^{(k)} \oplus x_{n-2}^{(k-1)}; & x_{n-3}^{(k)} := (s_{n-3}^{(k)} + x_{n-2}^{(k-1)}) - s_{n-2}^{(k)}; \\
s_{n-1}^{(k)} := s_{n-2}^{(k)} \oplus x_{n-1}^{(k-1)}; & x_{n-2}^{(k)} := (s_{n-2}^{(k)} + x_{n-1}^{(k-1)}) - s_{n-1}^{(k)}; \\
s_n^{(k)} := s_{n-1}^{(k)} \oplus x_n^{(k-1)}; & x_{n-1}^{(k)} := (s_{n-1}^{(k)} + x_n^{(k-1)}) - s_n^{(k)}; \\
 & x_n^{(k)} := s_n^{(k)};
\end{array} \tag{7}$$

for $k = 1, 2, 3, \ldots$. Then the sequence $\{x^{(k)}\}_{k=1,2,3,\ldots}$ has the following properties:

(a) $\bigwedge_{k \in N} \sum_{i=1}^{n} x_i^{(k)} = \sum_{i=1}^{n} x_i^{(0)}$,

(b)
$$\begin{array}{ll}
k = 1: & x_{n-1}^{(1)} \prec x_n^{(1)} \\
k = 2: & x_{n-2}^{(2)} \prec x_{n-1}^{(2)} \prec x_n^{(2)} \\
k = 3: & x_{n-3}^{(3)} \prec x_{n-2}^{(3)} \prec x_{n-1}^{(3)} \prec x_n^{(3)} \\
\vdots & \\
k = n-1: & x_1^{(n-1)} \prec x_2^{(n-1)} \prec \cdots \prec x_{n-1}^{(n-1)} \prec x_n^{(n-1)}.
\end{array}$$

*Proof*: (a) is an immediate consequence of (7).

(b): By Lemma 6.1(b) we obtain

$$\bigwedge_{k=1,2,3,\ldots} \bigwedge_{p=1(1)n-1} x_p^{(k)} \prec s_{p+1}^{(k)}. \tag{8}$$

We now proceed to prove the properties (b) of this lemma, step by step. The labels (c) and (d) used in the following argument refer to the corresponding properties of Lemma 6.1.

$k = 1$: Since $s_n^{(k)} = x_n^{(k)}$, setting $p = n - 1$ in (8) gives

$$\bigwedge_{k=1,2,3,\ldots} x_{n-1}^{(k)} \prec x_n^{(k)}. \tag{9}$$

This proves (b) for $k = 1$.

$k = 2$: Setting $p = n - 2$ in (8) gives

$$\bigwedge_{k=1,2,3,\ldots} x_{n-2}^{(k)} \prec s_{n-1}^{(k)}. \tag{$\alpha$1}$$

By (9) we obtain

$$\bigwedge_{k=2,3,4,\ldots} (x_{n-1}^{(k-1)} \prec x_n^{(k-1)} \underset{(d)}{\Rightarrow} x_{n-2}^{(k)} \prec x_n^{(k-1)}), \tag{$\beta$1}$$

$$(\alpha 1), (\beta 1) \Rightarrow \bigwedge_{k=2,3,4,\ldots} x_{n-2}^{(k)} \underset{(c)}{\prec} x_{n-1}^{(k)} \underset{(9)}{\prec} x_n^{(k)}. \tag{10}$$

This proves (b) for $k = 2$.

$k = 3$: Setting $p = n - 3$ in (8) gives

$$\bigwedge_{k=1,2,3,\ldots} x_{n-3}^{(k)} \prec s_{n-2}^{(k)}. \tag{$\alpha$2}$$

By (10) we obtain

$$\bigwedge_{k=3,4,5,\ldots} x_{n-1}^{(k-1)} \prec x_{n-1}^{(k-1)} \underset{(d)}{\Rightarrow} x_{n-3}^{(k)} \prec x_{n-1}^{(k-1)}, \tag{$\beta$2}$$

$$(\alpha 2), (\beta 2) \Rightarrow \bigwedge_{k=3,4,5,\ldots} x_{n-3}^{(k)} \underset{(c)}{\prec} x_{n-2}^{(k)} \underset{(10)}{\prec} x_{n-1}^{(k)} \prec x_n^{(k)}. \tag{11}$$

This proves (b) for $p = n - 3$.

$k = 4$: This case is proved analogously by using (8) and (11). Similarly for $k = 5, \ldots, n - 1$. ■

Lemma 6.2 asserts that all of the sums

$$\sum_{i=1}^{n} x_i^{(k)}$$

for $k = 0, 1, 2, \ldots$ are equal and that with increasing $k$ the $x_i^{(k)}$ become progressively ordered with respect to the relation $\prec$. Thus the iteration process is finitely convergent and may be terminated after at most $n - 1$ steps.

The algorithm shown in Figs. 47 and 48 applies the iteration method of Lemma 6.2 in $S_{b,2l}$ to the computation of the sum of $n$ double-length

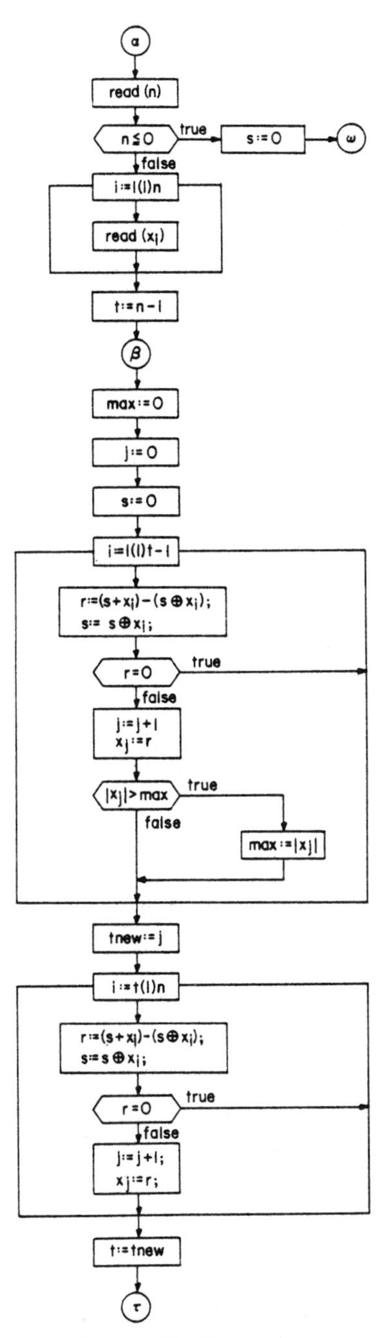


FIGURE 47. Summation

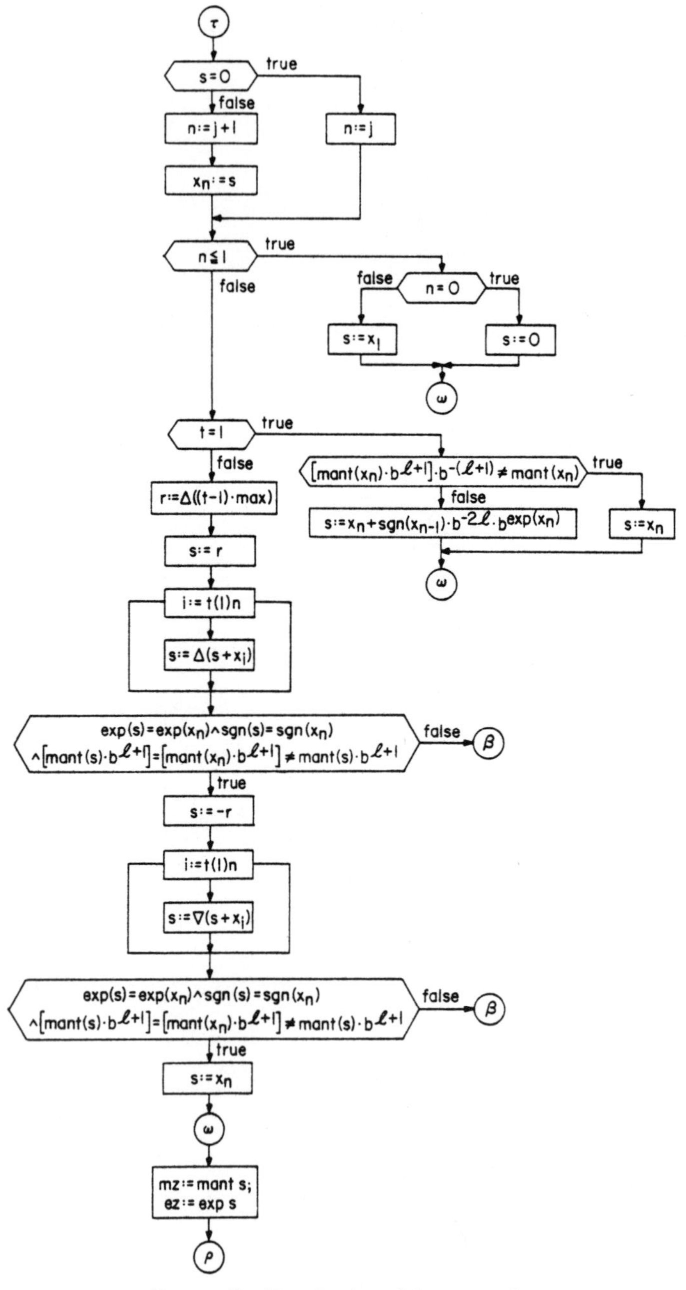


FIGURE 48. Termination of the summation.

floating-point numbers. The variable $t$ in this algorithm plays the role of $n - k$ in property (b) of Lemma 6.2.

The third statement in the summation algorithm (Fig. 47) is described by a symbol that has not yet been used. This symbol denotes a for statement and has the meaning

$$\text{for} \quad i := 1(1)n \qquad \text{do} \quad \text{read}(x_i).$$

Other for-statements that occur in the algorithms of Figs. 47 and 48 are described by the same symbol, and mean: for $i := \ldots$, do the statements contained in the following rectangle. A path (arrow) leading to this rectangle signifies a return to the next loop of the for-statement with the index $i$ increased by unity.

To permit a compact description of the algorithms, all variables are assumed to be either integers or floating-point numbers. In practice during the actual execution, the latter have to be decomposed into mantissa and exponent. In the algorithms, the operation signs $+, -, \cdot$ denote precise real (resp. integer) operations. Rounded floating-point operations are denoted as usual by enclosure with the corresponding rounding symbol $\bigcirc, \bigtriangledown$, or $\bigtriangleup$.

The second for-statement in Fig. 47 determines the sum of all the $x_i$, which are not yet ordered with respect to $\lessdot$. It also determines the maximum of these elements. The next for-statement in Fig. 47 continues the summation with these $x_i$, which are already ordered. Errors with the value zero are eliminated in both of these for-statements.

The first statement in the second and third for-statements of Fig. 47 represents the actual summation step.

Since all of the $x_i$, $i = 1(1)n$, are assumed to be floating-point numbers with a mantissa of length $2l$, the addition in question can be performed either by using an accumulator of $4l + 1$ digits of base $b$ after the point and an additional binary digit in front of the point or by using an accumulator of $2l + 2$ digits of base $b$ and an additional binary digit after the point as well as a binary digit in front of the point.

As we have noted above, the execution of the addition $s := s \oplus x_i$ automatically determines the corresponding $r$.

The statement $r := \bigtriangleup((t - 1) \cdot \max)$ in Fig. 48 determines an upper bound of the $t - 1$ summands $x_i$, which are not yet ordered, and max is their maximal value.

The next for-statement estimates the error. It determines the sum of those $x_i$ that are already ordered and the upper bound of the as yet unordered summands. The next for-statement shows a corresponding computation of a lower bound.

The iteration stops as soon as the sum $\square \sum x_i$ is calculable for all roundings of the set $\{\bigtriangledown, \bigtriangleup, \square_\mu, \mu = 0(1)b\}$.

The last statement in Fig. 48 symbolizes the decomposition of the sum into exponent and mantissa. This permits a direct transfer to the rounding algorithms described above.

The following Theorem 6.3 shows that the algorithm leads to (1), the desired result.

**Theorem 6.3:** Let $S = S_{b,l}$ be a floating-point system with $l \geq 3$ and $x_i \in S_{b,2l}$, $i = 1(1)n$, $n$ floating-point numbers of double length. Let $\triangledown, \triangle, \square_\mu, \mu = 0(1)b: \boldsymbol{R}^* \to S^*$ be roundings to single-length floating-point numbers. Then the algorithm of Figs. 47 and 48 computes an approximation,

$$s = \widetilde{\sum_{i=1}^{n} x_i} \in S_{b,2l}$$

of the sum $\sum_{i=1}^{n} x_i$, with the property

$$\bigwedge_{(x_i)\in S_{b,2l}^n} \quad \bigwedge_{\square \in \{\triangledown, \triangle, \square_\mu, \mu=0(1)b\}} \square \sum_{i=1}^{n} x_i = \square s. \tag{12}$$

*Proof*: Lemma 6.2 establishes that the $x_i$ are progressively ordered with respect to the relation $\prec$, while their sum remains unchanged. Thus the algorithm stops after at most $n - 1$ iteration steps with $t = 1$.

In order to prove (12), we show that $s$ ultimately equals $\sum x_i$ in sign, exponent, and the first $l + 1$ digits of the mantissa and that the remaining $l - 1$ digits of $s$ are zero if and only if $s = \sum x_i$. We distinguish the following three cases with which the algorithm can stop:

**Case 1:** $n \leq 1 \Rightarrow s = \sum x_i \Rightarrow (12)$.

**Case 2:** We suppose that $n > 1$. If the algorithm stops because $t = 1$, then by Lemma 6.2(b) all $x_i$ are ordered with respect to the relation $\prec$:

$$\bigwedge_{x_i \neq 0} x_1 \prec x_2 \prec \cdots \prec x_n.$$

This means by Lemma 6.1(b) that

$$\exp(x_n) - \exp(x_{n-1}) \geq 2l.$$

Consequently, in general $x_n$ is the result $s$ unless the last $l - 1$ digits of the mantissa of $x_n$ are all zeros. In this case a bit has to be added to (resp. subtracted from) the last digit of the mantissa of $x_n$ to take into account the influence of $x_{n-1}$ on the result. Since $l \leq 3$, then one of the last $l - 1$ digits of $s$ differs from zero.

**Case 3:** We suppose that $n > 1 \wedge t > 1$. If the algorithm stops because the summands $x_1, x_2, \ldots, x_{n-1}$ have no significant influence on $x_n$, i.e., if they cannot change the sign, the exponent, or the first $l+1$ digits of the mantissa of $x_n$ and if they do not change the last $l-1$ digits of $x_n$ into zeros, then $x_n$ is the result $s$. ■

## REMARKS AND COMMENTS

1. In practice the result is often already available with the required accuracy after the first summation. Then iteration is not required, and the running time is proportional to $n$.

2. If the exponents of the summands $x_i \in S_{b,2l}$, $i = 1(1)n$, are bounded by integers $e1$ and $e2$, i.e., if

$$\bigwedge_{i=1(1)n} \exp x_i \in \{e1, e1+1, \ldots, e2\},$$

then an exponent range $\{e1 - 2l + 1, \ldots, e2\lceil \log_b(n) \rceil\}$ is needed† for the intermediate values of $s$ and of the $x_i$. In a typical computer, this enlarged exponent range may be accommodated by decomposing the $x_i$ and treating the signed mantissas $mx_i$ and the exponents $ex_i$ as integers separately.

3. The summation algorithm was developed primarily for the computation of scalar products. If only the sum of $n$ single precision floating-point numbers is needed, some storage space is wasted by the algorithm. This could be avoided through use of more complicated termination criteria.

4. In using the summation algorithm described above for the computation of scalar products $\sum_{i=1}^{n} a_i b_i$, we presume that the products $a_i \cdot b_i$, $i = 1(1)n$, are previously calculated. If $a_i, b_i \in S_{b,l}$, the multiplication leads to a floating-point numbers of length $2l$. Thus the summation algorithm is developed for $n$ floating-point numbers of length $2l$. The actual summation then can be performed with the accumulators described above, where the longer accumulator, which uses $4l+1$ digits of base $b$, is the more convenient. The length of this accumulator can be shortened in an essential way by use of the following measures:

Instead of precalculating the product $a_i \cdot b_i$, we observe that multiplication of two floating-point numbers $x$ and $y$ with the mantissas,

$$mx = 0.x[1]x[2]\ldots x[l] \qquad \text{and} \qquad my = 0.y[1]y[2]\ldots y[l],$$

by means of the formula

$$mx \cdot my = mx \cdot y[1] \cdot b^{-1} + mx \cdot y[2] \cdot b^{-2} + \cdots + mx \cdot y[l] \cdot b^{-l}$$

† $\lceil x \rceil$ denotes the least integer greater than or equal to $x$.

also includes a summation process. Thus this summation can be performed by the Bohlender algorithm. Instead of $n$ summands of length $2l$, we then obtain $ln$ summands of length at most $l + 1$ nonzero digits of base $b$. Then the entire summation process can conveniently be performed with an accumulator of $2l + 3$ digits of base $b$ after the point plus one binary digit in front of the point.

## 11. AN ALTERNATIVE ALGORITHM FOR COMPUTING THE SUM OF *n* FLOATING-POINT NUMBERS

In this section we discuss an alternative algorithm for computing the sum of $n$ floating-point numbers

$$\square \sum_{i=1}^{n} x_i \tag{1}$$

for all roundings $\square \in \{\triangledown, \triangle, \square_\mu, \mu = 0(1)b\}$. This alternative actually predates the preceding algorithm, see [43, 40]. It proceeds by sorting the numbers $x_i$ according to their exponents. For large $n$ this sorting process requires a time proportional to $n\lceil \log_2 n \rceil$. In practice the exponent is often represented by only two decimal digits or eight binary digits (one byte), and the sorting of eight binary digits can be a very fast process. Thus, whether the Bohlender algorithm or the present alternate is the faster may depend on the characteristics of the computer being used.

We present this algorithm in its original form as given in [43]. However, we stress that the sorting process can and should be made efficient, for example, by avoiding the return to storage of summands. Indeed, our description of this process, through use of pointers, indicates how this may be accomplished.

The algorithm to be developed here will be applied to the computation of scalar products of the form

$$\square\left(\sum_{i=1}^{n} a_i b_i\right). \tag{2}$$

Here $a_i$ and $b_i$, $i = 1(1)n$, are floating-point numbers with $l$ digit mantissas. The exact product $a_i b_i$ can then be generated in an accumulator of $L = 2l$ digits. Let us suppose that this is done. Then (2) can be generated if we implement the sum

$$z := \square\left(\sum_{i=1}^{n} x_i\right) = \square\left(\overbrace{\sum_{i=1}^{n} x_i}\right) \tag{3}$$

on the computer, where the $x_i$, $i = 1(1)n$, denote floating-point numbers of $L = 2l$ digit mantissas and $z$ is an $n$ digit floating-point number. Note that the algorithm to be developed can also be used to produce a floating-point number $z$ defined by (3) of $n$, $n + 1, \ldots, L = 2n$ digits by rounding the intermediate result $\widetilde{\sum_{i=1}^{n} x_i}$ to other lengths.

The entire algorithm for implementing (3) can in principle be separated into the following nine steps:

1. decomposition of the $x_i, i = 1(1)n$, i.e., separation of the $x_i$ into exponent part $e_i$ and mantissa $m_i$;
2. elimination of those $x_i, i = 1(1)n$, that are zero;
3. ordering the exponents: $e_1 \geq e_2 \geq \cdots \geq e_n$;
4. ordering the exponents: $e_1 > e_2 > \cdots > e_n$;
5. addition from left to right;
6. addition from right to left;
7. normalization;
8. rounding;
9. composition, i.e., combination of the resulting exponent and mantissa into a floating-point number.

Figure 49 gives a diagram of these nine steps. Between these steps, labels contained in circles are used to separate these steps into explicit flow diagrams, which we discuss below. The algorithm uses an accumulator composed as follows: one digit, which can be a binary digit in front of the point, $L + 2$ digits of base $b$, and an additional binary digit after the point. If $l$ denotes the number of digits of the floating-point mantissa, then $L = 2l$.

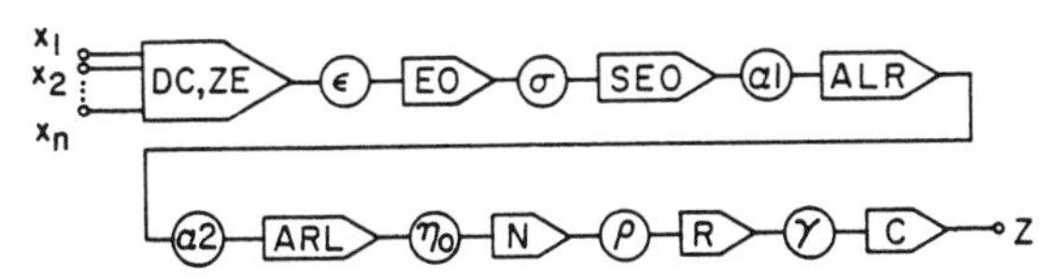


$x_i = m_i \cdot b^{e_i}$; $z = m \cdot b^e$; DC: decomposition; ZE: zero elimination; EO: exponent order; SEO: strong exponent order; ALR: addition from left to right; ARL: addition from right to left; N: normalization; R: rounding; C: composition.

accumulator length:

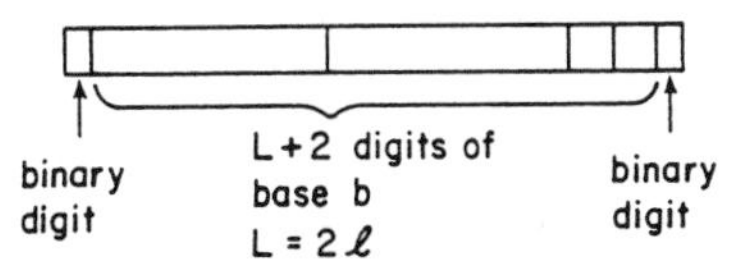


FIGURE 49. Flow diagram of the summation $z = \square(\widetilde{\sum_{i=1}^{n} x_i}) = \square(\sum_{i=1}^{n} x_i)$ and composition of the accumulator.

In the entire process, care must be taken that the formulas (RG), (R1), (R2), and (R4) (resp. (R3)) are strictly realized. With these requirements, further reduction of accumulator length seems difficult (resp. impossible) from any practical point of view.

The following algorithms show how (3) is implemented for the roundings $\square \in \{\nabla, \triangle, \square_\mu, \mu = 0(1)b\}$. The nine steps mentioned above are mutually independent. This means, in particular, that the result $\widetilde{\sum_{i=1}^n x_i}$ can be chosen independently of the rounding, which is later applied to it. In the diagram of Fig. 49, therefore, instead of $R$, any of the roundings $\square \in \{\nabla, \triangle, \square_\mu, \mu = 0(1)b\}$ can be substituted, and we obtain the result defined by that rounding without changing any other part of the entire algorithm. Explicit algorithms for the steps 8 (rounding) and 9 (composition) are not given in this section since they were already discussed in Section 5. In the following diagrams for-statements are described by the same symbol as in Section 5. At any stage of this nine-step process, we use the symbols $x_i$ and $z$ generically. The $x_i$, $i = 1(1)n$, are input variables and are denoted by $x_i = m_i \cdot b^{e_i}$; the output is denoted by $z = m \cdot b^e$. The $k$th digit of $m_i$ (resp. $m$) is denoted by $m_i[k]$ (resp. $m[k]$).

The working of the algorithm is similar to the algorithm for the *execution of the addition with the short accumulator* (Fig. 45) with the replacement there of $n$ by $L$. Since the sum of two floating-point numbers of length $L$ is replaced by the sum of $n$ of them, we first order the $x_i$ with respect to their exponents. Figures 50, 51, 52 give the algorithms for the decomposition and zero elimination, ordering (Fig. 51), and strict ordering of the exponents (Fig. 52).

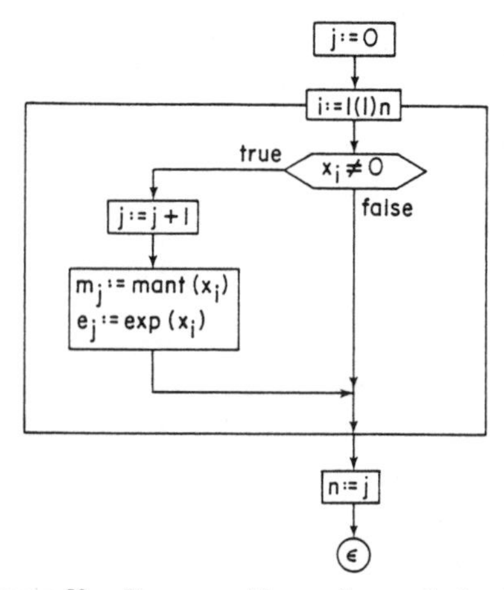


FIGURE 50. Decomposition and zero elimination.

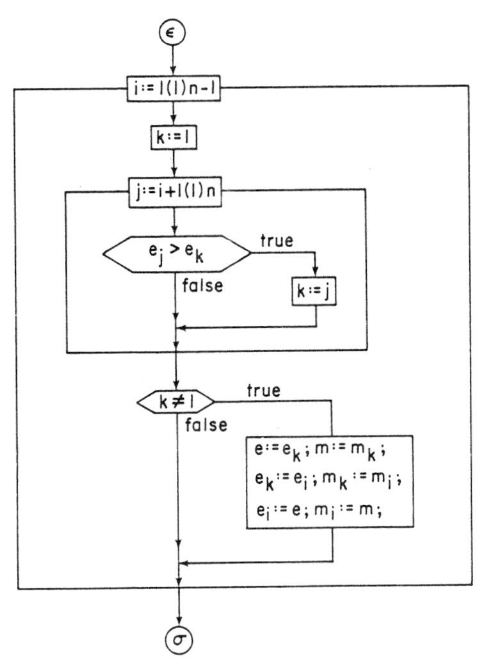


FIGURE 51. Execution of an exponent ordering $e_1 \geq e_2 \geq \cdots \geq e_n$.

After these steps the $x_i$, $i = 1(1)n$, are ordered as follows:

$$e_1 > e_2 > \cdots e_n. \tag{4}$$

The mantissas $m_i$ of the $x_i, i = 1(1)n$, may be denormalized: they may have leading zeros, but they do not have carry bits.

During the ordering process, summand storage should be avoided since this is a time wasting procedure. This ordering is conveniently and rapidly performed on computers in which record and pointer types are available. A floating-point number then can be represented as a record in which the first component represents the exponent and the second component the mantissa. A pointer variable in the third record component can then be used in order to create a new linking of the summands in a list with strictly decreasing exponents.

We now consider the summation process: $n$ elements $x_i$, $i = 1(1)n$, with the property (4) could be added from right to left, as is customary, without accumulation of rounding errors. However, the entire sum $\sum_{i=1}^{n} x_i$ could be affected by catastrophic cancellation. Since in general the magnitude of the

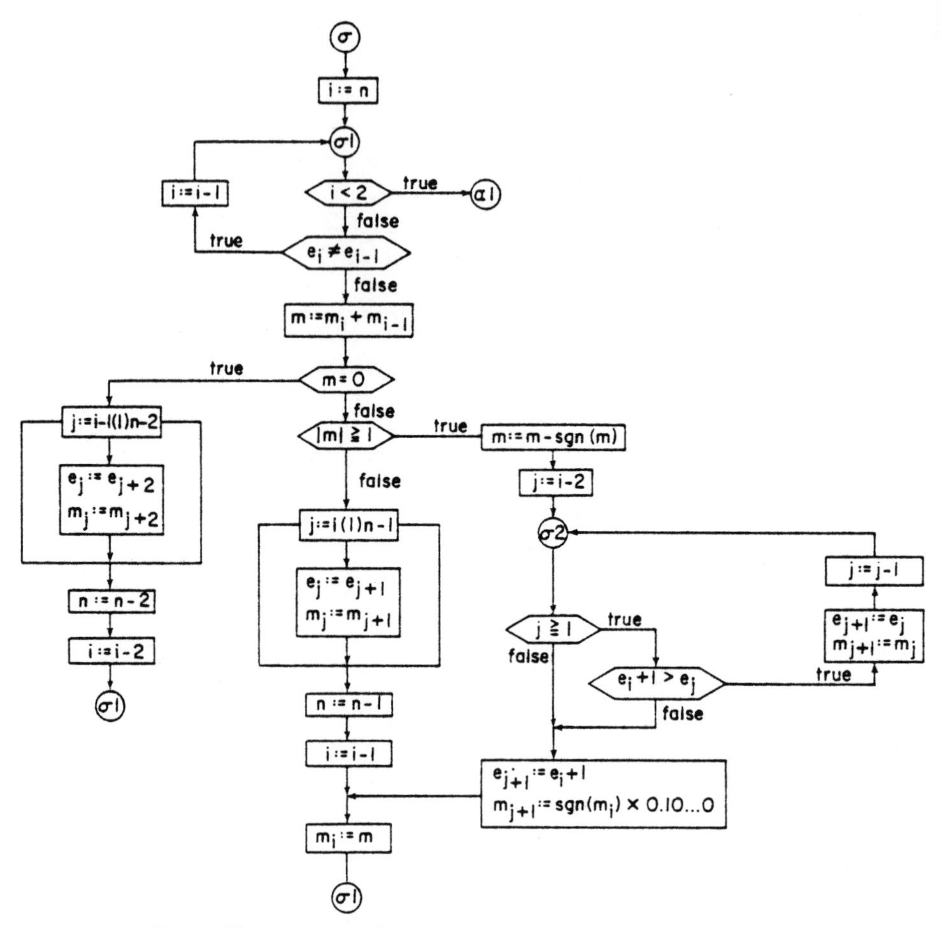


FIGURE 52. Execution of a strong exponent ordering $e_1 > e_2 > \cdots > e_n$.

result $z$ in (3) is unknown in advance, it is difficult to estimate the number of digits over which this addition would have to be carried out. Therefore, we begin the entire summation process with an addition from left to right as described in Fig. 53. It delivers the magnitude of the result $z$ and continues as long as it may do so with complete precision. When the latter is no longer possible, the algorithm of Fig. 53 terminates. The output mantissa $m$ has no leading zeros, but it may possibly have a carry and $L + 2$ digits of base $b$ on the right-hand side of the point. Since the accumulator has $L + 2$ digits (cf. Fig. 49), then

$$e - e_k \geq 3, \tag{5}$$

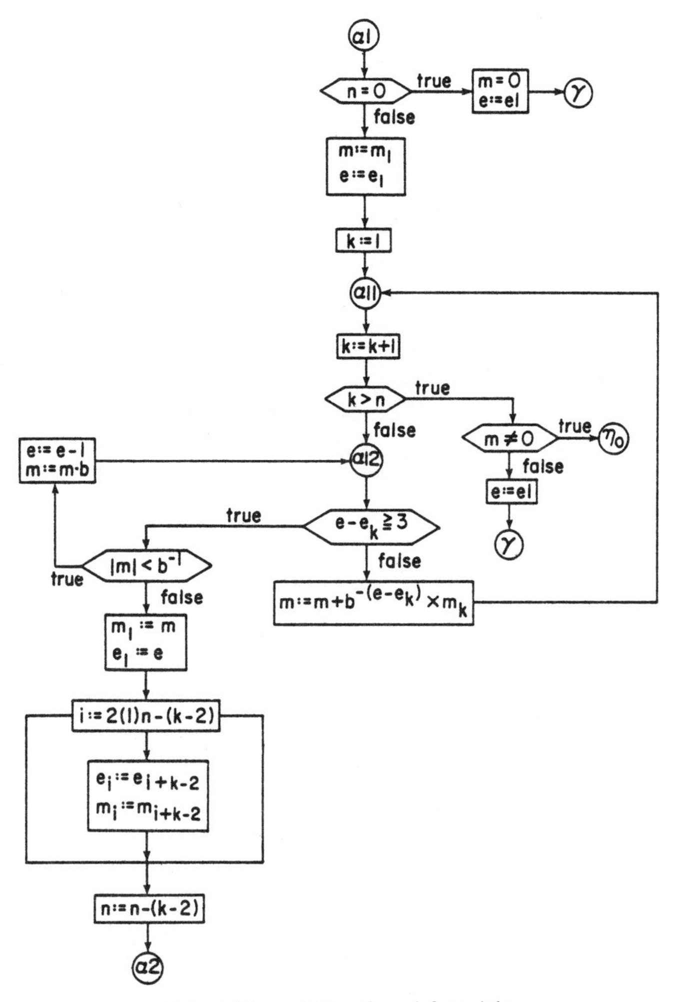


FIGURE 53. Addition from left to right.

where $k - 1$ is the number of elements already added at termination. Then

$$\left|mb^e + \sum_{i=k}^{n} m_i b^{e_i}\right| \geq b^{e-1} - \sum_{i=k}^{n} b^{e_i} \underset{(4)}{\geq} b^{e-1} - b^{e_k} \sum_{i=0}^{n-k} b^{-i}$$

$$> b^{e-1} - b^{e_k} \sum_{i=0}^{\infty} b^{-i} \underset{(5)}{\geq} b^{e-1} - b^{e-3} \frac{1}{1 - b^{-1}}$$

$$> b^{e-1} - b^{e-2} > b^{e-2}.$$

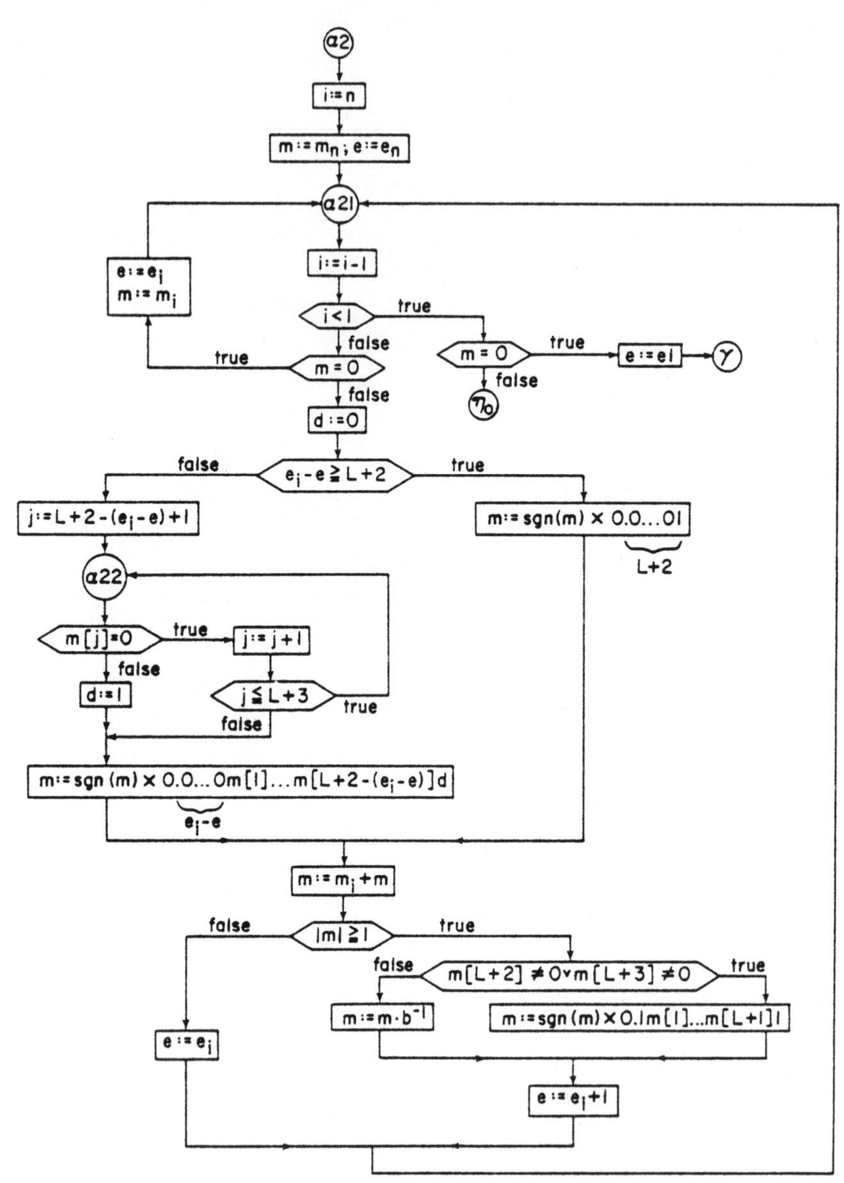


FIGURE 54. Addition from right to left.

Thus in the entire sum a cancellation of at most one additional digit can occur, and the remaining summands, therefore, can be added from right to left as described in Fig. 54. For this purpose we again use the accumulator that was described by Fig. 49. In order to allow a simple description in the algorithm of Fig. 54, we have assumed that the $(L + 3)$rd digit is a full digit of base $b$. It could, however, be represented by a single binary digit.

The algorithm in Fig. 54 adheres to the strategy that after every addition the intermediate or final result $m$ is correct to within the first $L + 2$ digits to the right of the point, as well as the carry digit, if the latter occurs. The $(l + 3)$rd digit to the right of the point carries the information that is required in order to obtain the correct result when one of the roundings $\nabla, \triangle, \square_0$, or $\square_b$ is used.

Since after the addition from left to right $m$ may possess $L + 2$ digits of base $b$, the case $e_i - e \leq 2$ cannot be treated in a separate and simpler branch as was done for the addition of two elements of length $L$ (cf. Fig. 45).

Figure 55 gives the algorithm for the normalization.

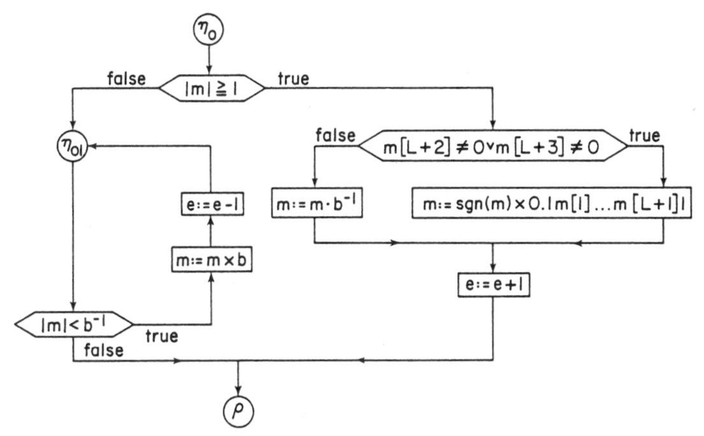


FIGURE 55. Normalization.

## 12. THE LEVEL 3 ROUTINES

This section contains a brief discussion of the level 3 arithmetic. The routines associated with this arithmetic are supposed to furnish all operations defined in the sets of Fig. 1 as well as between these sets. The basic ideas of these operations were already extensively studied in Chapters 3 and 4. Thus our objective is simply to summarize the definition of these operations and to point out that they all can be performed by using the level 2 operations.

We don't derive algorithms for the level 3 arithmetic because they are simple. In principle, they present no difficulties provided the level 3 operations are clearly defined and the level 2 routines are available. Once more we consider the five basic data sets (types) $\boldsymbol{Z}$, $S$, $\boldsymbol{CS}$, $\boldsymbol{IS}$, $\boldsymbol{ICS}$, already mentioned in the Introduction. In an appropriate programming language they may be called *integer*, *real*, *complex*, *real interval*, *complex interval*. We assume here that the integer arithmetic in $\boldsymbol{Z}$ is available in the computer hardware. We now define the arithmetic in the remaining basic data sets, beginning with $S$.

(a) The arithmetic in $S$ (*real*). We assumed throughout the preceding chapters that the floating-point arithmetic in $S$ is defined by semimorphism, i.e., by means of the formula

$$\text{(RG)} \quad \bigwedge_{a,b \in S} a \boxast b := \square(a * b) \qquad \text{for} \quad * \in \{+, -, \cdot, /\},$$

where $\square$ denotes one of the monotone and antisymmetric roundings $\square_\mu$, $\mu = 0(1)b$. All of these roundings, as well as the operations (RG), are already included in the level 2 arithmetic.

(b) The arithmetic in $\boldsymbol{CS}$ is also defined by semimorphism,

$$\text{(RG)} \qquad \bigwedge_{(a,b),(c,d) \in \boldsymbol{CS}} (a, b) \boxast (c, d) := \square((a, b) * (c, d)), \qquad * \in \{+, -, \cdot, /\},$$

where the rounding $\square : \boldsymbol{C} \to \boldsymbol{CS}$ is defined by

$$\bigwedge_{(a,b) \in \boldsymbol{C}} \square(a, b) := (\square a, \square b).$$

Once again the rounding on the right-hand side denotes one of the roundings $\square_\mu, \mu = 0(1)b$. This leads to the following explicit representation of the operations for all couples of elements $(a, b), (c, d) \in \boldsymbol{CS}$:

$$(a, b) \boxplus (c, d) = (a \boxplus c, b \boxplus d),$$
$$(a, b) \boxminus (c, d) = (a \boxminus c, b \boxminus d),$$
$$(a, b) \boxdot (c, d) = (\square(ac - bd), \square(ad + bc)),$$
$$(a, b) \boxslash (c, d) = \left(\square\left(\frac{ac + bd}{c^2 + d^2}\right), \square\left(\frac{bc - ad}{c^2 + d^2}\right)\right).$$

Division is not defined if $c = d = 0$.

Thus addition and subtraction can be executed in terms of the corresponding operations in $S$. The products occurring on the right-hand side of the multiplication formula lead to floating-point numbers of length $2l$. The summation and the rounding appearing in the formula for multiplication can, for instance, be performed by the Bohlender algorithm.

The quotient can be performed by another application of the Bohlender algorithm. In the quotient formula expressions of the form

$$q := (x_1 y_1 + x_2 y_2)/(u_1 v_1 + u_2 v_2)$$

occur with $x_i, y_i, u_i, v_i \in S_{b,l}$. This quotient is the only operation, the execution of which was not extensively studied so far. Let us, therefore, comment briefly on how it can be performed. See [19]. First compute the products $x_i y_i$ and $u_i v_i$ of length $2l$. Then compute $l + 8$ digit approximations $\tilde{n}$ and $\tilde{d}$ for the numerator and denominator of the quotient $q$. With these expressions, one gets an approximation $\tilde{q} := \tilde{n}/\tilde{d}$ of $l + 5$ digits for the exact quotient $q$. In the most cases it turns out that $\square\tilde{q} = \square q$ for all roundings $\square \in \{\bigtriangledown, \bigtriangleup, \square_\mu, \mu = 0(1)b\}$. If $\square\tilde{q} \neq \square q$, the correct result $\square q$ can be determined from $\tilde{q}$, and the residual

$$r := (u_1 \cdot v_1 \cdot \tilde{q} + u_2 \cdot v_2 \cdot \tilde{q} - x_1 \cdot y_1 - x_2 \cdot y_2).$$

Since cancellation occurs in the computation of this expression, the Bohlender algorithm has to be applied for its evaluation. See [19].

The order relation $\leq$ in $\boldsymbol{CS}$ is defined in terms of the order relation in $S$:

$$(a, b) \leq (c, d) :\Leftrightarrow a \leq c \wedge b \leq d.$$

(c) The arithmetic in $\boldsymbol{IS}$ is defined by the semimorphism

$$\text{(RG)} \qquad \bigwedge_{A,B \in \boldsymbol{IS}} A \mathbin{\Diamond\!\!\!\!\!\ast} B := \Diamond(A \boxast B)$$

with the monotone upwardly directed rounding $\Diamond: \overline{\boldsymbol{IR}} \to \overline{\boldsymbol{IS}}$.

In Chapter 4 we derived the following formulas for the execution of (RG) for all couples of intervals $[a, b], [c, d] \in \boldsymbol{IS}$:

$$\begin{aligned}
[a, b] \mathbin{\Diamond\!\!\!\!\!+} [c, d] &= [a \mathbin{\bigtriangledown\!\!\!\!\!+} c, b \mathbin{\bigtriangleup\!\!\!\!\!+} d], \\
[a, b] \mathbin{\Diamond\!\!\!\!\!-} [c, d] &= [a \mathbin{\bigtriangledown\!\!\!\!\!-} d, b \mathbin{\bigtriangleup\!\!\!\!\!-} c], \\
[a, b] \mathbin{\Diamond\!\!\!\!\cdot} [c, d] &= [\min\{a \mathbin{\bigtriangledown\!\!\!\!\cdot} c, a \mathbin{\bigtriangledown\!\!\!\!\cdot} d, b \mathbin{\bigtriangledown\!\!\!\!\cdot} c, b \mathbin{\bigtriangledown\!\!\!\!\cdot} d\}, \\
&\phantom{= [} \max\{a \mathbin{\bigtriangleup\!\!\!\!\cdot} c, a \mathbin{\bigtriangleup\!\!\!\!\cdot} d, b \mathbin{\bigtriangleup\!\!\!\!\cdot} c, b \mathbin{\bigtriangleup\!\!\!\!\cdot} d\}], \\
[a, b] \mathbin{\Diamond\!\!\!\!/} [c, d] &= [\min\{a \mathbin{\bigtriangledown\!\!\!\!/} c, a \mathbin{\bigtriangledown\!\!\!\!/} d, b \mathbin{\bigtriangledown\!\!\!\!/} c, b \mathbin{\bigtriangledown\!\!\!\!/} d\}, \\
&\phantom{= [} \max\{a \mathbin{\bigtriangleup\!\!\!\!/} c, a \mathbin{\bigtriangleup\!\!\!\!/} d, b \mathbin{\bigtriangleup\!\!\!\!/} c, b \mathbin{\bigtriangleup\!\!\!\!/} d\}].
\end{aligned}$$

The division is not defined if $0 \in [c, d]$. In the multiplication and division formulas, the minimum and maximum can be determined by using the Tables 3 and 4 in Section 4.5.

The order relations $\leq$ and $\subseteq$ in $\boldsymbol{IS}$ are defined by

$$\begin{aligned}
[a, b] \leq [c, d] &:\Leftrightarrow a \leq c \wedge b \leq d, \\
[a, b] \subseteq [c, d] &:\Leftrightarrow c \leq a \wedge b \leq d.
\end{aligned}$$

Here the relation $\leq$ on the right-hand side denotes the order relation in $S$.

In addition to the arithmetic operations in $\boldsymbol{IS}$, the intersection $\cap$ and the convex hull $\bar{\cup}$ have to be made available:

$$[a,b] \cap [c,d] :\Leftrightarrow [\max\{a,c\}, \min\{b,d\}],$$
$$[a,b] \bar{\cup} [c,d] :\Leftrightarrow [\min\{a,c\}, \max\{b,d\}].$$

The intersection is not defined if $b < c$ or $d < a$.

(d) The arithmetic in $\boldsymbol{ICS}$ is also defined by semimorphism

$$\bigwedge_{\Phi,\Psi \in \boldsymbol{ICS}} \Phi \mathbin{\Diamond\!\!\!\!\ast} \Psi := \Diamond (\Phi \boxed{*} \Psi),$$

where $\Diamond : \overline{\boldsymbol{IC}} \to \overline{\boldsymbol{ICS}}$ denotes the monotone upwardly directed rounding. We showed in Section 7 of Chapter 4 that these operations are isomorphic to the operations in $\boldsymbol{CIS}$ as far as addition, subtraction, and multiplication are concerned. See also Fig. 27. The division in $\boldsymbol{CIS}$ delivers upper bounds for the quotient in $\boldsymbol{ICS}$.

Thus for the execution of the above semimorphism, we obtain the following formulas for all couples of elements $(A,B),(C,D) \in \boldsymbol{CIS}$ with $A = [a_1, a_2], B = [b_1, b_2], C = [c_1, c_2], D = [d_1, d_2] \in \boldsymbol{IS}$:

$$(A,B) \mathbin{\Diamond\!\!\!\!+} (C,D) = (A \mathbin{\Diamond\!\!\!\!+} C, B \mathbin{\Diamond\!\!\!\!+} D) = ([a_1 \mathbin{\nabla\!\!\!\!+} c_1, a_2 \mathbin{\triangle\!\!\!\!+} c_2], [b_1 \mathbin{\nabla\!\!\!\!+} d_1, b_2 \mathbin{\triangle\!\!\!\!+} d_2]),$$
$$(A,B) \mathbin{\Diamond\!\!\!\!-} (C,D) = (A \mathbin{\Diamond\!\!\!\!-} C, B \mathbin{\Diamond\!\!\!\!-} D) = ([a_1 \mathbin{\nabla\!\!\!\!-} c_2, a_2 \mathbin{\triangle\!\!\!\!-} c_1], [b_1 \mathbin{\nabla\!\!\!\!-} d_2, b_2 \mathbin{\triangle\!\!\!\!-} d_1]),$$

$$(A,B) \mathbin{\Diamond\!\!\!\!\cdot} (C,D)$$
$$= (\Diamond(A \boxdot C \boxminus B \boxdot D), \Diamond(A \boxdot D \boxplus B \boxdot C))$$
$$= \left( \left[ \nabla \left\{ \min_{i,j=1,2} (a_i \cdot c_j) - \max_{i,j=1,2} (b_i \cdot d_j) \right\}, \triangle \left\{ \max_{i,j=1,2} (a_i \cdot c_j) - \min_{i,j=1,2} (b_i \cdot d_j) \right\} \right], \right.$$
$$\left. \left[ \nabla \left\{ \min_{i,j=1,2} (a_i \cdot d_j) + \min_{i,j=1,2} (b_i \cdot c_j) \right\}, \triangle \left\{ \max_{i,j=1,2} (a_i \cdot d_j) + \max_{i,j=1,2} (b_i \cdot c_j) \right\} \right] \right),$$

$$(A,B) \mathbin{\Diamond\!\!\!\!/} (C,D) = (\Diamond((A \boxdot C \boxplus B \boxdot D) \boxslash (C \boxdot C \boxplus D \boxdot D)),$$
$$\Diamond((B \boxdot C \boxminus A \boxdot D) \boxslash (C \boxdot C \boxplus D \boxdot D))).$$

The division is defined only if $0 \notin C \boxdot C \boxplus D \boxdot D$. The operations $\boxed{*}$, $* \in \{+, -, \cdot, /\}$, on the right-hand side of the last two formulas denote operations in $\boldsymbol{IR}$. For the execution of $\boxdot$ and $\boxslash$, see Chapter 4, Tables 1 and 2, and Corollary 4.10.

The order relations $\leq$ and $\subseteq$ in $\boldsymbol{CIS}$ are defined by

$$(A,B) \leq (C,D) :\Leftrightarrow A \leq C \wedge B \leq D,$$
$$(A,B) \subseteq (C,D) :\Leftrightarrow A \subseteq C \wedge B \subseteq D.$$

Here the relations $\leq$ and $\subseteq$ on the right-hand side are those in $\boldsymbol{IS}$.

The intersection $\cap$ and the convex hull $\bar{\cup}$ in $\boldsymbol{CIS}$ are defined componentwise:

$$(A, B) \cap (C, D) :\Leftrightarrow (A \cap C, B \cap D),$$
$$(A, B) \bar{\cup} (C, D) :\Leftrightarrow (A \bar{\cup} C, B \bar{\cup} D),$$

Here the operations $\cap$ and $\bar{\cup}$ on the right-hand side are those in $\boldsymbol{IS}$.

We have now completed our discussion that defines inner arithmetic operations and order relations for all the five basic data sets (types) $\boldsymbol{Z}$, $S$, $\boldsymbol{CS}$, $\boldsymbol{IS}$, $\boldsymbol{CIS}$. We still have to define operations and relations between elements of different sets, whenever these may occur.

In case of addition, subtraction, and multiplication, the types of the result of the operation are given in the table of Fig. 56. Figure 56 also describes the result of the division with the one exception that the quotient of two integers is defined to be of type $S$.

| $a$ \ $b$ | $\boldsymbol{Z}$ | $S$ | $\boldsymbol{CS}$ | $\boldsymbol{IS}$ | $\boldsymbol{CIS}$ |
|---|---|---|---|---|---|
| $\boldsymbol{Z}$ | $\boldsymbol{Z}$ | $S$ | $\boldsymbol{CS}$ | $\boldsymbol{IS}$ | $\boldsymbol{CIS}$ |
| $S$ | $S$ | $S$ | $\boldsymbol{CS}$ | — | — |
| $\boldsymbol{CS}$ | $\boldsymbol{CS}$ | $\boldsymbol{CS}$ | $\boldsymbol{CS}$ | — | — |
| $\boldsymbol{IS}$ | $\boldsymbol{IS}$ | — | — | $\boldsymbol{IS}$ | $\boldsymbol{CIS}$ |
| $\boldsymbol{CIS}$ | $\boldsymbol{CIS}$ | — | — | $\boldsymbol{CIS}$ | $\boldsymbol{CIS}$ |

FIGURE 56. Resulting type of operations among the basic data sets.

The general rule for performing the operations upon operands of different types is to lift the operand of the simpler type into the type of the remaining operand by means of a type transfer. Then the operation can be executed as one of the inner operations all of which we have already dealt with. If, for instance, a multiplication $a \cdot b$ has to be performed, where $a \in S$ and $b \in \boldsymbol{CS}$, we transfer $a$ into an element of $\boldsymbol{CS}$ by adjoining an imaginary part that is zero. Then we multiply the two elements of $\boldsymbol{CS}$. Or if, for instance, an addition $a + b$ has to be performed with $a \in \boldsymbol{Z}$ and $b \in \boldsymbol{IS}$, we first transfer $a$ into a floating-point number of $S$ and then this number into the interval $[a, a] \in \boldsymbol{IS}$. Then the addition in $\boldsymbol{IS}$ can be used in order to execute the operation.

A dash in the table of Fig. 56 means that it is not reasonable to define the corresponding operation $a * b$, a priori. According to our point of view, a floating-point number of *S*, for instance, is an approximate representation of a real, while an interval is a precisely defined object. The product of the two, which ought to be an interval, may then not be precisely specified. However, if the user does indeed need, for instance, to multiply a floating-point number of *S* and a floating-point interval of ***IS***, he may do so by employing a transfer function—which may be predefined—which transforms the floating-point operand of *S* into a floating-point interval of ***IS***. However, in doing so, he should be aware of the possible loss of meaning of the interval as a precise bound. This requirement on the part of the user to invoke explicitly the transfer function in these cases is intended to alert him to questions of accuracy in the arithmetic.

The table given in Fig. 57 shows how to perform the intersection and convex hull between operands of different basic data types whenever this is reasonable. If necessary, the type lifting is also performed by automatic type transfer.

| $\cap$ \ $\bar{\cup}$ | *IS* | *CIS* |
|---|---|---|
| *IS* | *IS* | *CIS* |
| *CIS* | *CIS* | *CIS* |

FIGURE 57. Table of intersections and convex hulls between the basic interval types.

We have now completed our definition and discussion of the inner and mixed operations for the five basic data sets. We are now going to define operations for the product sets (matrices and vectors) of these sets. All arithmetic operations are again defined by semimorphisms.

Let $T$ be one of the sets (types) $\boldsymbol{Z}$, $S$, $CS$, $\boldsymbol{IS}$, $\boldsymbol{CIS}$. We denote the set of $m \times n$ matrices with elements in $T$ by

$$M_{mn}T := \{(a_{ij}) \,|\, a_{ij} \in T \text{ for } i = 1(1)m \wedge j = 1(1)n\}.$$

In $M_{mn}T$ we define operations $*: M_{mn}T \times M_{mn}T \to M_{mn}T$ by

$$\bigwedge_{(a_{ij}),(b_{ij}) \in M_{mn}T} (a_{ij}) * (b_{ij}) := (a_{ij} * b_{ij}).$$

Here the operation sign $*$ on the right-hand side denotes certain operations, depending on the following three cases. In

Case $T = \boldsymbol{Z}$: one of the integer operations $+, -$;
Case $T = S$ or $CS$: one of the rounded operations $\boxplus, \boxminus$;
Case $T = \boldsymbol{IS}$ or $\boldsymbol{CIS}$: one of the operations $\diamond\!\!\!+, \diamond\!\!\!-, \cap, \bar{\cup}$.

In addition to these operations, we define outer multiplications $\cdot : T \times M_{mn}T \to M_{mn}T$ by

$$\bigwedge_{a \in T} \bigwedge_{(b_{ij}) \in M_{mn}T} a \cdot (b_{ij}) := (a \cdot b_{ij}).$$

Here the multiplication sign $\cdot$ on the right-hand side has three cases of realization also. In

Case $T = \boldsymbol{Z}$: integer multiplication;
Case $T = S$ or $\boldsymbol{CS}$: rounded multiplication $\boxdot$;
Case $T = \boldsymbol{IS}$ or $\boldsymbol{ICS}$: multiplication $\diamond$.

Finally, we consider the product of matrices. It is a mapping $\cdot : M_{mn}T \times M_{np}T \to M_{mp}T$. These are five cases for definition of this multiplication:

Case $T = \boldsymbol{Z}$:

$$\bigwedge_{(a_{ij}) \in M_{mn}\boldsymbol{Z}} \bigwedge_{(b_{jk}) \in M_{np}\boldsymbol{Z}} (a_{ij}) \cdot (b_{jk}) := \left( \sum_{j=1}^{n} a_{ij} \cdot b_{jk} \right).$$

Case $T = S$:

$$\bigwedge_{(a_{ij}) \in M_{mn}S} \bigwedge_{(b_{jk}) \in M_{np}S} (a_{ij}) \boxdot (b_{jk}) := \left( \square \sum_{j=1}^{n} a_{ij} \cdot b_{jk} \right).$$

Here $\square$ denotes one of the roundings $\square_\mu$, $\mu = 0(1)b$. If the $a_{ij}$ and $b_{ij}$ are floating-point numbers of length $l$, the products $a_{ij} \cdot b_{jk}$ are of length $2l$. The rounded scalar products can be calculated by the Bohlender algorithm.

Case $T = \boldsymbol{CS}$:

$$\bigwedge_{((a_{ij}, b_{ij})) \in M_{mn}\boldsymbol{CS}} \bigwedge_{((c_{jk}, d_{jk})) \in M_{np}\boldsymbol{CS}} ((a_{ij}, b_{ij})) \boxdot ((c_{jk}, d_{jk}))$$

$$:= \left( \left( \square \sum_{j=1}^{n} (a_{ij}c_{jk} - b_{ij}d_{jk}), \square \sum_{j=1}^{n} (a_{ij}d_{jk} + b_{ij}c_{jk}) \right) \right).$$

Here $\square$ denotes one of the roundings $\square_\mu$, $\mu = 0(1)b$. If the $a_{ij}$, $b_{ij}$, $c_{jk}$, and $d_{jk}$ are floating-point numbers of length $l$, the components of the product matrix can be calculated, for example, by the Bohlender algorithm.

Case $T = \boldsymbol{IS}$:

$$\bigwedge_{([a_{ij}, b_{ij}]) \in M_{mn}\boldsymbol{IS}} \bigwedge_{([c_{ij}, d_{ij}]) \in M_{np}\boldsymbol{IS}} ([a_{ij}, b_{ij}]) \circledast ([c_{jk}, b_{jk}])$$

$$:= \left( \left[ \triangledown \sum_{j=1}^{n} \min\{a_{ij}c_{jk}, a_{ij}d_{jk}, b_{ij}c_{jk}, b_{ij}d_{jk}\}, \right. \right.$$

$$\left. \left. \triangle \sum_{j=1}^{n} \max\{a_{ij}c_{jk}, a_{ij}d_{jk}, b_{ij}c_{jk}, b_{ij}d_{jk}\} \right] \right).$$

If the $a_{ij}$, $b_{ij}$, $c_{jk}$, and $d_{jk}$ are floating-point numbers of length $l$, all the products in the curly brackets are of length $2l$ and can be determined exactly. However, it is not necessary to calculate all these products since in order to determine their minimum (resp. maximum), Table 3 in Section 4.5 is to be used. The sum then can be evaluated, for example, by applying the Bohlender algorithm.

Case $T = \boldsymbol{CIS}$: For all $(([a_{ij}^1, a_{ij}^2], [b_{ij}^1, b_{ij}^2])) \in M_{mn}\boldsymbol{CIS}$ and $(([c_{jk}^1, c_{jk}^2], [d_{jk}^1, d_{jk}^2])) \in M_{np}\boldsymbol{CIS}$, we have

$$
\begin{aligned}
&(([a_{ij}^1, a_{ij}^2], [b_{ij}^1, b_{ij}^2])) \mathbin{\Diamond\!\!\!\!\cdot\,} (([c_{jk}^1, c_{jk}^2], [d_{jk}^1, d_{jk}^2])) \\
&:= \Bigg(\Bigg(\Diamond \boxed{\sum_{j=1}^{n}} \{[a_{ij}^1, a_{ij}^2] \boxdot [c_{jk}^1, c_{jk}^2] \boxminus [b_{ij}^1, b_{ij}^2] \boxdot [d_{jk}^1, d_{jk}^2]\}, \\
&\qquad \Diamond \boxed{\sum_{j=1}^{n}} \{[a_{ij}^1, a_{ij}^2] \boxdot [d_{jk}^1, d_{jk}^2] \boxplus [b_{ij}^1, b_{ij}^2] \boxdot [c_{jk}^1, c_{jk}^2]\}\Bigg)\Bigg) \\
&= \Bigg(\Bigg(\Bigg[\nabla \sum_{j=1}^{n} \Bigg\{\min_{r,s=1,2} (a_{ij}^r \cdot c_{jk}^s) - \max_{r,s=1,2} (b_{ij}^r \cdot c_{jk}^s)\Bigg\}, \\
&\qquad \triangle \sum_{j=1}^{n} \Bigg\{\max_{r,s=1,2} (a_{ij}^r \cdot c_{jk}^s) - \min_{r,s=1,2} (b_{ij}^r \cdot c_{jk}^s)\Bigg\}\Bigg], \\
&\qquad \Bigg[\nabla \sum_{j=1}^{n} \Bigg\{\min_{r,s=1,2} (a_{ij}^r \cdot c_{jk}^s) + \min_{r,s=1,2} (b_{ij}^r \cdot c_{jk}^s)\Bigg\}, \\
&\qquad \triangle \sum_{j=1}^{n} \Bigg\{\max_{r,s=1,2} (a_{ij}^r \cdot c_{jk}^s) + \max_{r,s=1,2} (b_{ij}^r \cdot c_{jk}^s)\Bigg\}\Bigg]\Bigg)\Bigg).
\end{aligned}
$$

As before, the products occurring in this expression are of length $2l$.

To determine their minimum and maximum, Table 3 in Section 4.5 can be used. The sums (and differences) then can be evaluated by applying one of the algorithms for an $n$-fold sum.

If $T$ denotes one of the sets (types) $\boldsymbol{Z}$, $S$, $\boldsymbol{CS}$, $\boldsymbol{IS}$, $\boldsymbol{CIS}$, we denote the set of vectors with components in $T$ by

$$V_n T := \{(a_i) \mid a_i \in T \text{ for } i = 1(1)n\}$$

and the set of transposed vectors with components in $T$ by

$$V_n^{\mathrm{T}} T := \{(a_i)^{\mathrm{T}} \mid a_i \in T \text{ for } i = 1(1)n\}.$$

The operations defined above for matrices are directly extended to vectors if we identify the sets

$$V_n T \equiv M_{n1} T, \qquad V_n^{\mathrm{T}} T \equiv M_{1n} T, \qquad T = M_{11} T.$$

| $T = \mathbf{Z}$ : $+, -$ <br> $T = S$ or $\mathbf{CS}$ : $\boxplus, \boxminus$ <br> $T = \mathbf{IS}$ or $\mathbf{CIS}$ : $\diamond\!\!\!+, \diamond\!\!\!-$ <br> $\cap \bar{\cup}$ | $T$ | $M_{mn}T$ | $V_nT$ | $V_n^{\mathrm{T}}T$ |
|---|---|---|---|---|
| $T$ | $T$ | — | — | — |
| $M_{mn}T$ | — | $M_{mn}T$ | — | — |
| $V_nT$ | — | — | $V_nT$ | — |
| $V_n^{\mathrm{T}}T$ | — | — | — | — |

| $T = \mathbf{Z}$ : $\cdot$ <br> $T = S$ or $\mathbf{CS}$ : $\boxdot$ <br> $T = \mathbf{IS}$ or $\mathbf{CIS}$ : $\diamond$ | $T$ | $M_{np}T$ | $V_nT$ | $V_n^{\mathrm{T}}T$ |
|---|---|---|---|---|
| $T$ | $T$ | $M_{np}T$ | $V_nT$ | — |
| $M_{mn}T$ | — | $M_{mp}T$ | $V_mT$ | — |
| $V_nT$ | — | — | — | $M_{nn}T$ |
| $V_n^{\mathrm{T}}T$ | — | $V_p^{\mathrm{T}}T$ | $T$ | — |

FIGURE 58. Type of results for matrix and vector operations

The relevant operators, as well as the types of the results, are shown in the two tables of Fig. 58.

Transposed vectors are only used in order to perform scalar or dyadic products.

The intersection and convex hull of matrices and vectors with components in $\mathbf{IS}$ or $\mathbf{CIS}$ are taken componentwise.

The order relation $\leq$ for matrices of $M_{mn}T$ with $T \in \{\mathbf{Z}, S, \mathbf{CS}, \mathbf{IS}, \mathbf{CIS}\}$ is defined by

$$\bigwedge_{(a_{ij}),(b_{ij}) \in M_{mn}T} (a_{ij}) \leq (b_{ij}) :\Leftrightarrow \bigwedge_{i=1(1)n} \bigwedge_{j=1(1)n} a_{ij} \leq b_{ij}$$

and

$$\bigwedge_{a,b \in M_{mn}T} a < b :\Leftrightarrow a \leq b \wedge a \neq b.$$

In $M_{mn}\mathbf{IS}$, inclusion $\subseteq$ is defined by

$$\bigwedge_{(a_{ij}),(b_{ij}) \in M_{mn}\mathbf{IS}} (a_{ij}) \subseteq (b_{ij}) :\Leftrightarrow \bigwedge_{i=1(1)m} \bigwedge_{j=1(1)n} a_{ij} \subseteq b_{ij}$$

and

$$\bigwedge_{a,b \in M_{mn}IS} a \subset b :\Leftrightarrow a \subseteq b \wedge a \neq b.$$

In $M_{mn}CIS$, inclusion is defined by

$$\bigwedge_{((a_{ij},b_{ij})),((c_{ij},d_{ij})) \in M_{mn}CIS} ((a_{ij}, b_{ij})) \subseteq ((c_{ij}, d_{ij}))$$

$$:\Leftrightarrow \bigwedge_{i=1(1)m} \bigwedge_{j=1(1)n} a_{ij} \subseteq c_{ij} \wedge b_{ij} \subseteq d_{ij}$$

and

$$\bigwedge_{a,b \in M_{mn}CIS} a \subset b :\Leftrightarrow a \subseteq b \wedge a \neq b.$$

We have now completed specification of matrix and vector operations in the cases in which the components of both operands are of the same type for all of the basic data sets (types) $\boldsymbol{Z}$, $S$, $CS$, $IS$, $CIS$. We now comment on the situation in which the components of both operands are of different type.

As before, the general rule for performing the matrix and vector operations for operands of different component type is to lift the operand with the simpler component type into the type of the remaining operand by means of a type transfer. Then the operation can be executed as one of the inner operations all of which we have already dealt with. If, for instance, a matrix

| $+ \ - \ \cap \ \bar{\cup}$ | $T_2$ | $M_{mn}T_2$ | $V_nT_2$ | $V_n^TT_2$ |
|---|---|---|---|---|
| $T_1$ | $T_3$ | — | — | — |
| $M_{mn}T_1$ | — | $M_{mn}T_3$ | — | — |
| $V_nT_1$ | — | — | $V_nT_3$ | — |
| $V_n^TT_1$ | — | — | — | — |

| | $T_2$ | $M_{np}T_2$ | $V_nT_2$ | $V_n^TT_2$ |
|---|---|---|---|---|
| $T_1$ | $T_3$ | $M_{np}T_3$ | $V_nT_3$ | — |
| $M_{mn}T_1$ | — | $M_{mp}T_3$ | $V_mT_3$ | — |
| $V_nT_1$ | — | — | — | $M_{nn}T_3$ |
| $V_n^TT_1$ | — | $V_p^TT_3$ | $T_3$ | — |

FIGURE 59. Type of the result of matrix and vector operations.

$a \in M_{nn}S$ has to be multiplied by a vector $b \in V_n\boldsymbol{C}S$, we transfer $a$ into an element of $M_{nn}\boldsymbol{C}S$ by adding zero imaginary parts to all of the components of $a$. Then we multiply this matrix of $M_{nn}\boldsymbol{C}S$ with the vector $b \in V_n\boldsymbol{C}S$. Or if, for instance, a matrix $a \in M_{mn}\boldsymbol{Z}$ has to be multiplied by a vector $b \in V_n\boldsymbol{I}S$, we first transfer $a$ into a floating-point matrix of $M_{mn}S$ and then into an interval matrix of $M_{mn}\boldsymbol{I}S$ whose components are all point intervals. Then we multiply the matrix $a \in M_{mn}\boldsymbol{I}S$ by the vector $b \in V_n\boldsymbol{I}S$.

Now let $T_1$, $T_2$, and $T_3$ each denote one of the basic data sets (types) $\boldsymbol{Z}$, $S$, $\boldsymbol{C}S$, $\boldsymbol{I}S$, $\boldsymbol{CI}S$. We consider the set of matrices $M_{mn}T_i$, vectors $V_nT_i$, and transposed vectors $V_n^{\mathrm{T}}T_i$, $i = 1(1)3$, whose components are chosen from the basic data types. Among other operations, elements of these sets can be multiplied. The second table in Fig. 59 displays the types of such products. In this table, $T_3$ is to be replaced by the resulting type of Fig. 56 if the components of the operands are of the type $T_1$ and $T_2$, respectively.

A corresponding table for matrix and vector addition, subtraction, convex hull, and intersection is also given in Fig. 59.

To conclude we emphasize that the formulas and the discussion in this section show that all of the level 3 operations and relations can be performed on a computer if the level 2 operations are available.

# Chapter 7 / COMPUTER ARITHMETIC AND PROGRAMMING LANGUAGES

**Summary:** In the preceding chapters we considered five basic data types, which we variously called integer, real, complex, real interval, and complex interval. We also considered the sets of vectors and matrices over these data types. Then we discussed the arithmetic operations that are defined for all of these types. In this chapter we show how to embed all these data types and operations into existing higher programming languages. Most such languages provide only the two numerical data types: integer and real. To accept the new data types and their arithmetic operations in the form of operators, the syntax and semantics of these languages have to be extended. It turns out, in particular, that this affects the input–output statements, as well as the syntax variable for expression. We give a general description of the language extension; we describe the additional standard functions required for the new data types, and we develop the syntax for the additional numerical data types and operations in the form of easily readable syntax diagrams. We do this in the context of PASCAL, *one of the most encompassing of the commonly used programming languages.* Corresponding language extensions for other programming languages can be developed quite analogously. See [78] for the extension of FORTRAN.

## 1. INTRODUCTION

Higher programming languages such as ALGOL, FORTRAN, PASCAL, BASIC, and PL/1 usually only provide the two numerical data types *integer* and *real* as well as the associated operators. On occasion an additional data type, *complex*, is also available. To accommodate all other data types, as well as their operations (as displayed in Fig. 1 and which are discussed in the preceding chapters), requires simulation and packaging into arrays, records, and procedures. This requirement often causes loss of speed and accuracy as well as loss of certain convenient arithmetic properties. Moreover, a complicated notation for arithmetic operations and expressions is required in programs. Each numerical operation has to be written in the form of a procedure call. This makes even simple programs extremely long and difficult to read. Cognizant of such difficulties, we demanded at the outset of this text that *in a good programming system these operations should be available as operators for all admissible data types.*

In this chapter we present an extension of the syntax of programming languages, which accepts operands and variables of all sets or data types displayed in Fig. 1 and which allows for an operational notation of the arithmetic operations defined in these sets. For the sake of clarity and of precision in detail, we do this for a special choice of programming language: PASCAL. PASCAL is one of the most encompassing of the commonly used programming languages. It turns out, in particular, that the concepts of INPUT–OUTPUT STATEMENT and EXPRESSION have to be generalized. The generalization of these syntax variables and their interaction with the remaining syntax of the language are described by the syntax diagrams given below. Corresponding extensions of other programming languages can be developed quite analogously [78].

Along with the integers $\boldsymbol{Z}$, the spaces occurring in numerical computations are displayed in Fig. 1. This leads to the five basic numerical data types $\boldsymbol{Z}$, $S$, $\boldsymbol{CS}$, $\boldsymbol{IS}$, $\boldsymbol{ICS}$ already described in the Introduction above. In an appropriate programming language, they may be called integer, real, complex, real interval, or simply interval and complex interval. For a still simpler notation for these data types in the syntax diagrams to follow, we rename them somewhat and in an obvious way. Thus we obtain the following five basic data types:

| *Symbol* | *Name* | *Meaning* |
|---|---|---|
| $Z$ | integer | subset of integers |
| $R$ | real | real floating-point numbers |
| $C$ | complex | complex floating-point numbers |
| $I$ | interval | floating-point intervals |
| $CI$ | complex interval | complex floating-point intervals |

In addition to the data types $T \in \{Z, R, C, I, CI\}$, we consider the structured types

$VT$ vectors with components of $T$,
$V^{T}T$ transposed vectors with components of $T$,
$MT$ matrices with components of $T$.

In the preceding chapters, we developed two different ways for defining the arithmetic operations within and among these sets. These were called the vertical and the horizontal methods, respectively. As far as the syntax of a programming language is concerned, it is inessential by which such method the operations are defined. The interpretation of the operations is part of the semantics of the language.

However, we once again point out that we strongly favor the horizontal method. It leads to an optimal arithmetic in many respects: accuracy, theoretical describability, closedness of the theory, applicability. Furthermore, the entire arithmetic package as prescribed by Fig. 1 and its context can be defined within the programming language axiomatically *for all rows of Fig. 1* by the simple concept of a semimorphism: Let $N \subseteq M$ be a set–subset pair of Fig. 1 and $\square : M \to N$ a mapping. Then all inner and outer operations in $N$ are concisely defined by the following axioms:

(RG) $\bigwedge_{a,b} a \boxast b := \square(a * b)$,

(R1) $\bigwedge_{a \in N} \square a = a$,

(R2) $\bigwedge_{a,b \in M} (a \leq b \Rightarrow \square a \leq \square b)$,

(R4) $\bigwedge_{a \in M} \square(-a) = -\square a$.

The directed roundings and corresponding operations can also be axiomatically defined in the programming language by (RG), (R1), (R2), and

(R3) $\bigwedge_{a \in M} \nabla a \leq a$ or $\bigwedge_{a \in M} a \leq \Delta a$.

The simplicity of these rules for all arithmetic operations of a programming language as well as the fact that they produce ordered or weakly ordered ringoids or vectoids in all rows of Fig. 1 should be very useful for correctness proofs of arithmetic properties in numerical algorithms and programs.

The inclusion of all such data types and their associated operations into a programming language results in several advantages: Programs become much shorter; they are much easier to read: they allow an improved documentation, and error correction and verification become easier.

## 2. GENERAL DESCRIPTION OF THE LANGUAGE EXTENSION

We now discuss the question of how a programming language has to be extended in order that it accept expressions composed of symbols and operators chosen from within and between the different sets. By this we mean that the syntax of the language should, for instance, allow assigments to be written in the form

$$z := ((a + b) * x - c) * y$$

not only for reals and integers but for all of our data types: complex numbers, intervals, matrices, complex matrices, and so on.

The tables in Figs. 56–59 of Chapter 6 show which operations are to be permitted and which are to be excluded by the syntax. In particular, they prescribe the resultant type of those operations when the operands are of differing types. In such cases we assume that the compiler automatically performs a type conversion from the simpler to the more complicated type.

Thus, we see from Fig. 56 that the complier has to be able to convert integers into reals, complex numbers, intervals or complex intervals, as well as reals into complex numbers and intervals into complex intervals. The same conversions have to be performable for matrices and vectors.

In the Introduction we made a count of the number of operations occurring in computer arithmetic as tabulated in Fig. 1 and its context. We found that there are, in principal, 264 different multiplications, 33 different divisions, 99 different additions and subtractions, and 12 different intersections and takings of the convex hull.

With the automatic type conversions performed by the compiler, these numbers are reduced considerably, and a new count shows that there remain 9 different multiplications, divisions, additions, and subtractions in Fig. 2. This leads to a total of $72(=8 \times 9)$ different multiplications, 9 different divisions, $27(=3 \times 9)$ different additions and subtractions, and $6(=3 \times 2)$ different intersections and takings of the convex hull.

We already observed that these numbers of operations—which are still large—can be reduced to the few level 2 routines discussed in Chapter 6.

All type conversions under discussion are also provided for comparisons $(x \leq y)$ if both operands are of different types as well as for assignment statements ($y := x$) whenever the variable on the left hand-side is of a type into which the type of the expression on the right-hand side is convertible.

As an example, let us consider the product of an integer vector by an interval matrix. According to Fig. 59, the result is an interval vector. Before performing the multiplication, the compiler automatically converts the integer vector into an interval vector in which the lower and upper bounds of each component are identical to the corresponding components of the original integer vector. Then the multiplication of an interval vector by an interval matrix is performed.

Compared with the standard definition of PASCAL, the extended language contains three additional data types: complex ($C$), interval ($I$), and complex interval ($CI$). They are considered as unstructured types, although they may be represented as records or arrays of reals.

Vectors and matrices of the five basic numerical data types are considered to be arrays with one or two scalar index types. For all these data types the operations described above, which are summarized in the tables of Figs. 58 and 59, are to be permitted. Comparisons are also permitted for all these types or between different types if one operand is convertible to the higher level.

In case of intervals, interval matrices, interval vectors, complex intervals, complex interval vectors, and complex interval matrices, the underlying order relation is the set inclusion.

The new data types are also permitted as the output of functions (function procedures).

In input statements, variables of these types, and in output statements, expressions of these types are permitted.

For real and complex numbers, matrices, and vectors, the specification of $\bigtriangledown$ or $\bigtriangleup$ as rounding is permitted in expressions, input–output statements, and constants. If no rounding is specified, the rounding to the nearest number or another monotone and antisymmetric rounding as the standard rounding of the computer is employed.

In input–output statements for real and complex numbers, vectors, and matrices, the rounding selection is written as part of the format specification: For instance, for a real variable $x$, read $(x:\bigtriangleup)$ indicates that $x$ is to be rounded upwardly by the input procedure, and write $(x:\bigtriangledown)$ indicates that $x$ is to be rounded downwardly by the output procedure. For a complete definition of the syntax, see the syntax diagrams to follow. For intervals the lower bound is always rounded downwardly and the upper bound is always rounded upwardly.

Complex numbers are read and written in the form $(x, y)$, intervals in the form $[x, y]$, and complex intervals in the form $([u, v], [x, y])$. Here $u$, $v$, $x$, $y$ are real floating-point numbers.

During the reading of an interval, the relation $x \le y$ is automatically checked. In output statements integer expressions can be specified which control the format of the output; a standard output is used in the case of a default.

Vectors and matrices are read and written componentwise. The format specifications and roundings are applied componentwise too.

In input statements the conversion of integers into reals, complex numbers, intervals, and complex intervals, of reals into complex numbers, and of intervals into complex intervals is permitted. For vectors and matrices these conversions are applied componentwise.

With respect to the order of execution of the operations within an expression, the multiplicative operations ($*, /, \text{div}, \text{mod}$) always have higher priority than the additive operations ($+, -$). Otherwise, operations are always executed from left to right. Therefore, as indicated in the syntax diagrams scalar products are placed in parentheses when they follow /, div, mod.

For real constants and for real and complex operations, a rounding can be specified that indicates whether the constant and the result of the operation shall be rounded upwardly ($\triangle$) or downwardly ($\triangledown$), respectively. If no rounding is specified, the constant and the result shall be rounded by the standard rounding of the computer, which might be the rounding to the nearest number. For constants the rounding is appended to the number (e.g., $0.12345e20\triangle$), for operations it is appended to the operation (e.g., $+\triangle$, $+\triangledown$). Thus in the notation of the preceding chapters, $0.12345e20\triangle$ means $\triangle 0.12345e20$ and $+\triangledown$ (resp. $+\triangle$) means $\mathbin{\triangledown\!\!\!\!\!+}$ (resp. $\mathbin{\triangle\!\!\!\!\!+}$). This convention is adopted for convenience and ease of compilation as it is currently practiced.

For interval operations the lower and upper bounds are always rounded downwardly and upwardly, respectively. The operations $\overline{\cup}$ and $\cap$ denote the hull of the union and the intersection of intervals. On a conventional keyboard, they could be expressed by $++$ and $**$, respectively. If the intersection is empty, the result of the operation $\cap$ is not defined. The operation $\cap$ has lower priority than arithmetic operations and higher priority than $\overline{\cup}$.

## 3. ARITHMETIC STANDARD FUNCTIONS OF THE LANGUAGE EXTENSION

Table 5 shows additional standard functions that are useful if the programming language PASCAL is extended to the new arithmetic data types just discussed. The actual notational form of each such function is given by the syntax diagrams; the argument, result, and meaning are given in Table 5.

TABLE 5

NEW STANDARD FUNCTIONS

| Function | Argument | Result | Meaning |
|---|---|---|---|
| re/im | complex<br>complex interval | real<br>interval | real/imaginary part of the argument |
| inf/sup | interval<br>complex interval | real<br>complex | infimum/supremum of the interval |
| intval | 2 reals<br>2 complex numbers | interval<br>complex interval | composes interval out of two (real/complex) numbers |
| compl | 2 reals<br>2 intervals | complex<br>complex interval | composes a complex number/interval out of two reals/intervals |
| conj | complex<br>complex interval | complex<br>complex interval | conjugate complex number<br>conjugate complex interval |
| transp/herm | vector/matrix | transp. vector/ matrix | constructs the transpose of a vector/matrix, Hermitian in complex case (transpose of a vector only as intermediate result in expressions) |
| disj | 2 intervals<br>2 complex intervals<br>2 interval vectors/matrices | Boolean | determines whether the two intervals/complex intervals/interval vectors/ interval matrices have empty intersection |

In addition to these new standard functions, the traditional standard functions abs, sqr, sqrt, sin, cos, arctan, exp, ln are to be extended to complex and interval arguments.

## 4. SYNTAX OF THE LANGUAGE EXTENSION

To describe the syntax of the language extension, we use syntax diagrams that employ a new notation. This notation was recently introduced in [84] in order to describe the syntax of several programming languages in a unified and easily readable style. Most of the syntax extension described in this section was developed in [10]. For the sake of brevity, we omit a large number of diagrams dealing with TYPE DEFINITION, COMPARISON, ACTUAL ARGUMENT LIST (ACT ARG L), and so on. It is clear how the new data types can be inserted into the diagrams for these syntax variables. Apart from the types *CI*, *VCI*, and *MCI* and some other minor differences, the full set of syntax diagrams for the extended language PASCAL is given in [10]. The diagrams are to be read and interpreted in the following manner: solid lines are to be traversed from left to right and from top to bottom. Dotted lines are to be traversed oppositely, i.e., from right to left and from bottom to top. Syntax variables are written in upper-case letters. Basic symbols (i.e., the so-called terminals of formal language theory) are enclosed in circles. Names for standard functions or word symbols (i.e., sequences of basic symbols) are enclosed in ovals. The first one, two, or three letters of a syntax variable often indicate the relevant numerical space (e.g., $Z, R, C, I, CI, MZ, VZ, \ldots, MC, VC, \ldots, CI, MCI, VCI$). Furthermore, we use the abbreviations E for EXPRESSION, T for TERM, O for OPERAND, SFE for STANDARD FUNCTION EXPRESSION and ID for IDENTIFIER. These among other constructs of the language will be specified in the syntax diagrams to follow.

We note especially that syntax variables for integers are denoted by $Z$, while $I$ stands for interval. Several names for standard functions (trunc, round, ord, . . .) occur in the following syntax diagrams for standard function expressions, but are not explained in this section. Their explanation is part of the definition of PASCAL. One of these the standard function round, is not to be confused with RD (ROUNDING). The latter is explained in the syntax diagram 3 and is the rounding repeatedly referred to in this chapter.

The syntax diagrams that follow are organized and displayed in a pattern, awareness of which is useful in reading them: Diagrams 1 and 2 are the input and output diagrams, respectively. Diagram 3 is the diagram for ROUNDING and diagram 47 is the diagram for EXPRESSION. The intervening syntax diagrams 4–46 are organized into five consecutive sets, which deal, respectively, with the data types $Z$, $R$, $C$, $I$, and $CI$. In each case the first few diagrams define expressions for that type, while the remaining syntax diagrams define expressions for matrices and for vectors with components of that type.

The next section displays a list of all syntax variables used in the diagrams in alphabetical order. A dash in the first column means that no abbreviation is used for this syntax variable. The number in the third column denotes the syntax diagram in which that syntax variable is defined. A dash in the last column indicates that this syntax variable is already defined by the syntax of customary PASCAL.

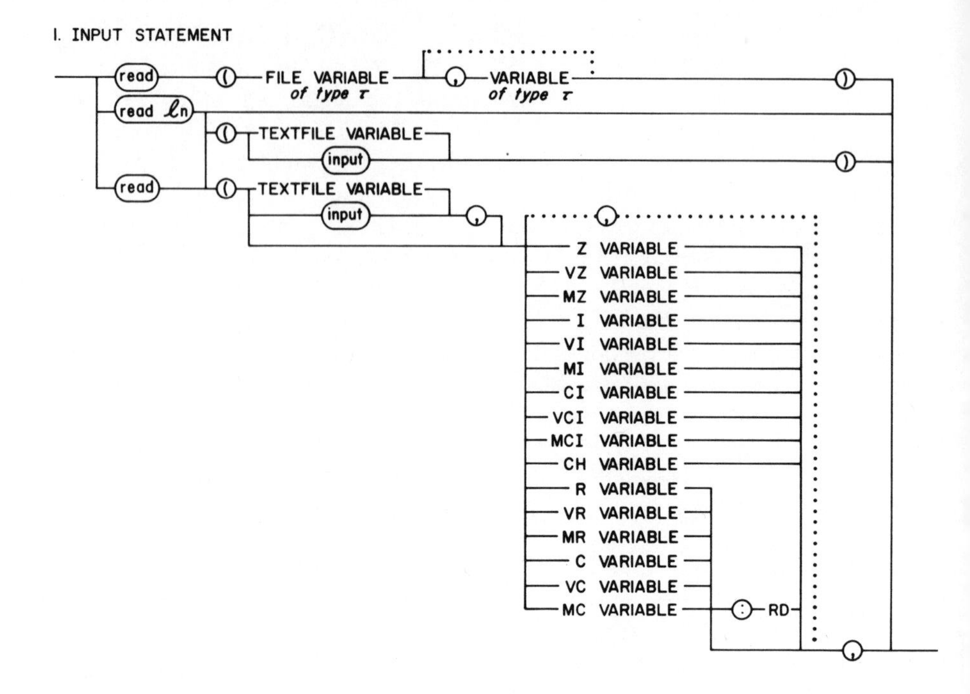
1. INPUT STATEMENT
read
read ℓn
read
FILE VARIABLE
of type τ
VARIABLE
of type τ
TEXTFILE VARIABLE
input
TEXTFILE VARIABLE
input
Z VARIABLE
VZ VARIABLE
MZ VARIABLE
I VARIABLE
VI VARIABLE
MI VARIABLE
CI VARIABLE
VCI VARIABLE
MCI VARIABLE
CH VARIABLE
R VARIABLE
VR VARIABLE
MR VARIABLE
C VARIABLE
VC VARIABLE
MC VARIABLE
RD

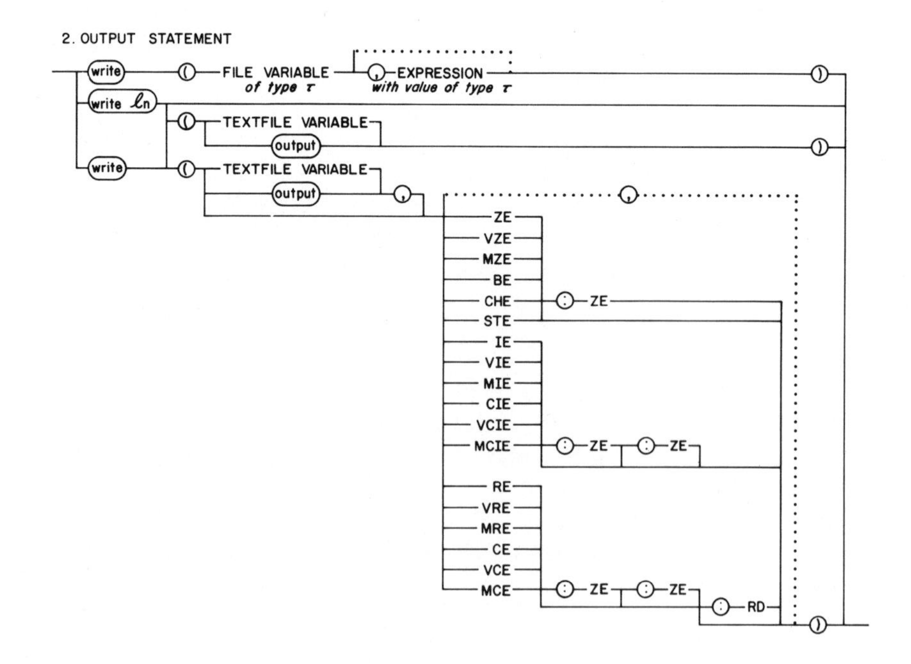
2. OUTPUT STATEMENT
write
write ℓn
write
FILE VARIABLE
of type τ
EXPRESSION
with value of type τ
TEXTFILE VARIABLE
output
TEXTFILE VARIABLE
output
ZE
VZE
MZE
BE
CHE
STE
ZE
IE
VIE
MIE
CIE
VCIE
MCIE
ZE
ZE
RE
VRE
MRE
CE
VCE
MCE
ZE
ZE
RD

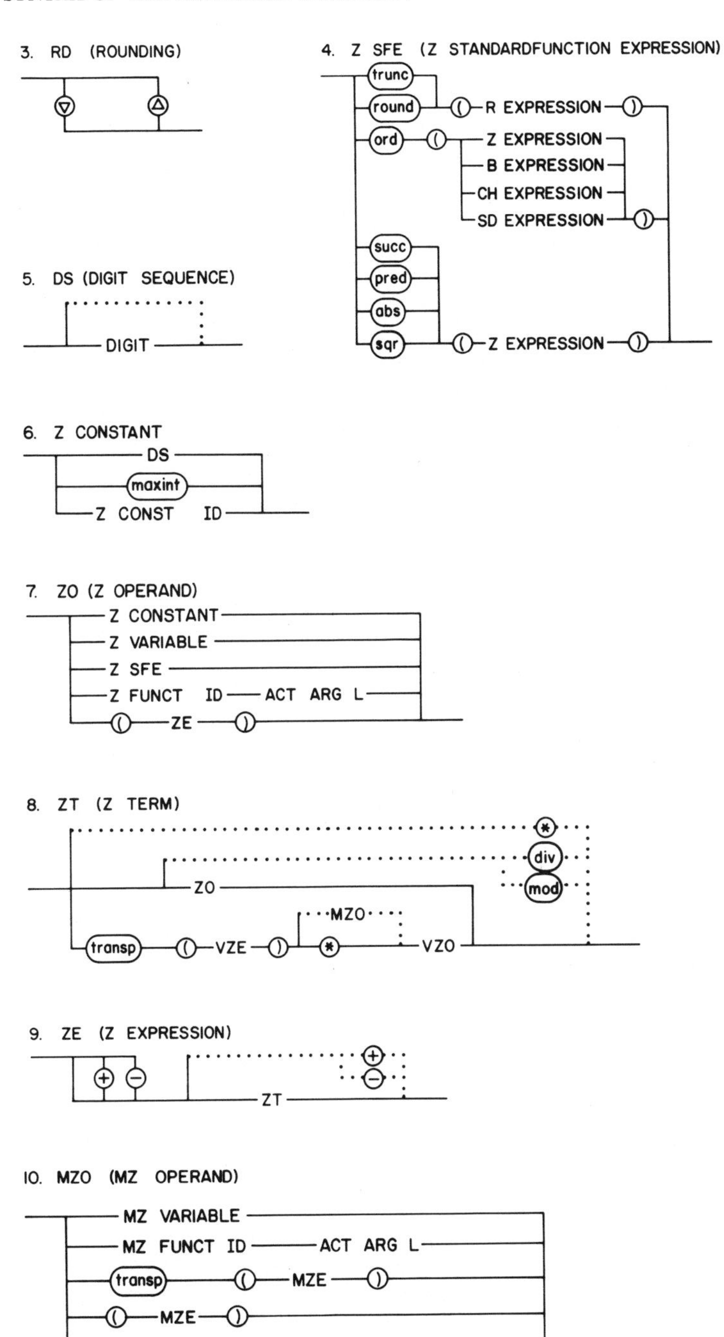
3. RD (ROUNDING)
4. Z SFE (Z STANDARDFUNCTION EXPRESSION)
trunc
round
( R EXPRESSION )
ord
( Z EXPRESSION
B EXPRESSION
CH EXPRESSION
SD EXPRESSION )
succ
pred
abs
sqr
( Z EXPRESSION )
5. DS (DIGIT SEQUENCE)
DIGIT
6. Z CONSTANT
DS
maxint
Z CONST ID
7. ZO (Z OPERAND)
Z CONSTANT
Z VARIABLE
Z SFE
Z FUNCT ID ACT ARG L
( ZE )
8. ZT (Z TERM)
*
div
mod
ZO
MZO
transp ( VZE ) * VZO
9. ZE (Z EXPRESSION)
+ −
ZT
10. MZO (MZ OPERAND)
MZ VARIABLE
MZ FUNCT ID ACT ARG L
transp ( MZE )
( MZE )
VZO * transp ( VZE )

11. MZE (MZ EXPRESSION)

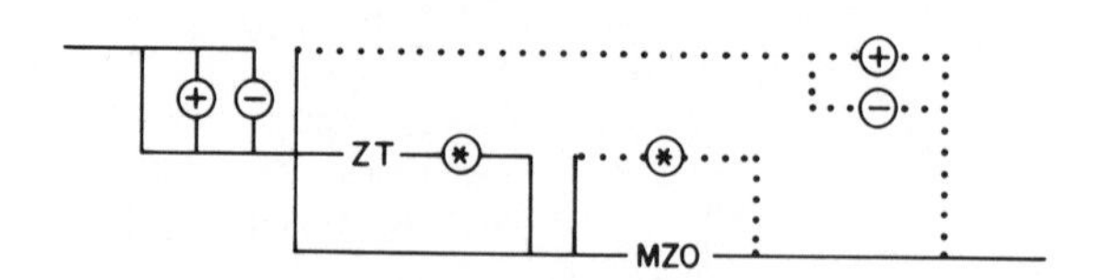


12. VZO (VZ OPERAND)

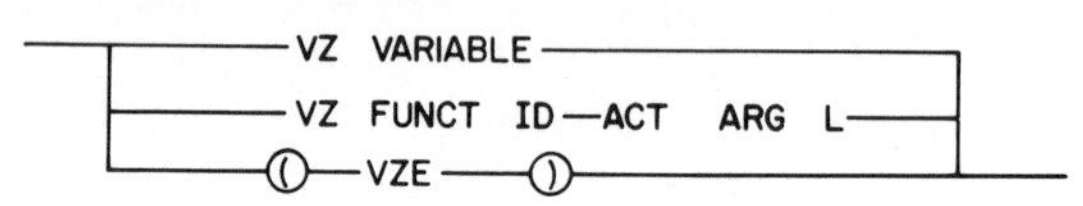


13. VZE (VZ EXPRESSION)

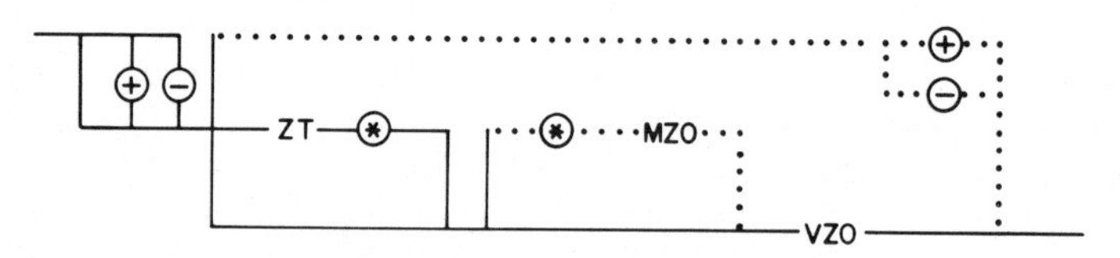


14. R SFE (R STANDARDFUNCTION EXPRESSION)

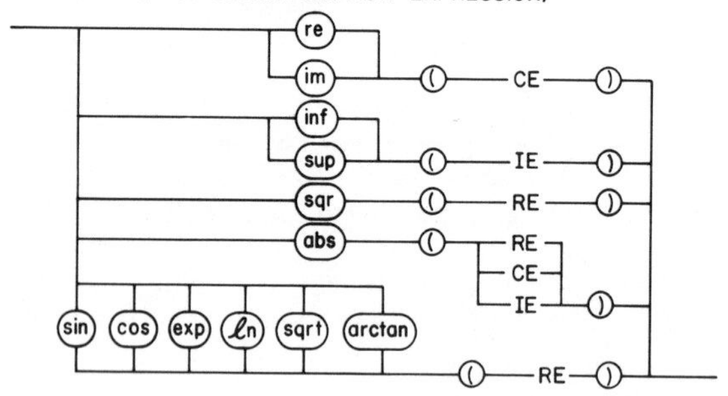


15. R CONSTANT

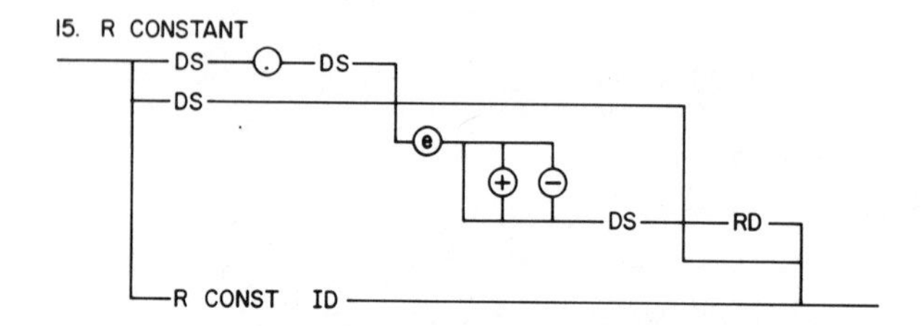

16. RO (R OPERAND)

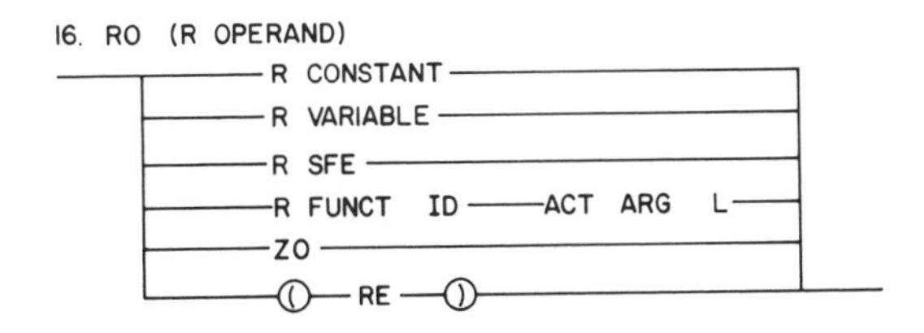


17. RT (R TERM)

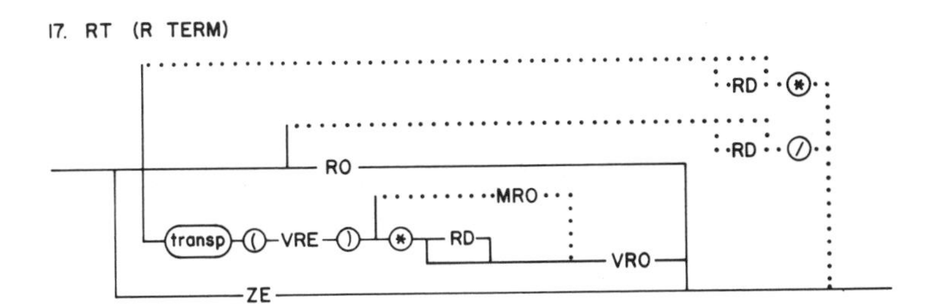


18. RE (R EXPRESSION)

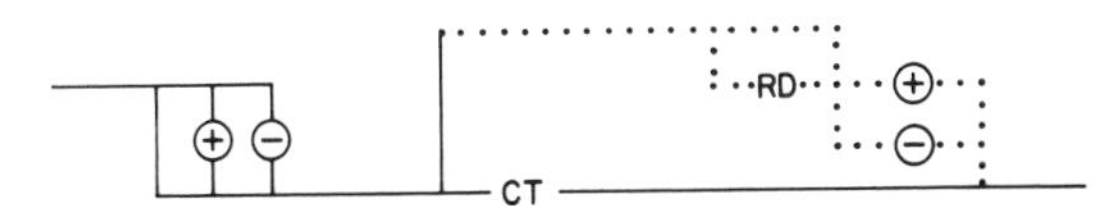


19. MRO (MR OPERAND)

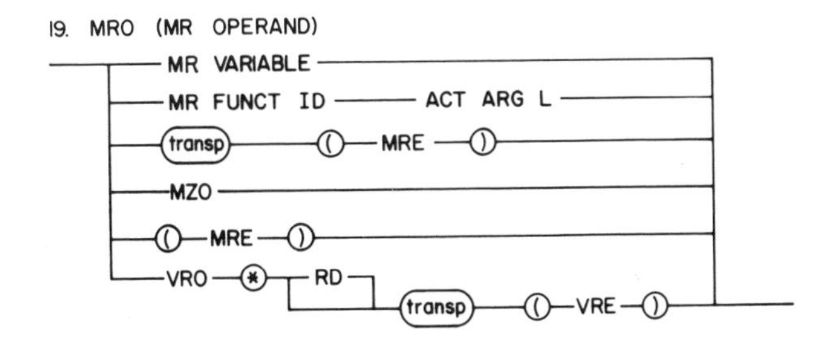


20. MRE (MR EXPRESSION)

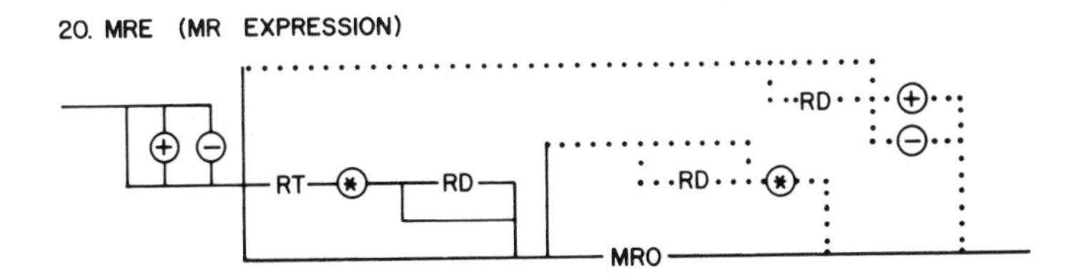

21. VRO (VR OPERAND)

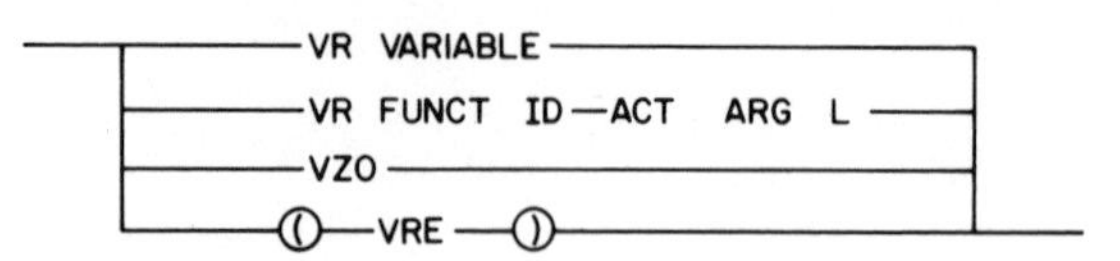


22. VRE (VR EXPRESSION)

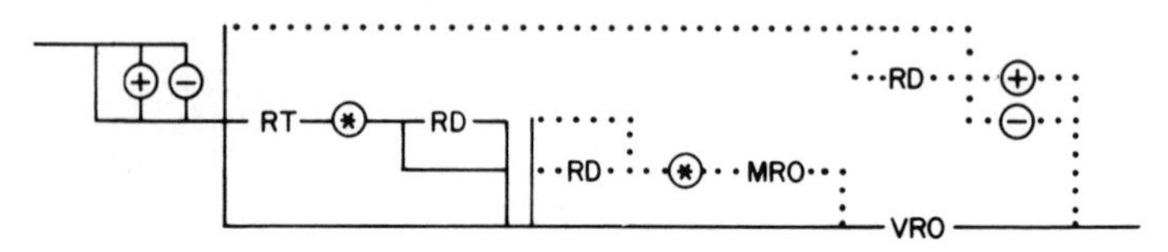


23. C SFE (C STANDARDFUNCTION EXPRESSION)

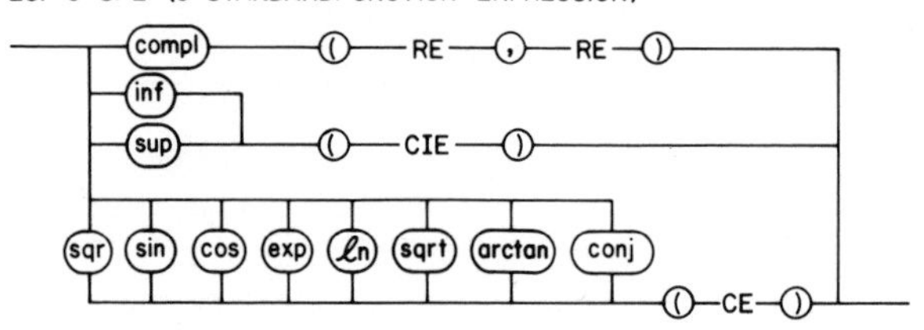


24. CO (C OPERAND)

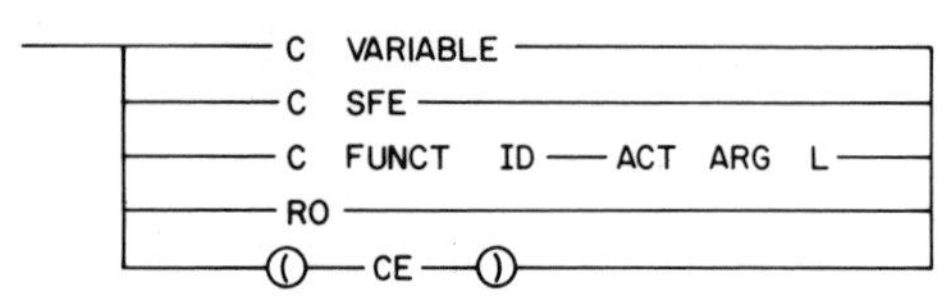


25. CT (C TERM)

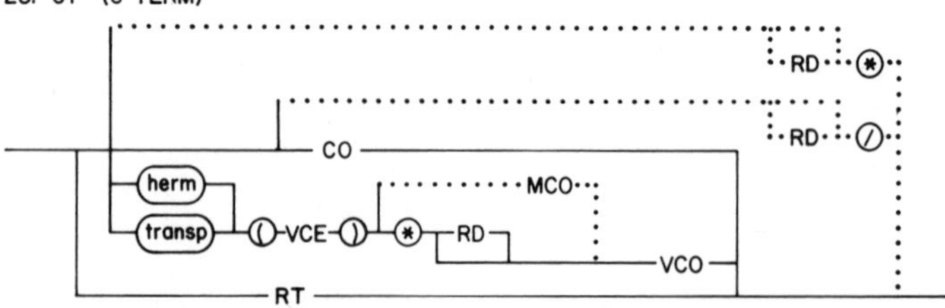


26. CE (C EXPRESSION)

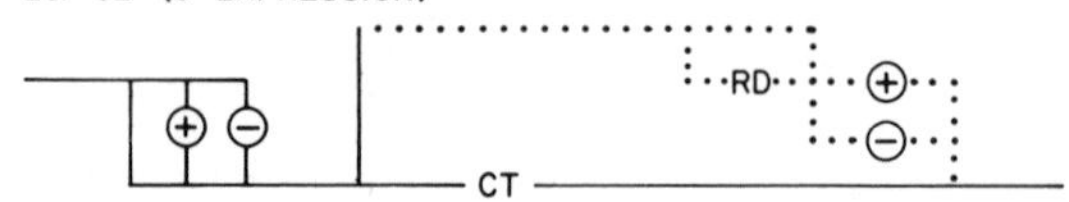

27. MCO (MC OPERAND)

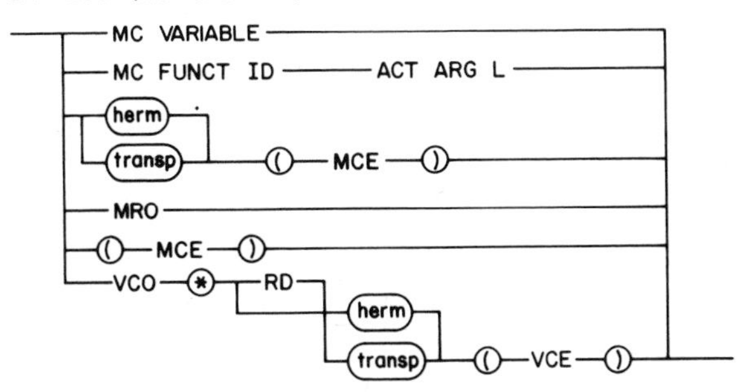


28. MCE (MC EXPRESSION)

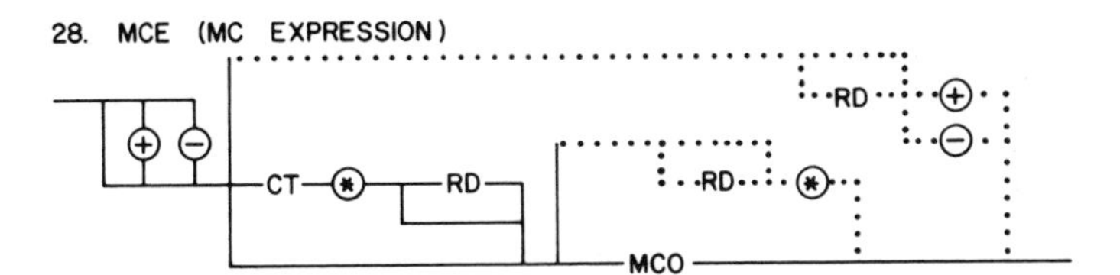


29. VCO (VC OPERAND)

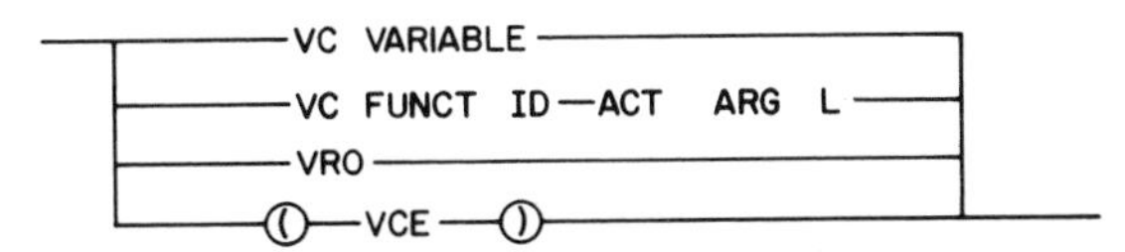


30. VCE (VC EXPRESSION)

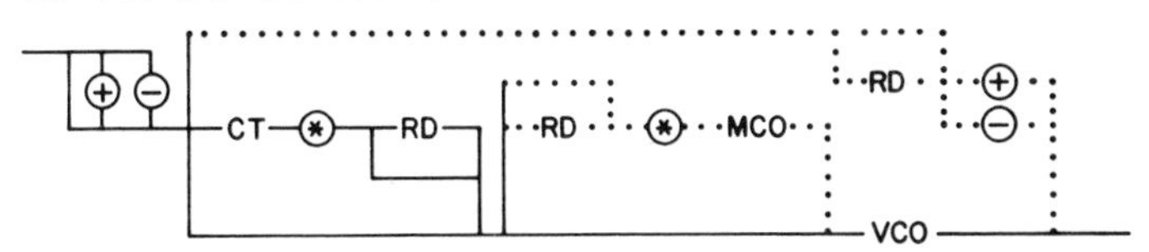


31. I SFE (I STANDARDFUNCTION EXPRESSION)

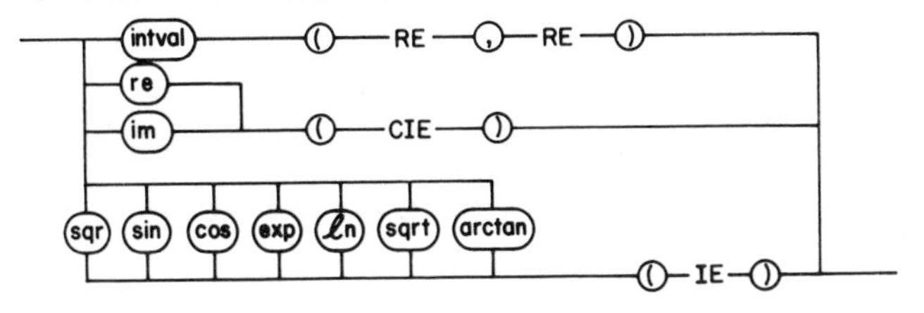

32. IO (I OPERAND)

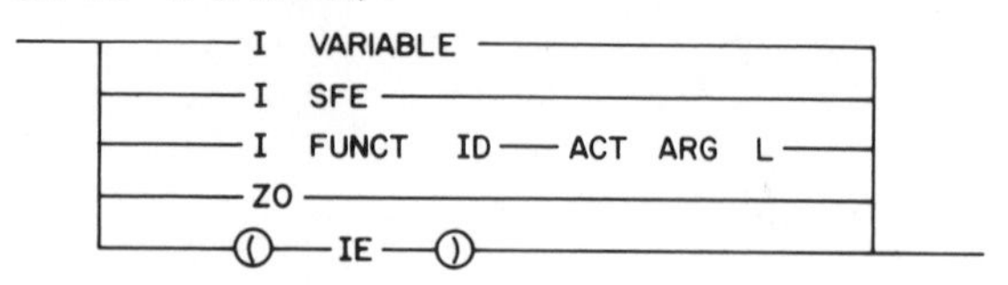


33. IT (I TERM)

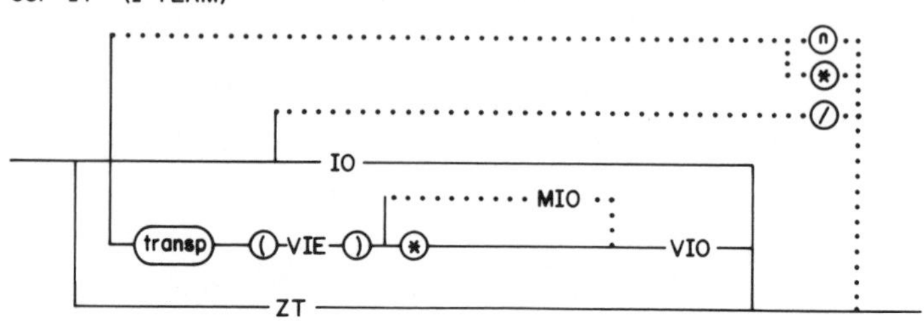


34. IE (I EXPRESSION)

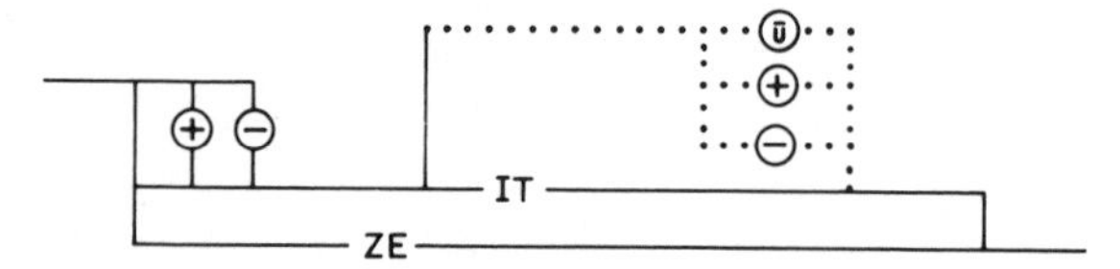


35. MIO (MI OPERAND)

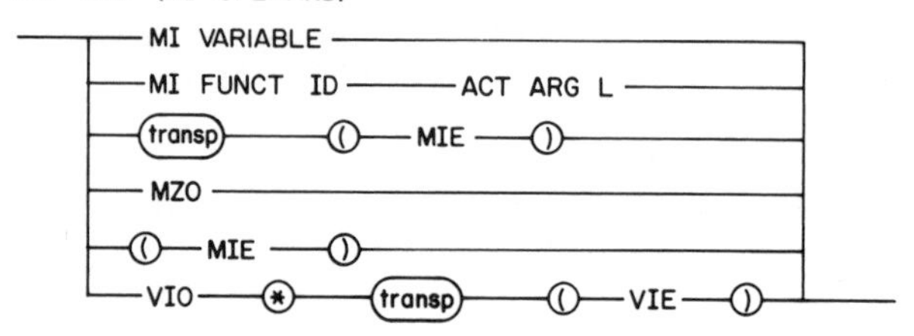


36. MIE (MI EXPRESSION)

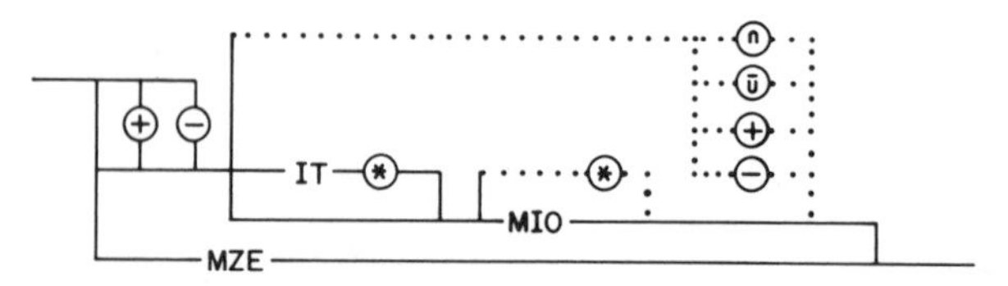


37. VIO (VI OPERAND)

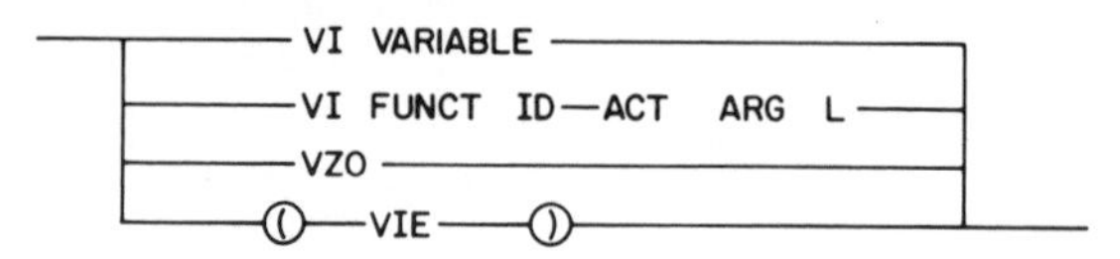

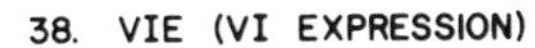

38. VIE (VI EXPRESSION)

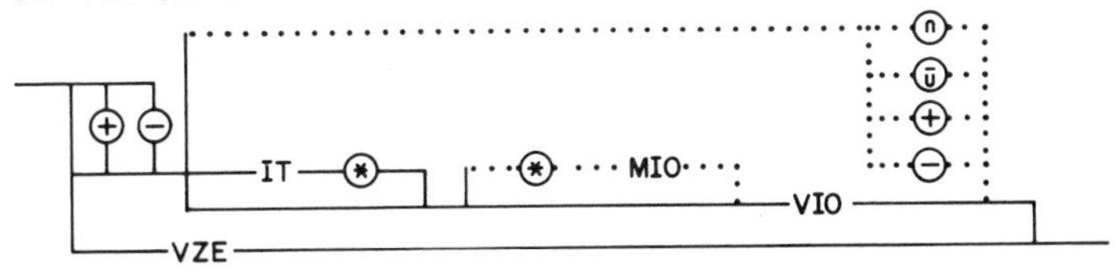


39. CI SFE (CI STANDARDFUNCTION EXPRESSION)

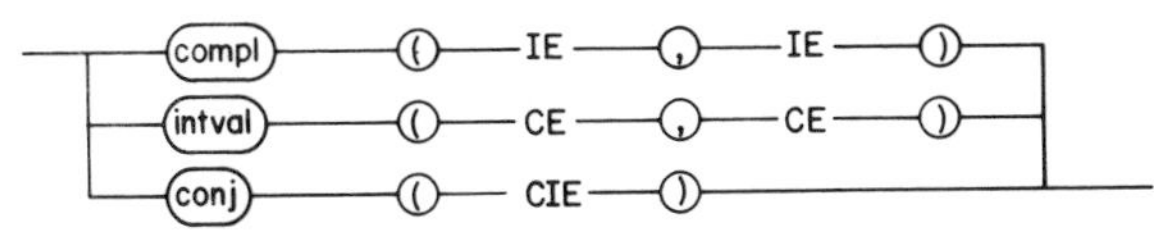


40. CIO (CI OPERAND)

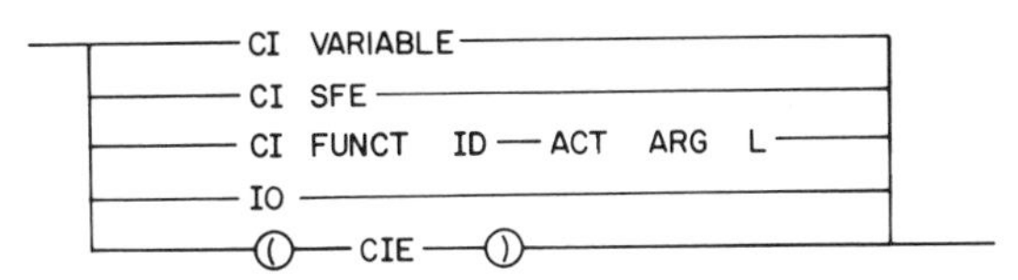


41. CIT (CI TERM)

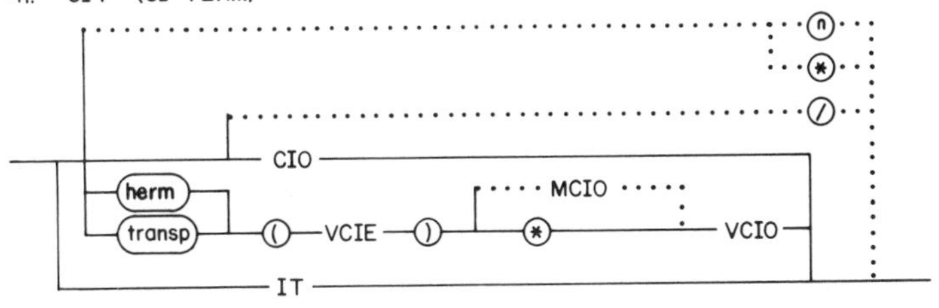


42. CIE (CI EXPRESSION)

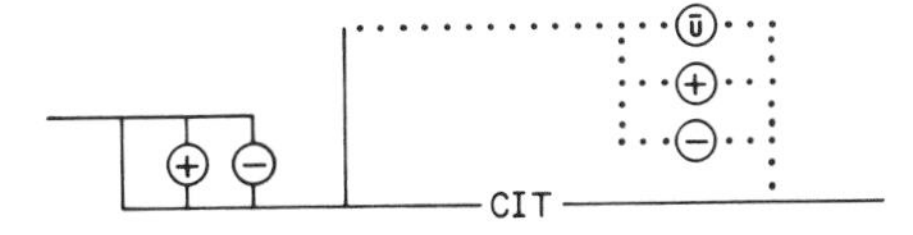


43. MCIO (MCI OPERAND)

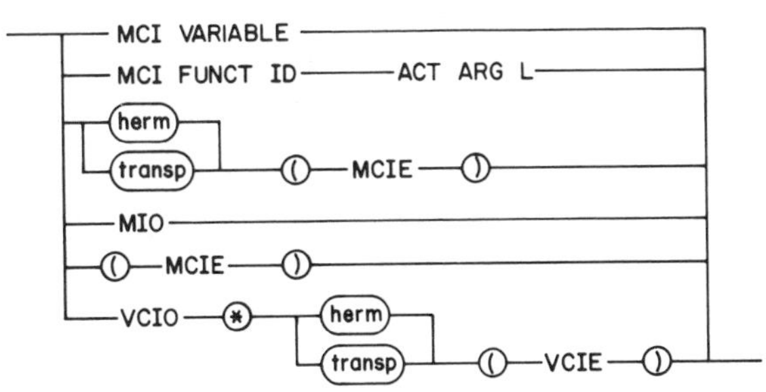

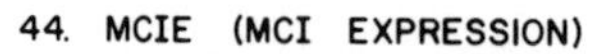
44. MCIE (MCI EXPRESSION)

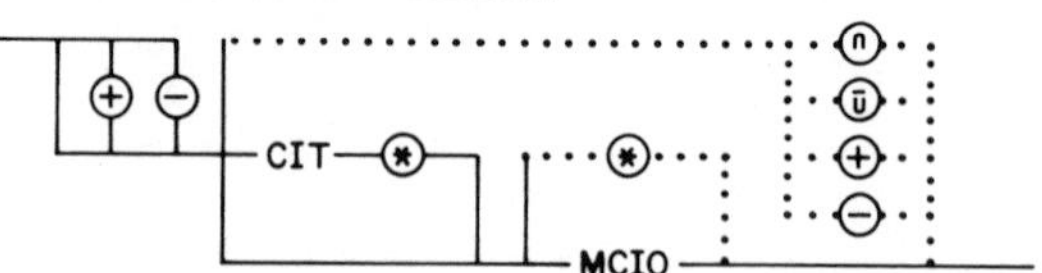


45. VCIO (VCI OPERAND)

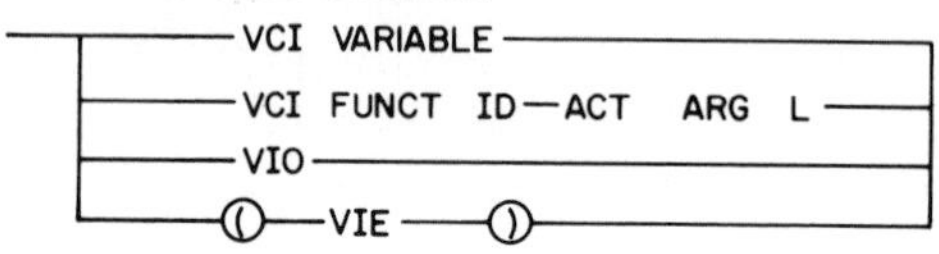


46. VCIE (VCI EXPRESSION)

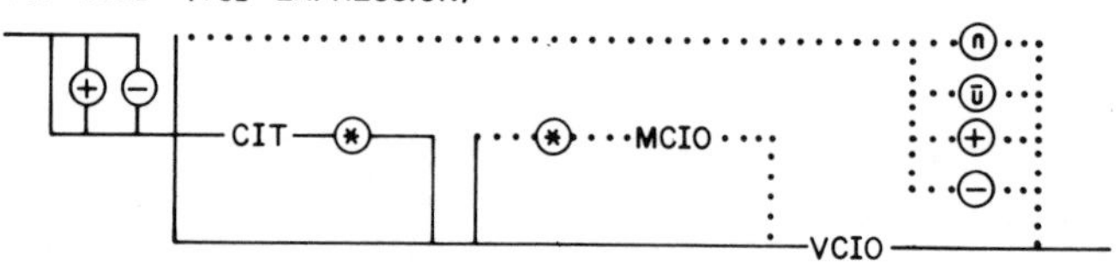


47. EXPRESSION

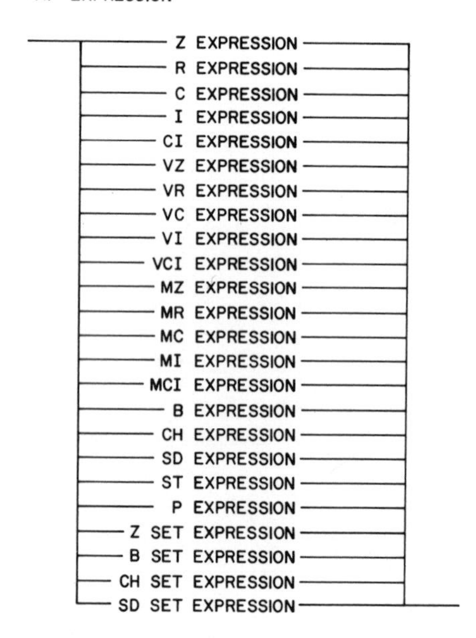

## 5. SYNTAX VARIABLES USED IN THE DIAGRAMS

| Abbreviation | Full Denotation | Defined in Diagram Number |
|---|---|---|
| ACT ARG L | ACT ARGUMENT LIST | — |
| BE | BOOLEAN EXPRESSION | — |
| B SET EXPRESSION | BOOLEAN SET EXPRESSION | — |
| CE | C EXPRESSION | 26 |
| C FUNCT ID | C FUNCTION IDENTIFIER | — |
| CHE | CHAR EXPRESSION | — |
| CH SET EXPRESSION | CHAR SET EXPRESSION | — |
| CH VARIABLE | CHAR VARIABLE | — |
| CIE | CI EXPRESSION | 42 |
| CI FUNCT ID | CI FUNCTION IDENTIFIER | — |
| CIO | CI OPERAND | 40 |
| CI SFE | CI STANDARD FUNCTION EXPRESSION | 39 |
| CIT | CI TERM | 41 |
| — | CI VARIABLE | — |
| CO | C OPERAND | 24 |
| C SFE | C STANDARD FUNCTION EXPRESSION | 23 |
| CT | C TERM | 25 |
| — | C VARIABLE | — |
| DS | DIGIT SEQUENCE | 5 |
| — | EXPRESSION | 47 |
| — | FILE VARIABLE | — |
| IE | I EXPRESSION | 34 |
| I FUNCT ID | I FUNCTION IDENTIFIER | — |
| — | INPUT STATEMENT | 1 |
| IO | I OPERAND | 32 |
| I SFE | I STANDARD FUNCTION EXPRESSION | 31 |
| IT | I TERM | 33 |
| — | I VARIABLE | — |
| MCE | MC EXPRESSION | 28 |
| MC FUNCT ID | MC FUNCTION IDENTIFIER | — |
| MCIE | MCI EXPRESSION | 44 |
| MCI FUNCT ID | MCI FUNCTION IDENTIFIER | — |
| MCIO | MCI OPERAND | 43 |
| — | MCI VARIABLE | — |
| MCO | MC OPERAND | 27 |
| — | MC VARIABLE | — |
| MIE | MI EXPRESSION | 36 |
| MI FUNCT ID | MI FUNCTION IDENTIFIER | — |
| MIO | MI OPERAND | 35 |
| — | MI VARIABLE | — |
| MRE | MR EXPRESSION | 20 |
| MR FUNCT ID | MR FUNCTION IDENTIFIER | — |
| MRO | MR OPERAND | 19 |
| — | MR VARIABLE | — |
| MZE | MZ EXPRESSION | 11 |

| Abbreviation | Full Denotation | Defined in Diagram Number |
|---|---|---|
| MZ FUNCT ID | MZ FUNCTION IDENTIFIER | — |
| MZO | MZ OPERAND | 10 |
| — | MZ VARIABLE | — |
| — | OUTPUT STATEMENT | 2 |
| P EXPRESSION | POINTER EXPRESSION | — |
| — | R CONSTANT | 15 |
| R CONST ID | R CONSTANT IDENTIFIER | — |
| RD | ROUNDING | 3 |
| RE | R EXPRESSION | 18 |
| R FUNCT ID | R FUNCT IDENTIFIER | — |
| RO | R OPERAND | 16 |
| R SFE | R STANDARD FUNCTION EXPRESSION | 14 |
| RT | R TERM | 17 |
| — | R VARIABLE | — |
| SD EXPRESSION | EXPRESSION OF SELFDEFINED TYPE | — |
| SD SET EXPRESSION | EXPRESSION OF SELFDEFINED SET TYPE | — |
| STE | STRING EXPRESSION | — |
| — | TEXTFILE VARIABLE | — |
| — | VARIABLE | — |
| VCE | VC EXPRESSION | 30 |
| VC FUNCT ID | VC FUNCTION IDENTIFIER | — |
| VCIE | VCI EXPRESSION | 46 |
| VCI FUNCT ID | VCI FUNCTION IDENTIFIER | — |
| VCIO | VCI OPERAND | 45 |
| VCO | VC OPERAND | 29 |
| — | VCI VARIABLE | — |
| — | VC VARIABLE | — |
| VIE | VI EXPRESSION | 38 |
| VI FUNCT ID | VI FUNCTION IDENTIFIER | — |
| VIO | VI OPERAND | 37 |
| VRE | VR EXPRESSION | 22 |
| VR FUNCT ID | VR FUNCTION IDENTIFIER | — |
| VRO | VR OPERAND | 21 |
| — | VR VARIABLE | — |
| VZE | VZ EXPRESSION | 13 |
| VZ FUNCT ID | VZ FUNCTION IDENTIFIER | — |
| VZO | VZ OPERAND | 12 |
| — | VZ VARIABLE | — |
| — | Z CONSTANT | 16 |
| Z CONST ID | Z CONSTANT IDENTIFIER | — |
| ZE | Z EXPRESSION | 9 |
| Z FUNCT ID | Z FUNCTION IDENTIFIER | — |
| ZO | Z OPERAND | 7 |
| — | Z SET EXPRESSION | — |
| Z SFE | Z STANDARD FUNCTION EXPRESSION | 4 |
| ZT | Z TERM | 8 |
| — | Z VARIABLE | — |

# REFERENCES


[1] Albrecht, R., Grundlagen einer Theorie algebraischer Verknüpfungen in topologischen Vereinen, *Computing Suppl.* **1**, 1–14 (1977).

[2] Alefeld, G., Intervallrechnung über den komplexen Zahlen und einige Anwendungen. Dissertation, Universität Karlsruhe, 1968.

[3] Alefeld, G., and Herzberger, J., "Einführung in die Intervallrechnung," Reihe Informatik, **12**. Wissenschaftsverlag des Bibliographischen Instituts Mannheim, 1974.

[4] Apostolatos, N., Christ, H., Santo, H., and Wippermann, H., Rounding Control and the Algorithmic Language ALGOL-68. Report Universität Karlsruhe, Rechenzentrum, Juli 1968.

[5] Bauer, F. L., and Samelson, K., Optimale Rechengenauigkeit bei Rechenanlagen mit gleitendem Komma, *Z. Angew. Math. Phys.* **4**, 312–316 (1953).

[6] Behnke, H., *et al.*, "Grundzüge der Mathematik," **1** und **2**. Vandenhoeck u. Rupprecht, Göttingen, 1962.

[7] Birkhoff, G., "Lattice Theory." Amer. Math. Soc., Providence, Rhode Island, 1967.

[8] Bohlender, G., Floating-point computation of functions with maximum accuracy, IEEE Trans. Compute. **C-26**, No. 7, 621–632 (1977).

[9] Bohlender, G., Genaue Summation von Gleitkommazahlen, *Computing Suppl.* **1**, 21–32 (1977).

[10] Bohlender, G., Genaue Berechnung mehrfacher Summen, Produkte und Wurzeln von Gleitkommazahlen und allgemeine Arithmetik in Höheren Programmiersprachen. Dissertation, Universität Karlsruhe, 1978.

[11] Claudio, D. M., Beiträge zur Struktur der Rechnerarithmetik. Dissertation, Universität Karlsruhe, 1979.

[12] Christ, H., Realisierung einer Maschinenintervallarithmetik auf beliebigen ALGOL-60 Compilern, *Elektron. Rech.* **10**, 217–222 (1968).

[13] Collatz, L., "Funktionalanalysis und Numerische Mathematik," Springer-Verlag, Berlin and New York, 1968.

[14] Forsythe, G. E., Pitfalls in computation, or why a math book isn't enough, Tech. Rep. No. CS147, pp. 1–43. Computer Science Department, Stanford University, Stanford, California, 1970.

[15] Forsythe, G. E., and Moler, C. B., "Computer Solution of Linear Algebraic Systems," Prentice-Hall, Englewood Cliffs, New Jersey, 1967.

[16] Gericke. H., "Theorie der Verbände." Bibliographisches Institut Mannheim, 1967.

[17] Grüner, K., Fehlerschranken für lineare Gleichungssysteme, *Computing Suppl.* **1**, 47–55 (1977).

[18] Grüner, K., Allgemeine Rechnerarithmetik und deren Implementierung. Dissertation, Universität Karlsruhe, 1979.

[19] Haas. H. Ch., Implementierung der komplexen Gleitkommaarithmetik mit maximaler Genauigkeit. Diplomarbeit, Institut für Angewandte Mathematik, Universität Karlsruhe, 1975.

[20] Heinhold, J., and Riedmüller, B., "Lineare Algebra und Analytische Geometrie." Carl Hanser Verlag, München, 1971.

[21] Hermes, H., "Einführung in die Verbandstheorie." Springer-Verlag, Berlin and New York, 1967.

[22] Herzberger, J., Metrische Eigenschaften von Mengensystemen und einige Anwendungen. Dissertation, Universität Karlsruhe, 1969.

[23] Kahan, W., and Parlett, B. N., Können Sie Sich auf Ihren Rechner verlassen?, "Jahrbuch Überblicke Mathematik 1978." Wissenschaftsverlag des Bibliographischen Instituts Mannheim, pp. 199–216, (1978).

[24] Kaucher, E., Über metrische und algebraische Eigenschaften einiger beim numerischen Rechnen auftretender Räume. Dissertation, Universität Karlsruhe, 1973.

[25] Kaucher, E., Algebraische Erweiterungen der Intervallrechnung unter Erhaltung der Ordnungs-und Verbandsstrukturen, *Computing Suppl.* **1**, 65–79 (1977).

[26] Kaucher, E., Über Eigenschaften und Anwendungsmöglichkeiten der erweiterten Intervallrechnung und des hyperbolischen Fastkörpers über $\boldsymbol{R}^*$, *Computing Suppl.* **1**, 81–94 (1977).

[27] Kaucher, E., Über eine Überlaufarithmetik auf Rechenanlagen und deren Anwendungsmöglichkeiten, *Z. Angew. Math. Mech.* **57**, T286–T287 (1977).

[28] Klatte, R., Zyklisches Enden bei Iterationsverfahren. Dissertation, Universität Karlsruhe, 1975.

[29] Klatte, R., and Ullrich, Ch., Consequences of a properly implemented computer arithmetic for periodicities of iterative methods, IEEE Computer Society, Symposium on Computer Arithmetic, Dallas, 24–32 (1975).

[30] Knuth, D., "The Art of Computer Programming," Vol. 2. Addison-Wesley. Reading, Massachusetts, 1962.

[31] Kulisch, U., Grundzüge der Intervallrechnung, Überblicke Mathematik **2**, Bibliographisches Institut Mannheim, 51–98 (1969).

[32] Kulisch, U., An axiomatic approach to rounded computations, TS Report No. 1020, Mathematics Research Center, University of Wisconsin, Madison, Wisconsin, 1969, and *Numer. Math*, **19**, 1–17 (1971).

[33] Kulisch, U., On the concept of a screen, TS Report No. 1084, Mathematics Research Center, University of Wisconsin, Madison, Wisconsin, 1970; and *Z. Angew. Math. Mech.* **53**, 115–119 (1973).

[34] Kulisch, U., Rounding invariant structures, TS Report No. 1103, Mathematics Research Center, University of Wisconsin, Madison, Wisconsin, 1970.

[35] Kulisch, U., Interval arithmetic over completely ordered ringoids, TS Report No. 1105, Mathematics Research Center, University of Wisconsin, Madison, Wisconsin, 1970.

[36] Kulisch, U., Implementation and formalization of floating-point arithmetics, Report RC 4608, IBM Thomas J. Watson Research Center, 1973, and *C. Caratheodory Symp., 1973*, 328–369.

[37] Kulisch, U., Formalization and implementation of floating-point arithmetic, *Computing* **14**, 323–348 (1975).

[38] Kulisch, U., Über die Arithmetik von Rechenanlagen, "Jahrbuch Überblicke Mathematik 1975," Wissenschaftsverlag des Bibliographischen Instituts Mannheim, 68–108 (1975).

[39] Kulisch, U., Mathematical foundation of computer arithmetic, IEEE Trans. Comput. **C-26**, No. 7, 610–621 (1977).

[40] Kulisch, U., "Grundlagen des Numerischen Rechnens-Mathematische Begründung der Rechnerarithmetik," Reihe Informatik, **19**. Wissenschaftsverlag des Bibliographischen Instituts Mannheim, 1976.

[41] Kulisch, U., Ein Konzept für eine allgemeine Theorie der Rechnerarithmetik, *Computing Suppl.* **1**, 95–105 (1977).

[42] Kulisch, U., Über die beim numerischen Rechnen mit Rechenanlagen auftretenden Räume, *Computing Suppl.* **1**, 107–119 (1977).

[43] Kulisch, U., and Bohlender, G., Formalization and implementation of floating-point matrix operations, *Computing* **16**, 239–261 (1976).

[44] Kulisch, U., and Miranker, W. L., Arithmetic operations in interval spaces, Report RC 7681, IBM Thomas J. Watson Research Center, 1979; *Computing Suppl.* **2**, 51–67 (1980).

[45] Lortz, B., Eine Langzahlarithmetik mit optimaler einseitiger Rundung. Dissertation, Universität Karlsruhe, 1971.

[46] Luxemburg, W. A., and Zaanen, A. C., "Riesz Spaces," Vol. 1. North-Holland Publ. Amsterdam, 1971.

[47] Matula, D. W., In-and-out conversions, *Comm. ACM* **11**, 47–50 (1968).

[48] Matula, D. W., Radix Arithmetic: Digital algorithms for computer architecture, Report. No. CS-74-3 Department of Applied Mathematics and Computer Science, Washington University, St. Louis, 1974.

[49] Mayer, O., Über die in der Intervallrechnung auftretenden Räume und einige Anwendungen. Dissertation, Universität Karlsruhe, 1968.

[50] Mayer, O., Algebraische und metrische Strukturen in der Intervallrechnung und einige Anwendungen, *Computing* **5**, 144–162 (1970).

[51] McShane, E. J., and Botts, T. A., "Real Analysis." Van Nostrand–Reinhold, Princeton, New Jersey, 1959.

[52] Mealy, G. H., Floating scale base conversion, Center for Research in Computing Technology, Harvard University, Cambridge, Massachusetts, 1968.

[53] Meschkowski, H., "Mathematisches Begriffswörterbuch." Bibliographisches Institut Mannheim, 1965.

[54] Moore, R. E., "Interval Analysis," Prentice-Hall, Englewood Cliffs, New Jersey, 1966.

[55] Perron, O., "Irrationalzahlen." de Gruyter, Berlin, 1960.

[56] Pichat, M., Correction d'une somme en arithmetic a virgule flottante, *Numer. Math.* **19**, 400–406 (1972).

[57] Reinsch, Ch., Die Behandlung von Rundungsfehlern in der Numerischen Analysis, "Jahrbuch Überblicke Mathematik 1979," Wissenschaftsverlag des Bibliographischen Instituts Mannheim, 43–62 (1979).

[58] Rudin. W., "Principles of Mathematical Analysis." McGraw-Hill, New York, 1953.

[59] Rutishauser, H., Eine Axiomatik des numerischen Rechnens und ihre Anwendung auf den Quotienten-Differenzen-Algorithmus. "Vorlesungen über Numerische Mathematik," **2**, 179–221. Birkhäuser-Verlag, Basel, 1976.

[60] Spaniol, O., "Arithmetik in Rechenanlagen." Teubner Verlag, Stuttgart, 1976.

[61] Stummel, F., Fehleranalyse numerischer Algorithmen, Universität Frankfurt a.M., Vorlesungsskriptum, Wintersemester 1977/78.

[62] Szàsz, G., "Introduction to Lattice Theory." Academic Press, New York, 1963.

[63] Ullrich, Ch., Rundungsinvariante Strukturen mit äusseren Verknüpfungen. Dissertation, Universität Karlsruhe, 1972.

[64] Ullrich, Ch., Uber die beim numerischen Rechnen mit komplexen Zahlen und Intervallen vorliegenden mathematischen Strukturen, Computing **14**, 51–65 (1975).

[65] Ullrich, Ch., Zum Begriff des Rasters und der minimalen Rundung, *Computing Suppl.* **1**, 129–134 (1977).

[66] Ullrich, Ch., Zur Konstruktion komplexer Kreisarithmetiken. *Computing Suppl.* **1**, 135–150 (1977).

[67] Ullrich, Ch., Über schwach zyklische Abbildungen in nicht linearen Produkträumen und einige Monotonieaussagen, *Apli. Mati.* **24**, 209–234 (1979).

[68] Wilkinson, J. H., "Rounding Errors in Algebraic Processes." Prentice-Hall, Englewood Cliffs, New Jersey, 1963.

[69] Wilkinson, J. H., "Rundungsfehler." Springer-Verlag, Berlin and New York, 1969.

[70] Wilkinson, J. H., "The Algebraic Eigenvalue Problem." Oxford Univ. Press (Clarendon), London and New York, 1965.

[71] Wippermann, H.-W., Realisierung einer Intervallarithmetik in einem ALGOL-60 System, *Electron. Rech.* **9**, 224–233 (1967).

[72] Wippermann, H.-W., Implementierung eines ALGOL-60 Systems mit Schrankenzahlen, *Electron. Datenverab.* **10**, 189–194 (1968).

[73] Wippermann, H.-W., Definition von Schrankenzahlen in TRIPLEX-ALGOL, *Computing* **3**, 99–109 (1968).

[74] Yohe, J. M., Interval bounds for square roots and cube roots, *Computing* **11**, 51–57 (1973).

[75] Yohe, J. M., Roundings in floating-point arithmetic, *IEEE Trans. Comput.* **C.12** No. 6, 577–586 (1973).

## SUPPLEMENTARY REFERENCES

[76] Albrecht, R., Roundings and approximation in ordered sets, *Computing Suppl*, **2**, 17–32 (1980).

[77] Bohlender, G., Embedding universal computer arithmetic in higher programming languages, *Computing* **24**, 149–160 (1980).

[78] Bohlender, G., Kaucher, E., Klatte, R., Kulisch, U., Miranker, W. L., Ullrich, Ch., and Wolff von Gudenberg, J., FORTRAN for contemporary numerical computation, Report RC 8348, IBM Thomas J. Watson Research Center, 1980.

[79] Claudio, D. M., Contributions to the structure of computer arithmetic, *Computing* **24**, 115–118, (1980).

[80] Grüner, K., Implementation of universal arithmetic with optimal accuracy, *Computing* **24**, 181–193 (1980).

[81] Heinhold, J., and Perron, O., "Jahrubuch Überblicke Mathematick 1980," Wissenschaftsverlag der Bibliographischen Instituts, Mannheim, 121–139 (1980).

[82] Kaucher, E., Interval analysis in the extended interval space IR, *Computing Suppl.* **2**, 33–50, (1980).

[83] Kaucher, E., Rump, S. M., Generalized iteration methods for bounds of the solution of fixed point operator equations, *Computing* **24**, 131–137 (1980).

[84] Kaucher, E., Klatte, R., and Ullrich, Ch., "Höhere Programmiersprachen ALGOL, FORTRAN, PASCAL in einheitlicher und übersichtlicher Darstellung," Reihe Informatik, **24**. Wissenschaftsverlag des Bibliographischen Instituts Mannheim, 1978.

[85] Netzer, N., Verallgemeinerte topologische Strukturen, "Jahrbuch Überblicke Mathematik, 1978." Wissenschaftsveralg des Bibliographischen Instituts Mannheim, 87–106, (1979).

[86] Nöbeling, G., "Grundlagen der analytischen Topologie." Springer-Verlag, Berlin and New York, 1954.

[87] Rump, S. M., and Kaucher, E., Small bounds for the solution of systems of linear equations, *Computing Suppl.* **2**, 157–164 (1980).

[88] Stummel, F., Rounding error analysis of elementary numerical algorithms, *Computing Suppl.* **2**, 169–196 (1980).

[89] Ullrich, Ch., Iterative methods in the spaces of rounded computations, *Computing Suppl.* **2**, 197–210 (1980).

[90] Woff von Gudenberg, J. Evaluation of the standard functions in generalized computer arithmetic, *Computing* (to be published).

# GLOSSARY OF SYMBOLS AND FORMULAS

| | |
|---|---|
| $\boldsymbol{N}$ | set of natural numbers |
| $\boldsymbol{Z}$ | set of integers |
| $\boldsymbol{R}$ | set of real numbers |
| $\boldsymbol{C}$ | set of complex numbers |
| $\boldsymbol{C}S$ | set of pairs over $S$ |
| $\boldsymbol{P}S$ | power set of the set $S$ |
| $\boldsymbol{I}S$ | set of intervals over the ordered set $\{S, \leq\}$ |

| | |
|---|---|
| (O1), (O2), (O3), (O4) | order properties, see Definition 1.1, p. 13 |
| (O5) | lattice property, see Definition 1.7, p. 18 |
| (S1), (S2) | properties of a screen see Definition 1.18, p. 26 |
| (S3) | symmetric screen, see Definition 3.2, p. 67 |
| (R1), (R2), (R3) | properties of roundings, see Definition 1.22, p. 30 |
| (R4) | antisymmetry of a rounding, see Definition 3.2, p. 67 |
| (R) | monotone directed roundings, see Theorem 1.21, p. 28 |
| (RG) | definition of operations on a screen, see Theorems 1.30 and 1.34, pp. 36 and 40, and Definition 3.2, p. 67 |

| | |
|---|---|
| (RG1), (RG2), (RG3) | properties of operations on a screen, see Definitions 1.29, 1.33, 3.6, and 3.7, pp. 36, 40, 71, and 72 |
| (D1), (D2), . . . , (D9) | properties of a ringoid, see Definition 2.1, p. 41ff |
| (OD1), (OD2), . . . , (OD6) | order properties of a ringoid, see Definition 2.1, p. 42ff |
| (V1), (V2), . . . , (V5) | properties of the inner operations of a vectoid, see Definition 2.7, p. 52 |
| (VD1), (VD2), . . . , (VD5) | properties of the outer operation of a vectoid, see Definition 2.7, p. 52ff |
| (OV1), (OV2), . . . , (OV5) | order properties of a vectoid, see Definition 2.7, p. 53ff |

# INDEX

**H**



**I**



**L**



**M**



**N**



**O**



**P**



**R**



**S**



**T**

## U



## U



## W

**Computer Science and Applied Mathematics**

A SERIES OF MONOGRAPHS AND TEXTBOOKS

Editor
**Werner Rheinboldt**
*University of Pittsburgh*

HANS P. KÜNZI, H. G. TZSCHACH, and C. A. ZEHNDER. Numerical Methods of Mathematical Optimization: With ALGOL and FORTRAN Programs, Corrected and Augmented Edition

AZRIEL ROSENFELD. Picture Processing by Computer

JAMES ORTEGA AND WERNER RHEINBOLDT. Iterative Solution of Nonlinear Equations in Several Variables

AZARIA PAZ. Introduction to Probabilistic Automata

DAVID YOUNG. Iterative Solution of Large Linear Systems

ANN YASUHARA. Recursive Function Theory and Logic

JAMES M. ORTEGA. Numerical Analysis: A Second Course

G. W. STEWART. Introduction to Matrix Computations

CHIN-LIANG CHANG AND RICHARD CHAR-TUNG LEE. Symbolic Logic and Mechanical Theorem Proving

C. C. GOTLIEB AND A. BORODIN. Social Issues in Computing

ERWIN ENGELER. Introduction to the Theory of Computation

F. W. J. OLVER. Asymptotics and Special Functions

DIONYSIOS C. TSICHRITZIS AND PHILIP A. BERNSTEIN. Operating Systems

ROBERT R. KORFHAGE. Discrete Computational Structures

PHILIP J. DAVIS AND PHILIP RABINOWITZ. Methods of Numerical Integration

A. T. BERZTISS. Data Structures: Theory and Practice, Second Edition

N. CHRISTOPHIDES. Graph Theory: An Algorithmic Approach

ALBERT NIJENHUIS AND HERBERT S. WILF. Combinatorial Algorithms

AZRIEL ROSENFELD AND AVINASH C. KAK. Digital Picture Processing

SAKTI P. GHOSH. Data Base Organization for Data Management

DIONYSIOS C. TSICHRITZIS AND FREDERICK H. LOCHOVSKY. Data Base Management Systems

JAMES L. PETERSON. Computer Organization and Assembly Language Programming

WILLIAM F. AMES. Numerical Methods for Partial Differential Equations, Second Edition

Arnold O. Allen. Probability, Statistics, and Queueing Theory: With Computer Science Applications

Elliott I. Organick, Alexandra I. Forsythe, and Robert P. Plummer. Programming Language Structures

Albert Nijenhuis and Herbert S. Wilf. Combinatorial Algorithms. Second edition.

James S. Vandergraft. Introduction to Numerical Computations

Azriel Rosenfeld. Picture Languages, Formal Models for Picture Recognition

Isaac Fried. Numerical Solution of Differential Equations

Abraham Berman and Robert J. Plemmons. Nonnegative Matrices in the Mathematical Sciences

Bernard Kolman and Robert E. Beck. Elementary Linear Programming with Applications

Clive L. Dym and Elizabeth S. Ivey. Principles of Mathematical Modeling

Ernest L. Hall. Computer Image Processing and Recognition

Allen B. Tucker, Jr., Text Processing: Algorithms, Languages, and Applications

Martin Charles Golumbic. Algorithmic Graph Theory and Perfect Graphs

Gabor T. Herman. Image Reconstruction from Projections: The Fundamentals of Computerized Tomography

Webb Miller and Celia Wrathall. Software for Roundoff Analysis of Matrix Algorithms

Ulrich W. Kulisch and Willard L. Miranker. Computer Arithmetic in Theory and Practice

*In preparation*

Louis A. Hageman and David M. Young. Applied Iterative Methods